ENERGY MANAGEMENT SYSTEMS

Edited by **P. Giridhar Kini**
and **Ramesh C. Bansal**

INTECHOPEN.COM

Energy Management Systems

Edited by P. Giridhar Kini and Ramesh C. Bansal

CBS, Edition 2016

Published by InTech

Janeza Trdine 9, 51000 Rijeka, Croatia

Publishing Process Manager Iva Simcic

Technical Editor Teodora Smiljanic

Cover Designer InTech Design Team

Image Copyright 2010. Used under license from Shutterstock.com

Additional hard copies can be obtained from orders@intechopen.com

Energy Management Systems, Edited by P. Giridhar Kini and Ramesh C. Bansal
p. cm.
ISBN 978-953-307-579-2

Contents

Preface

Energy management has become an important issue in recent times when many utilities around the world find it very difficult to meet energy demands which have led to load shedding and power quality problems. An efficient energy management in residential, commercial and industrial sector can reduce the energy requirements and thus lead to savings in the cost of energy consumed which also has positive impact on environment. Energy management is not only important in distribution system but it has great significance is generation system as well. Smart grid management and renewable energy integration are becoming important aspects of efficient energy management.

The management of energy technology and its applications in residential, commercial and industrial sector is a diversified topic and quite difficult task to document in a single book. This book tries to cover many important aspects of energy management, forecasting, optimization methods and their applications in selected industrial, residential, and generation system. This book comprises of 13 chapters which are arranged in two sections. Section one covers energy efficiency, optimization, forecasting, modelling and analysis and section two covers some of the diversified applications of energy management systems for buildings, renewable energy (photovoltaic system), design of cooling water systems, super capacitor for transportation systems, locomotive energy systems and smart grid management. Brief discussion of each chapter is as follows.

Chapter 1 looks into the energy audit and management requirements, alternate sources of energy, power quality issues, instrumentation requirements, financial analysis, energy policy framework and energy management information systems (EMIS) for an industrial utility.

Chapter 2 presents a detailed methodology for the development of energy management in industrial plants. The main aspects of energy management methodology are: energy cost and consumption data and their analysis, forecasting, sub-metering, tariff analysis, consumption control, budgeting and machines management optimization.

Chapter 3 presents an innovative methodology for the productive process quantification optimization in aluminum bar industry. Energy optimization has a high impact on service industry which has been discussed for a water supply company.

Chapter 4 discusses the online energy system optimization and demonstrates energy optimization application in thermal power generation sector. A detailed model of energy systems comprising of fuel system, boiler feed water, steam, electricity generation, and condensate network is built within energy management system (EMS) environment and it is continuously fed with real time data. Optimization is configured to minimize the total cost. Besides real time optimization, key performance indicators (KPIs) targets can also be set up. The chapter also discusses many examples in open and closed loop implementation in power generation sector.

Chapter 5 presents energy demand analysis and forecasting. The modeling results are interpreted by statistical tests and the focus of the investigation lies in the application of regression methods and neural networks for the forecast of the power and heat demand for cogeneration systems. The application of the proposed method is demonstrated by the heat and power demand forecast for a real district heating system containing different cogeneration units.

Chapter 6 discusses about intelligent buildings their automation and home automation networks. This chapter surveys appliance and lighting load energy management strategies that works to achieve the three goals of building energy management, i.e., reduction of energy consumption of building; reduction of electricity bills while increasing comfort and productivity to occupants; and the improvement of environmental stewardship. Smart grid security and   security threats that need to be addressed are also discussed in the chapter.

Chapter 7 reviews the parameters that affect PV systems' efficiency and diffuse of solar irradiance. The results of energy yield and gains by the optimal fixed azimuth and tilt angle are presented. The important results of the chapter are the contour plots with appropriate combination of tilt and azimuth angles for four typical locations in Slovenia.

Chapter 8 presents an optimization model and detailed design of cooling water systems. The cooling water structure embeds all possible combinations of series-parallel arrangements of heat exchanger units. The model is based on a mixed-integer nonlinear programming to determine the cooling water system design which minimizes the total annual cost. Two examples are demonstrated to show the savings which can be obtained with the proposed design.

Chapter 9 discusses the fundamental characteristics of super capacitor devices. Some preliminary consideration with respect to optimization methodologies are presented and light transportation systems modeling for both stationary storage systems and on-board are discussed. A numerical application is reported for a case study with two trains along double track dc electrified subway networks both for stationary and on-board applications.

Chapter 10 presents different types of locomotive energy saving systems which are used in aeroefficient optimized trains, energy management control, energy storage systems. New technologies of traction motors of increased energy efficiency at reduced volume and weight are discussed. The theoretical and practical possibilities of dc/dc, ac/dc, ac/ac traction system locomotive regenerative braking energy management are suggested. Catenary free system for trams, light rail vehicles, trolleybuses are presented. Energy saving and power supply optimization possibilities using regenerative braking energy are also discussed in the chapter.

Chapter 11 has developed an adaptive energy management system (A-EMS) for controlling energy consumption by converging heterogeneous networks such as power line communications (PLC), Wi-Fi networks, ZigBee, and future sensor networks. In this work a prototype system enables users to freely configure a cooperative network of sensors and home appliances from a mobile device. Experimental results demonstrate that the proposed system can easily detect waste electrical energy.

Chapter 12 discusses the eight priories of US smart grid, i.e., wide area awareness, demand response and consumer energy efficiency, energy storage, electric transportation, cyber security, network communications, advance metering infrastructure, and distribution grid management. This chapter discusses the importance of customer feedback loop in smart grid. Finally chapter discusses about customer load response, commercial and industrial dynamic power management strategies, distributed generation and industrial micro grids.

Chapter 13 presents discussion on demand management and use of wire sensor networks (WSN) in generation, transmission and distribution and in demand side management. Various types of demand management system, i.e., communication based, incentive based, real time and optimization based have also been discussed in the chapter.

Editors are grateful to a number of individuals who have directly or indirectly contributed to this book. In particular Editors would like to thank all authors for their contributions. Editors are indebted to all the reviewers for reviewing the book chapters which has improved the quality of book.

Editors would like to thank the authorities and staff members of Manipal University and The University of Queensland who have been very generous and helpful in maintaining a cordial atmosphere and extending all the facilities required for the book. Thanks are due to InTech - Open Access Publisher, especially to Ms. Viktorija Zgela and Ms. Iva Simcic, Publishing Process Manager for making sincere efforts in timely bringing out the book. Editors would like to express thanks and sincere regards to their family members who have provided great support for completion of this book.

P. Giridhar Kini
Dept of Electrical & Electronics Engg,
Manipal Institute of Technology, Manipal,
India

Ramesh C. Bansal
School of Information Technology & Electrical Engineering,
The University of Queensland,
Australia

Part 1

Energy Efficiency, Optimization, Forecasting, Modeling and Analysis

Energy Efficiency in Industrial Utilities

P. Giridhar Kini[1] and Ramesh C. Bansal[2]
[1]*Dept of Electrical & Electronics Engg, Manipal Institute of Technology, Manipal*
[2]*School of Information Technology & Electrical Engineering, The University of Queensland*
[1]*India*
[2]*Australia*

1. Introduction

The demand for electrical energy has increased tremendously over the last 25 years; its importance is such that it is now a vital component of any nation's economic progress. Increase in population has increased the energy requirements, coupled with the industrialization & socio-economic responsibilities; the energy supply has not kept pace with the demand. This has led to a bleak energy scenario whereby power generation and utilization from alternate energy sources has become very much a necessity. In the industrial context, electrical energy is used to produce a desired output, given the availability of all the relevant raw materials. For a nation to progress economically, the industrial sector has to be constantly encouraged and developed, but as the number of industrial units / sectors increase, energy demand and consumption also increases. The industrial sector is the largest consumer of all the electrical energy that is commercially generated for utilization, and process industries like fertilizer, cement, sugar, textile, aluminum, paper, etc.; are extremely energy intensive in nature. To reduce the increasing demand-supply gap, either more industrial energy has to be generated or the existing consumption of energy has to be brought down without any compromise on either product quality or quantity. The rate of growth in the domestic sector easily outnumbers the industrial sector. In the event of power shortage, industrial sector gains precedence over other sectors, thereby, domestic sector always bears the brunt of long or constant power cuts. The impact of rising energy cost has a disastrous impact on the day-to-day activities of industrial and domestic consumers wherein the prices of commodities, products and even essential services tend to cost more. One option is to improve the working efficiency of the process and systems. This will ensure the reduction in the product cost in addition to efficient energy management. The other option is the use of energy derived from non-conventional energy sources i.e.; ensure a balance between conventional and non-conventional energy sources in the process. With depleting coal reserves, it is very essential to develop the renewable energy technology so as to be an alternative option. Amongst the various renewable energy sources, solar energy is the best possible option and finds application in most of the domestic and industrial processes.

2. Energy audit

With the conventional fuel supplies becoming scarce and more expensive, and the initial investment for harnessing energy from renewable sources being too high; the concept of energy auditing and energy conservation / efficiency practices have gained significant importance especially in energy intensive industries. Energy audit programmes are inexpensive investments, as compared to the cost of energy utilization, and are an important tool in analyzing and controlling the demand-supply situation. An energy audit serves to identify and quantify all forms of energy usage. The main aim of energy audit is to maintain a proper balance between energy required and the energy actually utilized; while the main objectives of energy audit are to (i) analyze the energy consumed and wasted (ii) develop ways and means to utilize the available energy in the most efficient manner by the use of energy efficient devices (iii) adopt suitable operational strategies, and (iv) time scheduling (e) demand limiting. An industrial energy audit is the most effective tool in bringing out as well as pursuing an effective energy conservation programme. In the industrial context, energy auditing is the process of identifying energy dependent equipment in the system / sub-system processes, quantifying the amount of energy consumed by each of the individual equipment and then, analyzing the data obtained to identify energy conservation opportunities. As processes vary from plant to plant and from sector-to-sector, so do the use and type of equipment; the nature and type of energy audit also varies. Thus there cannot be a standard way of doing an energy audit, but it typically involves analyzing past and present energy consumption data, comparison of actual consumption to the standard consumption data, comparison of present consumption with other firms in the same industrial sector, checking working capacities and overall efficiencies of equipment, review of fuel storage and handling, development of energy use indices for performance comparison, analysis of energy saving incentives and reviewing the need for new energy saving techniques.

All processes and hence energy audit for the process can be divided into three general stages: input side, process side and the output side [1]. The input side energy audit involves an analysis of the fuels used in terms of quantity and quality. The process energy audit involves the analysis of the process in addition to the energy consuming equipment at various stages / sub-stages individually. The output side energy audit deals with the energy that is either rejected or lost out to the surrounding environment. As a number of stages and sub-stages are involved in every process, starting from the input to the final product, the type of the energy audit now becomes very important.

The type of energy audit can be classified into a walk-through audit and a comprehensive audit. A walk-through audit takes the least possible time and generally involves moving around the facility looking for simple possible steps to minimize energy wastage or improving the system process leading to better energy utilisation ie; walk-through audit covers only a general notation of the performance. A comprehensive audit involves a longer time frame and generally involves getting into the indepth details of process, i.e.; comprehensive audit covers specific information [2]. A well designed and properly executed extensive energy audit programme will reveal the various areas of energy wastage and process inefficiencies, thereby pin-pointing the areas of immediate improvement.

The energy audit whether walk-through or comprehensive should be carried out by an energy auditor specifically designated for the purpose. The energy auditor may be from within the organization or an expert / consultant not connected with the organization, but must be well versed with the process involved. The main responsibilities of the energy auditor are:

- Plan, direct and execute an extensive energy audit process wise and system wise
- Quantify the process and system energy needs
- Quantify the process and system efficiencies

- Compare the present data with historical data
- Identify the various energy saving measures, analyze their technical / financial feasibility
- Organize the energy saving measures into low / medium / high priority and also into proposals requiring small / large investments
- Documentation regarding the energy saving proposals
- Appraise / convince the management regarding the need for implementation of energy saving measures
- Follow up on the implementation with periodic monitoring / appraisal

Investment related energy saving proposals require the consent of the management. In such cases, the decision making does not rest with energy auditor. It therefore becomes important that the energy auditor must prepare a detailed report listing out the various options available, their technical feasibility, financial requirements, time frame needed and possible gains that can be achieved. Thus the role of the energy auditor becomes very important. An energy audit report for a work area within the plant facility should essentially cover the following and the same can be extended to other units with relevant changes so as to form an energy audit report for the plant.

- Company details: name, location, power, fuel & water demand, products manufactured
- Work area details: name, dimensions, working hours, power, fuel, water & process requirements, process description, work output
- Device details: devices used, nameplate details, accessories, metering, control parameters, age, assumptions, maintenance details
- Observations: input, process & output, loading pattern, inefficiencies, wastages, comparison with historical data, possible reasons for deviations, potential opportunities
- Operational difficulties: feedback from personnel, maintenance, housekeeping, maintenance records
- Recommendations: input, process & output changes, replacements, retrofits, investments, training
- Benefits: process & product improvements, power, material & monetary savings, economic analysis, payback periods, forecasting
- Options available: tariff, efficient devices, systems, vendors, rebates, subsidies
- Probable implementation plan: time period, priorities, training requirements, investments
- References list: technical reports, handbooks, manuals
- Team details: plant work area, energy audit

The energy audit report must be simple in presentation as it should be understood by all concerned be it the president of the company or the maintenance personnel. It is also essential that the report be written in a manner that even a non-technical person should be in a position to understand the underlying message. The report should be a blend of text, figures, tables, etc. so that it is not monotonous in nature. The assumptions made during the energy audit must be easily justifiable. All standard formulas and measuring units can be a part of the annexure for reference. There must be consistency in information flow, i.e.; same variables or notations must be not be used a second time that may lead to confusion. The energy audit report can be in two parts: first part being the overall summary that highlights the important details and second part being the indepth report. This will enable the top management to concentrate only on the first part and in case of need of discussions, second part will be handy. In addition, the report should be sensible from the technical & financial point of view, i.e.; it should be encouraging enough for the management to realize that implementation will lead to better prospects. Thus a reader friendly energy audit report is an important step in initiating the implementation of audit recommendations.

Successful completion of an energy audit identifies the areas for improvement, which leads to listing out a number of energy conservation proposals. Indepth discussion with the relevant people leads to an energy management strategy which is quite significant in bridging the gap between availability and requirement of electrical power.

3. Energy management

Energy management embodies engineering, design, applications, utilization, and to some extent the operation and maintenance of electric power systems to provide the optimal use of electrical energy [2]. The most important step in the energy management process is the identification and analysis of energy conservation opportunities, thus making it a technical and management function, the focus being to monitor, record, analyze, critically examine, alter and control energy flows through systems so that energy is utilized with maximum efficiency [1]. Every industrial facility in a particular location is unique in itself; hence a systematic approach is extremely necessary for reducing the power consumption, without adversely affecting the productivity, quality of work and working conditions. Thus, for any process, energy conservation methodologies can be categorized into (i) housekeeping measures (ii) equipment and process modifications (iii) better equipment utilization and (iv) reduction of losses in building shell [3]. Thus energy management involves consumption and optimization of energy usage at various stages in the plant process in the most efficient way.

Energy management is responsibility of all involved in the industrial process but there must be person(s) specifically designated to oversee the implementation of energy efficiency proposals. Thus the role of energy manager is equally important as that of the energy auditor. The energy manager should have upto date technical skills to understand intricate technicalities of the process and excellent managerial skills in order to plan, organize, direct and control the various energy requirements. This will ensure that competency of the energy manager will not be questioned at any point in time and also, the top management can rest assured that targets set will be easily achieved. The main responsibilities of the energy manager are:

* Setting up of an energy management cell with well-defined objectives
* Generate ideas for energy management to create / promote awareness
* Initiate regular training programmes for constant knowledge updation
* Initiate steps for appropriate monitoring and recording practices
* Set targets that are realistically achievable by all concerned in the process
* Proper implementation of the energy audit findings
* Ensure that all data related to unit / plant are maintained centrally and easily accessible
* Ensure coordination between top, middle and lower management personnel
* Associate with energy managers of related industries for information exchange
* Ensure easy information flow through proper communication

An energy manager's report for a work area within the plant facility should concentrate on the findings of the energy audit report, take into account the historical data and set realistic benchmarks / targets that contribute significantly towards energy efficiency. The reports prepared must be shared will all concerned especially with energy auditors. This will reassure the energy auditors that their reports are taken seriously and due importance / credit are attached to the work done. In short, the energy manager should be the bridge between the top management and unit personnel.

Demand side management (DSM) is an important policy issue in recent years in the context of energy management. DSM aims to (i) minimize energy consumption (ii) reduce maximum demand (iii) promote use of electricity to reduce green house gas emissions (iv) replacement of bio fuel by commercial energy to stop deforestation. The demand reduction management can be practiced by (i) efficient utilization of existing capacity (ii) reduction in transmission and distribution losses, and (iii) effective peak demand management. These methods are most appropriate for reducing the power bills and to meet the requirements of high quality of power. Implementation of DSM projects encourages introduction of energy efficient technology and equipment in all sectors.

4. Energy Management Strategies (EMS)

An energy management programme will be effective when the concerned persons are taken into confidence; awareness is created regarding the need / importance of the process and responsibility assigned so as to ensure team work. It is also important to understand that the EMS will not be same across the plant; rather it will be device specific. The first step in the development of an EMS is by forming a committee comprising of energy manager as the head, energy auditor, plant / unit manager, plant / unit personnel and maintenance personnel. The first meeting should basically discuss (a) energy auditor's findings (b) best operational procedures with unit and maintenance personnel (c) allotment of responsibilities (d) setting realistic benchmarks and targets (e) plan the work schedule. Subsequent meetings should discuss energy audit observations, targets set and targets achieved for energy management. A critical analysis of deviations in the targets set / achieved must be carried out and EMS reworked if necessary. The committee should meet on a regular basis perhaps every 10 days so as to ensure close monitoring. In addition to the above, it is very important to develop and motivate the concerned persons by upgrading their knowledge through regular workshops / training programmes. It is also important that competitions be conducted amongst the various units in the plant and incentives / awards be instituted for units / employees for best results achieved. In addition the EMS should be the basis for a comprehensive energy policy that will be implemented across the plant irrespective of the process involved.

Some of the commonly used equipment used in energy intensive processes are boiler systems, steam systems, refractories, furnaces, motor driven systems, compressed air systems, heating, ventilation and air conditioning systems, fans, blowers, pumping systems, cooling towers, illumination systems, diesel generators, etc. The energy management strategy should basically concentrate on:

- Boiler Systems: fuel, steam pressure, temperature, fans, blow down, ash handling, efficiency, heat loss, leaks, handling systems, dust collection, waste heat recovery, maintenance schedules, heat recovery, insulation requirements, feed water, piping, ventilation, economizers, air-preheaters, etc.
- Diesel Generators: fuel, quality, heat, exhaust, load, maintenance, etc.
- Electricity: magnitudes, frequency, quality, tariff, metering, power factor, load curve
- Energy Storage Systems: insulation, temperature, control, maintenance, etc.
- Furnaces: losses, conveyor, fixtures, storage, insulation, temperature, heaters, lining, ducts, coils, cover, temperature, slag, water, heat recovery, burners, etc.

- Heating Systems: temperature, ventilation, piping, controls, insulation, maintenance, load profiles, storage, etc.
- Illumination Systems: adequacy, luminaire, glare, sensors, standards, day lighting, control, maintenance, lamps, ballasts, etc.
- Instrumentation: analog, digital, calibration, panels, CTs, PTs, etc.
- Motor Systems: pumps, air compressors, fans, piping, volume, pressure, temperature, dust, control, ducts, leakage, nozzles, efficiency, loading, drive systems, class, instrumentation, etc.
- Refrigeration and Air Conditioning Systems: heat, load, windows, temperature, thermostats, air, illumination, insulation, ducts, piping, evaporators, condensers, heat exchangers, vapour, control, maintenance, etc.
- Steam Systems: pressure, temperature, superheating, piping, condensate recovery, leaks, steam traps, venting, maintenance, insulation, valves, etc.
- Ventilation Systems: air handling, thermal insulations, distribution, blockages, leakages, maintenance, control, heat recovery, etc.

All data have to be recorded and maintained for future reference. These facts and figures do give a fair idea about the pattern of energy consumption and its cost per unit of the finished product. As energy consumption is directly related to production rate, the energy consumed for every finished product can be used as a reference index. When sufficient amount of data has been built up over a period, the records then have to be converted into meaningful forms. Pictorial representations in the form of bar charts, pie charts and Sankey diagrams showing energy use and energy lost, process flow diagrams showing energy consumption at every stages of the operational process, etc. will go a long way in identifying the areas of high energy consumption, high costs of operation and in turn, the energy saving potential.

5. Renewable energy

In the previous century, the industrial revolution was powered by coal leading to setting up of large power plants as it was the only reliable source of energy available in abundance. Over the years, oil replaced coal as it was the cleaner form of fuel leading to increased industrialization. Due to increased usage of coal and oil in the name of economic development, environmental problem has started to put a lid on economic progress. The environmental concerns of fossil fuel power plants are due to sulfur oxides, nitrogen oxides, ozone depletion, acid rain, carbon dioxide and ash. The environmental concerns of hydroelectric power plants are flooding, quality, silt, oxygen depletion, nitrogen, etc. The environmental concerns of nuclear power plants are radioactive release, loss of coolant, reactor damage, radioactive waste disposal, etc. The environmental concerns of diesel power plants are noise, heat, vibrations, exhaust gases, etc. Finding and developing energy sources that are clean and sustainable is the challenge in the coming days.

Considering the depleting coal reserves, increasing power demand, cost of fuels and power generation, the power generating capacity can only be increased by involving renewable energy sources. The renewable energy source produce less pollution and are constantly replenished which is quite an advantage. Due to the future need of increasing power requirements, research has led to development of technology for efficient and reliable renewable energy systems. The various forms of renewable energy sources are solar, wind, biomass, tidal, fuel cells, geothermal, etc. The main advantages of renewable energy sources are sustainability, availability and pollution free. The disadvantages of renewable energy are variability, low density and higher cost of conversion. In order to sustain the present sources, the future energy will be mix of available energy sources utilised from multiple sources. This will ensure that the environment will be a lot less polluted. Renewable energy is the future from here on.

Among the various renewable energy sources, solar energy is the best usable source as the sun is the primary source of energy and the earth receives almost 90 % of its total energy from the sun. In one hour, the earth receives enough energy from the sun to meet its energy needs for almost a year. Solar energy can be converted through chemical, electrical or thermal processes. Solar radiation can be converted into heat and electricity using thermal and photovoltaic (PV) technologies. The thermal systems are used for hot water requirements, cooking, heating etc., while PV are used to generate electricity for standalone systems or fed into the grid. Solar energy has a lost economic, energy security and environmental benefits when compared to conventional energy for certain applications. Solar power is a cost effective solution to generate and supply power for a variety of applications, from small stand alone systems to large utility grid-tied installations. The conversion of solar energy requires certain equipment that have a relatively high initial cost but considering the lifetime of the solar equipment, these systems can be cost competitive as there are no major recurring cost and minimal maintenance cost. Even though solar energy systems have a reasonably high initial cost; they do not have fuel requirements and often require little maintenance. Hence the life cycle costs of a solar energy system should be understood for economic viability of the PV system. The important factors to be considered for a renewable energy system are power requirements, source availability, system type, system size, initial cost, operation cost, maintenance cost, depreciation, subsidies etc.

Grid connected PV system gives us the option to reduce the electricity consumption from the electricity grid and in some instances, to feed the surplus energy back into the electrical grid. The grid connected PV systems distinguish themselves through the lack of a need for energy storage device such as a battery. The basic building block of PV technology is the solar cell. Many solar cells can be wired together to form a PV module and many PV modules are linked together to form a PV array. A PV system usually consists of one or more PV modules connected to an inverter that changes the PV's DC to AC, not only to power our electrical devices that use alternating current (AC) but also to be compatible with the electrical grid.

Cogeneration is the conversion of energy into multiple usable forms. The cogeneration plant may be within the industrial facility and may serve one or more users. The advantages of cogeneration are fuel economy, lower capital costs, lower operational costs and better quality of supply.

6. Power quality

To overcome power shortage in addition to increasing power demand, industrial sectors are encouraged to adopt energy efficiency measures. Process automation involves extensive use of computer systems and adjustable speed drives (ASDs), power quality (PQ) has become a serious issue especially for industrial consumers. Power quality disturbances are a result of various events that are internal and external to industrial utilities. Because of interconnection of grid network, internal PQ problems of one utility become external PQ problems for the other.

The term power quality has been defined and interpreted in a number of ways: As per IEEE Std 1159, PQ refers to a wide variety of electromagnetic phenomena that characterize the voltage and current at a given time and at a given location on the power system [4]. As per

IEC 61000-1-1, electromagnetic compatibility is the ability of an equipment or system to function satisfactorily in its electromagnetic environment without introducing intolerable electromagnetic disturbances to anything in that environment [5]. In simple terms, power quality is considered to be a combination of voltage quality and current quality, and is mainly attributed to the deviation of these quantities from the ideal. Such a deviation is termed as power quality phenomena or power quality disturbance, which can be further divided into phenomenon: variations and events. Variations are small deviations away from the nominal or desired value involving voltage and current magnitude variations, voltage frequency variations, voltage and current unbalance, voltage fluctuations, harmonic voltage and current distortions, periodic voltage notching, etc. Events are phenomena that happen every once in a while involving interruption, under voltages, overvoltage, transients, phase angle jumps and three-phase unbalance [6].

The PQ problems can originate from the source side or the end user side. The source side of PQ disturbances involves events such as circuit breaker switching, reclosures, pf improvement capacitors, lightning strike, faults, etc., while the end user side of PQ disturbances involves non-linear loads, pf improvement capacitors, poor wiring & grounding techniques, electromagnetic interference, static electricity, etc. The effects of PQ disturbance depend upon the type of load and are of varied nature. Computers hang up leading to data loss, illumination systems often dim or flicker, measuring instruments give erroneous readings, communication systems experience noise, industrial process making use of adjustable speed drives inject harmonics as well as experience frequent shutdowns [7].

Industrial utilities need good PQ at all times as it vital to economic viability. The end users need standards that mainly set the limits for electrical disturbances and generated harmonics. The various organizations that publish power quality standards are ANSI (Steady State Voltage ratings), CENELEC (Regional Standards), CISPR (International Standards), EPRI (Signature newsletter on power quality standards), IEC (International Standards), IEEE (International and United States standards color book series).

There are generally two methods towards correction of PQ problems. The first method is load conditioning, wherein the balancing is done in such a manner that the equipments are made less sensitive to power disturbances and the other method is to install conditioning systems that either suppresses or opposes the disturbances. Active power filters offer an excellent solution towards voltage quality problem mitigation and can be classified into series active power filters and shunt active power filters. The selection of the type of active power filter to improve power quality depends on the type of the problem.

7. Economic analysis

With limited capacity addition taking place over the years, industrial utilities are forced to go for various energy management strategies. This may require additional financial commitment to achieve significant savings. The Life Cycle Cost (LCC) method is the most commonly accepted method for assessment of the economic benefits over their lifetime. The method is used to evaluate at least two alternatives for a given project of which only one alternative is selected for implementation based on the result of the economic analysis. In other words, LCC is the evaluation of a proposal over a reasonable time period considering all possible costs in addition to the time value of money. The initial investment made is called the capital cost while the equipment has a salvage value when it is sold. The additional investments exist in the form of recurring costs such as maintenance and energy usage. These costs are grouped as annual costs and expressed in a form that can be added directly to the capital cost. The capital cost can be segregated into two components: direct costs and indirect costs. Direct costs are monetary expenditures that can be directly assigned to the project such as material, labor for design and construction, start-up costs while indirect costs or overheads are expenditures that cannot be directly assigned to a project such as taxes, rent, employee benefits, management, corporate offices, etc. The capital cost now represents the total expenditure.

Economic analysis is an important step in the energy management process as they greatly influence decisions with regard to plant operations [2]. Though there are a number of economic models available for investment justification, LCC analysis is more advisable to be used as it takes into consideration the useful period of the equipment taking into account all costs and also the time value of money, and converting them to current costs. LCC is the evaluation of a proposal over a reasonable time period considering all pertinent costs and the time value of money, and is usually tailor made to suit specific requirement.

$$\text{As in [2], Total LCC} = PW_{CL} + PW_{OC} \tag{1}$$

PW_{CL} is the present worth of capital and installation cost given by

$$PW_{CL} = IC + (IC \times FWF \times PWF) \tag{2}$$

IC is the initial cost; FWF is the future worth factor; PWF is the present worth factor.

$$FWF = \text{future worth factor} = (1 + Inf)^N \tag{3}$$

Inf is the rate of inflation; N is the operating life in years.

$$PWF = \text{present worth factor} = 1/ (1+DR) \tag{4}$$

DR is the discount rate
As in [8], LCC can be represented in general mathematical form as

$$LCC (P_1, P_2, \ldots) = IC (P_1, P_2, \ldots) + ECC (P_1, P_2, \ldots) \tag{5}$$

IC is the initial cost of investment; ECC is the energy consumption cost; P_1, P_2, …are a set of design parameters.
As in [9], LCC can be mathematically expressed as for a specific case for motor options is

$$LCC = PP + [C \times N \times PWF] \times P_{LOSS} \tag{6}$$

PP is the purchase price; C is the power cost; N is the annual operating time; PWF is the cumulative present worth factor; P_{LOSS} is the evaluated loss
As in [10], LCC can also be expressed as

$$LCC = C_{IC} + C_{IN} + C_E + C_O + C_M + C_S + C_{ENV} + C_D \tag{7}$$

C_{IC} is the initial cost; C_{IN} is the installation and commissioning cost; C_E is the energy cost; C_O is the operating cost; C_M is the maintenance and repair cost; C_S is the down time cost; C_{ENV} is the environmental cost and C_D is the disposal cost.

LCC is the total discounted cost of owning, operating, maintaining, and disposing of equipment over a period of time. Thus the various components of LCC are:

a. Initial & Future Expenses: Initial expenses are all costs incurred prior to occupation of the facility while future expenses are all costs incurred after occupation of the facility.

b. Residual Value: Residual value is the net worth of a building at the end of the study period.
c. Study Period: The study period is the period of time over which ownership and operations expenses are to be evaluated.
d. Real Discount Rate: The discount rate is the rate of interest reflecting the investor's time value of money. Discount rates can be further separated into two types: real discount rates and nominal discount rates. The difference between the two is that the real discount rate excludes the rate of inflation and the nominal discount rate includes the rate of inflation.
e. Present Value: Present value is the time-equivalent value of past, present or future cash flows as of the beginning of the base year. The present value calculation uses the discount rate and the time a cost was or will be incurred to establish the present value of the cost in the base year of the study period.
f. Capital Investment: The amount of money invested in a project or a piece of equipment (this includes labor, material, design, etc.)

The LCC process involves the following steps:

1. Define cost analysis goals: This involves analysis objectives, identification of critical parameters and the various problems in analysis.
2. Identify guidelines and constraints: This involves evaluation of the available resources, determination of schedule constraints, management policy and technical constraints involved.
3. Identify feasible alternatives: This involves identification of all available options, practical and non-practical options.
4. Develop cost breakdown structure: This involves identification of all LCC elements, cost categories and their break downs.
5. Select / develop cost models: This involves identification of available cost models and construction of new models if necessary.
6. Developing cost estimating relationships: This involves identification of the input and supporting data.
7. Develop Life Cycle Cost profile: This involves identification of all present and future based cost related activities taking into consideration the inflationary effects.
8. Perform sensitivity analysis: This involves analysis of important parameters and its impact on overall cost and LCC.
9. Select best value alternatives: This involves choosing the best alternative that maximizes reliability with minimal cost.

Thus the life cycle cost is now written for specific situations taking into consideration all possible relevant parameters that need to support economic decisions regarding the various possible energy management options.

8. Energy Management Information Systems (EMIS)

EMIS is an IT based specialized software application solution that enables regular energy data gathering and analysis, used as a tool for continuous energy management. The main advantage of an EMIS application is the possibility of data collection, processing, maintenance, analysis and display on a continuous basis. A modern EMIS is integrated into an organization's systems for online process monitoring and control. An EMIS provides sensitive information to manage energy use in all aspects and is therefore an important element of an energy management programme. The nature of the EMIS will depend on company, inputs, process, products, cost incurred, instrumentation, control systems, historical data, reporting systems, etc. The EMIS should provide a breakdown of energy use and cost by product / process at various levels to improve process, systems and achieve cost control. The information generated by an EMIS enables actions that create financial value through proper energy management and control. An EMIS can be effectively used for benchmarking energy usage to achieve cost control. Benchmarking can be defined as a systematic approach for comparing the performance of processes in the present state with the best possible results without reduction in quality or quantity. It is a positive step in achieving targets that would ensure process improvement. The various steps involved in benchmarking are:

1. For the similar process, obtain the best possible result from various sources and set as reference
2. Compare the working result with the reference result and analyze them for deviations
3. Present the findings to the personnel involved and discuss the options for sustained improvement
4. Develop action plans and assign responsibilities
5. Implement plans with regular monitoring

The success of EMIS depends upon management, policies, systems, project, investment, etc. Implementation of an EMIS should lead to early detection of early detection of deviations from historical energy usages thereby identification of energy management proposals, budgeting, implementation schedules, etc. It is important to recognize that the EMIS brings process and system benefits in addition to financial benefits.

9. Energy policy

An organization should show its commitment to energy management by having a well-defined energy policy. The energy policy should of some purpose and should be motivating enough for all employees to contribute towards achieving the organizational goals. The energy policy should essentially contain the following:

- Energy policy statement of purpose
- Objectives of the energy policy
- Commitment and involvement of employees
- Action plan with targets for every process and systems
- Budget allocation for various activities
- Responsibility and accountability at all levels

The policy should take into account the nature of the work, process, systems in use in addition to the work culture of the organization. The draft policy should be circulated amongst the employees for their inputs. Having taken all the employees into the process of energy policy formulation, the final version of the document should be approved by the top management and circulated within the organization for implementation. The above energy policy may be a summarized version and a detailed version. The summarized version should be displayed at various important locations while the detailed version should be filed as a hard copy in the various departments / units and sent as a email to all employees.

It is important to understand that the goals and objectives defined in the energy policy must be achievable. The energy policy implementation must be periodically reviewed and the expected outcomes compared with the results achieved. Wide deviations in the results should lead to a review of the process and systems in place in addition to the energy policy.

10. Conclusions

With increasing energy prices directly impacting the product prices in addition to widening energy demand-supply gap, industries are encouraged to go in for energy saving in addition to use of multiple energy sources. This can be accurately gauged by having an appropriate energy audit. A good and comprehensive energy audit will lead to a list of energy saving options that can be adopted. A detailed discussion on the audit findings leads to an energy management program. Some of the energy saving options requires additional investment. For major investments, life cycle cost (LCC) analysis is a useful tool as it evaluates a proposal over a reasonable time period considering all pertinent costs and the time value of money. It is also important to remember that introducing renewable energy sources into the process needs additional systems that concerns power quality issues. Energy management information system (EMIS) is an IT based specialized software application solution that enables regular energy data gathering and analysis used as a tool for continuous energy management. An EMIS provides sensitive information to manage energy use in all aspects and is therefore an important element of an energy management programme. All organization should show its commitment to energy management by having a well-defined energy policy. The energy policy should be definitive, straight-forward and motivating enough for all employees to contribute towards achieving the organizational goals. Thus energy management in industrial utilities is the identification and implementation of energy conservation opportunities, making it a technical and management function, thus requiring the involvement of all employees so that energy is utilized with maximum efficiency

11. References

[1] P. O'Callaghan, "Energy management: A comprehensive guide to reducing costs by efficient energy use", *McGraw Hill*, London, UK, 1992.

[2] IEEE Std. 739-1995, *IEEE Recommended practice for energy management in industrial and commercial facilities.*

[3] W. Lee and R. Kenarangui, "Energy management for motors, systems, and electrical equipment", *IEEE Transactions on Industry Applications*, vol. 38, no. 2, Mar./Apr. 2002, pp. 602-607.

[4] IEEE Recommended Practice for Monitoring Electric Power Quality, IEEE Std 1159-1995.

[5] IEC Standards on Electromagnetic Compatibility, IEC 61000-1-1.

[6] M.H.J. Bollen, *Understanding Power Quality Problems*, IEEE Press, New York, 2000, ISBN: 81-86308-84-9.

[7] B. Kennedy, Power Quality Primer, McGraw Hill, New York, 2000, ISBN: 0-07-134416.

[8] Canova, F. Profumo and M. Tartaglia, "LCC design Criteria in electrical plants oriented to energy saving", *IEEE Trans. Industry Applications*, vol. 39, no. 1, Jan./Feb. 2003, pp. 53-58.

[9] P.S. Hamer, D.M. Lowe, and S.E. Wallace, "Energy efficient induction motors performance characteristics and life cycle cost comparisons for centrifugal loads", *IEEE Trans. Industry Applications*, vol. 33, no. 5, Sept./Oct. 1997, pp. 1312–1320.

[10] Pump Life cycle cost: A guide to LCC analysis for pumping systems, Executive Summary, The Hydraulic Institute, New Jersey USA.

Methodology Development for a Comprehensive and Cost-Effective Energy Management in Industrial Plants

Capobianchi Simona[1], Andreassi Luca[2], Introna Vito[2],
Martini Fabrizio[1] and Ubertini Stefano[3]
[1]*Green Energy Plus Srl*
[2]*University of Rome "Tor Vergata"*
[3]*University of Naples "Parthenope"*
Italy

1. Introduction

Energy management can be defined as "the judicious and effective use of energy to maximise profits and to enhance competitive positions through organisational measures and optimisation of energy efficiency in the process " (Cape, 1997). Profits maximization can be also achieved with a cost reduction paying attention to the energy costs during each productive phase (in general the three most important operational costs are those for materials, labour and electrical and thermal energy) (Demirbas, 2001). Moreover, the improvement of competitiveness is not limited to the reduction of sensible costs, but can be achieved also with an opportune management of energy costs which can increase the flexibility and compliance to the changes of market and international environmental regulations (Barbiroli, 1996). Energy management is a well structured process that is both technical and managerial in nature. Using techniques and principles from both fields, energy management monitors, records, investigates, analyzes, changes, and controls energy using systems within the organization. It should guarantee that these systems are supplied with all the energy that they need as efficiently as possible, at the time and in the form they need and at the lowest possible cost (Petrecca, 1992).

A comprehensive energy management programme is not purely technical, and its introduction also implies a new management discipline. It is multidisciplinary in nature, and it combines the skills of engineering, management and maintenance. In literature there are many authors that approaching the different aspects of energy management in industries. For sake of simplicity, identifying the main issues of the energy management procedure in energy prices, energy monitoring, energy control and power systems optimal management and design, in Table 1, for every branch the most significant scientific results are listed.

Concerning energy price in the new competitive environment due to the energy markets liberalization, many authors face up the risks emerged for market participants, on either side of the market, unknown in the previous regulated area. Long-term contracts, like futures or forwards, traded at power exchanges and bilaterally over-the-counter, allow for price risk management by effectively locking in a fixed price and therefore avoiding uncertain future spot prices. In fact, electricity spot prices are characterised by high volatility and occasional spikes (Cesarotti et al., 2007), (Skantze et al., 2000), (Weron, 2008). Moreover finding the best tariff for an industrial plant presents great difficulties, in particular due to the necessity of a predictive consumption model for adapting the bids to the real consumption trends of the plants.

Energy management Areas	Main Issues	Bibliography
Energy costs	Forecasting price of energy Renewal of contracts	(Cesarotti et al., 2007), (Skantze et al., 2000), (Weron, 2008)
Energy budgeting	Forecasting consumption Monitoring and analyzing deviations from the energy budget	(Farla & Blok, 2000), (Worrel at al., 1997), (Kannan & Boie, 2003), (Cesarotti et al., 2009)
Energy consumption control	Design and implementing monitoring system Forecasting and control consumption of specific users	(Brandemuel & Braun, 1999), (Elovitz, 1995), (Krakow et al., 2000), (Di Silvio et al., 2007)
Optimization of power systems	Defining the equipments optimal set points Increasing the overall system efficiency	(Sarimveis et al., 2003), (Arivalgan et al., 2000), (Von Spakovsky et al., 1995), (Frangopoulos et al., 1996), (Puttgen & MacGregor, 1996), (Tstsaronis & Winhold, 1985), (Temir & Bilge, 2004), (Tstsaronis & Pisa, 1994)

Table 1. Energy management open issues

Several studies on energy monitoring by using physical indicators to analyse energy efficiency developments in the manufacturing industry (especially the energy-intensive manufacturing industry) highlight the close relationship with the concept of specific energy consumption (energy use at the process level) and the international comparability of the resulting energy efficiency indicators as arguments advocating the use of physical indicators in the manufacturing industry (Farla & Blok, 2000), (Worrel at al., 1997). Moreover in (Kannan & Boie, 2003) the authors illustrate the methodology of energy management that was introduced in a German bakery with a clear and consistent path toward introducing energy management. Finally in (Cesarotti et al., 2009) the authors provide a method for planning and controlling energy budgets for an industrial plant. The developed method aims to obtain a very high confidence of predicted electrical energy cost to include into the estimation of budget and a continuous control of energy consumption and cost.

The energy control for specific systems is mainly focused on implementing one energy management control function at a time with or without optimal control algorithms (Brandemuel & Braun, 1999), (Elovitz, 1995), (Krakow et al., 2000). In (Di Silvio et al., 2007) a method for condition-based preventive maintenance based on energy monitoring and control system is proposed. The methodology supports to identify maintenance condition through energy consumption characterization, predicting and control (Cesarotti et al., 2010). In (Sarimveis et al., 2003) an example of power systems management optimization through mathematical programming tools is presented. In other terms, the availability of optimization tools for the energy plant operation (i.e. the possibility of optimally determining when boilers, turbines, chillers or other types of machinery shall be set on or off or partialized) may lead to energetic, economic and environmental savings. In scientific literature, several criteria for the optimization of combined cooling, heating and power systems in industrial plants are available based on different management hypotheses and objective functions. The goal of the models is to optimize the operation of the energy system to maximize the return on invested capital. Many of these models do account for load operations but use simple linear relationships to describe thermodynamic and heat transfer process that can be inherently non-linear. In (Arivalgan et al., 2000) a mixed-integer linear programming model to optimize the operation of a paper mill is presented. It is demonstrated that the model provides the methods for determining the optimal strategy that minimize the overall cost of energy for the process industry. In (Von Spakovsky et al., 1995) the authors use a mixed integer linear programming approach which balances the competing costs of operation and minimizes these costs subject to the operational constraints placed on the system. The main issue of the model is the capability to predict the best operating strategy for any given day. Nevertheless, the model validity is strictly dependent on the linear behaviour of the plant components. In (Frangopoulos et al., 1996) the authors have employed linear programming techniques to develop an optimization procedure of the energy system supported by a thermoeconomic analysis of the system and modelling of the main components performance. In (Puttgen & MacGregor, 1996) a linear programming based model maximizing the total revenue subject to constraints due to conservation of mass, thermal storage restrictions and shiftable loads requirement is developed. Finally, thermoeconomics offers the most comprehensive theoretical approach to the analysis of energy systems where costs are concerned. It is based on the assumption that exergy is the only rational basis to assign cost. In other terms, the main issue is that costs occur and are directly related to the irreversibility taking place within each component. Accordingly thermoeconomics could represent a reliable approach to the optimisation of energy plants operation involving thermodynamic and economical aspects (Tstsaronis & Winhold, 1985), (Temir & Bilge, 2004), (Tstsaronis & Pisa, 1994).

However, these studies have paid little attention in integrating the different individual energy management functions into one overall system. From this point of view, in this chapter we provide a comprehensive integrated methodology for implementing an automated energy management in an industrial plant.

2. Background and motivation

In the last decade, energy management has undergone distinct phases representing different approaches (Piper, 2000):

- Quick fixes: facing with rapidly escalating costs and the prospect of closings resulting from energy shortages, facility managers responded by implementing a round of energy conservation measures.
- Energy projects: once a fairly wide range of quick fixes had been implemented, facility managers came to realize that additional savings would require the implementation of

energy conservation activities, which are expensive and time-consuming. The emphasis shifted from quick fixes to energy projects.

- Energy management system: to fight these rising costs, organizations developed more comprehensive approaches to energy management moving from simply reducing energy consumption to managing energy use.

Organizations (both national governments and industrial companies) are recognizing the value and the need of energy management. If they are to be successful, they must understand what worked in the past and why, and what did not work and why it failed. In the last few years, some energy management models have been developed inspired by quality and environment management systems (ISO 9001). For this purpose, in 2005, the ANSI set up and published the first regulation concerning energy management system: the MSE (management system for energy), published by the American National Standard Institute. The objective of this standard is the definition of a reliable model which can be used in different scenarios, to promote the reduction of the energy costs/product unit ratio. The model/standard has to manage all kind of energy costs, in each step of the energy supply chain: supply, transformation, delivery and use. In other words, the application of this standard means setting up programs to manage energy use, instead of randomly funding energy saving projects. In this scenario the energy saving should be performed through a systematic approach operating on energy costs, energy budget preparation, measure and control of power consumption and energy production and conversion. Energy saving can be, in fact, realized through different actions on both the utilization and the production sides. However, it is really a complex task as many factors influence energy usage, conversion and consumptions. Moreover, these factors are strictly interconnected. For example, when evaluating an action on the energy consumption/production, one should take care of the interactions, as one measure influences the saving effect of the other measures. Therefore, it is important to highlight that each element of this systematic approach is strictly connected to the others, as explained in the following.

First of all in the process of renewing the energy supply contract, it is necessary to compare several rate proposals, as in the electricity and fuel market there are a number of different suppliers. This comparison is quite difficult for two reasons. Firstly, the energy rate depends on numerous factors and is usually made up of many different voices. Secondly, although the rate per kWh may be disguised in the electric bill, it varies in function of time and/or power request. This means that the consumption profile has to be known in order to make a prevision on what one is going to pay.

As making this consumption profile on historical data may lead to wrong predictions and non-economic actions and, considering that the annual energy cost is significantly affected by the chosen rate, an energy consumption model should be built. This means modeling the industrial plant energy consumption in function of its major affecting factors (i.e. energy drivers), as production volume, temperature, daylight length etc. This model should give the expected consumption in function of time and the time-step should be as small as possible in order to have reliable predictions. By this way it could be possible to distinguish the plant consumption and the energy drivers variation within the time bands of the energy rate. This could be done by installing a measuring system to record energy consumption and energy drivers. The meters position within the plant is particularly important to correlate the energy consumption to the energy drivers (i.e. different production lines). Therefore, a preliminary analysis based, for example, on the nominal power and the utilization factor of the single machines should be performed in order to build a meters tree.

A reliable energy budget formulation is needed, not only as a part of the whole plant budget, but also to define possible future investments on the energy sector. The present methodology allows to build the energy budget on the predicted energy profile and not only on the historical data basis, as usually done, thus taking into account the possible variation of the energy drivers and of the energy price. The latter could be optimized as described in the previous paragraph and correlated to indexes, as for example the oil market price. Moreover, if an energy system is present in the plant, the budget could not be built on the basis of the previous consumption profile, as the quantity of electricity drawn from the public network could vary as the self-production varies in function of the utilization of the energy system itself (i.e. the optimization of the energy system management as a part of the present methodology).

As far as the possible investments on energy saving are concerned, a correct measure and control of energy consumption is crucial. First of all the energy use measurement alone is not enough, as the predictive model requires correlating energy consumption with several energy drivers that should be accurately and frequently collected, making different measures in different plant areas. This would allow, in fact, to better correlate the consumption to the production on one hand and to undertake energy saving operations specifically designed in each zone on the other. Besides, it is worth to note that the predicted consumption should be compared to reference values in order to understand if the industrial plant is efficient or not.

Finally, an optimal energy management methodology should take into account the management of the energy system machines of the industrial plant, which means setting the load of the energy conversion equipments (i.e. boilers, air-conditioning systems and refrigerators, thermal engines) that optimizes energy cost with a given energy consumption profile (both electrical and thermal). Usually these small energy systems are operated simply switching on and off the machines for long time intervals (i.e. night and day, winter and summer). However, the machines typically used in these systems have small thermal inertia, thus allowing quick load variation, and may be operated under partial load. As demonstrated by the authors in (Andreassi et al., 2009), the energy system model together with the energy consumption one may lead to an optimal management of the power plant thus reducing energy costs. This, again requires a detailed energy consumption profile and then an accurate data collection system.

Besides, on the wake of the previous models, the CEN-CENELEC elaborated the EN 16001, published in July 2009, with the reference standards for the Energy Management System. The rule covers the phases of purchasing, storing and use of the energy resources in different type of organizations (industrial, commercial, tertiary). As the ISO 9001 and ISO 14001, the rule is based on Deming Cycle and the Plan-Do-Check-Act approach.

The EN 16001 has the aim of specifying the requirements of an Energy Management System. The adoption and the maintaining of this standard demonstrates a concrete commitment for the rationalization and the "intelligent" management of the energy resources.

Moreover the ISO Project Committee ISO/PC242 is working to publish an International Standard for Energy Management named ISO 50001. Probably this will be the more important standard for Energy Management for the next years. By now the final version of ISO 50001 is due to be released in the third quarter 2011.

Starting from these critical issues, in this chapter, a methodology considering energy management in a comprehensive manner is provided. A method for energy efficiency based on a systematic approach for energy consumption/cost reduction, which could simultaneously keep into proper account all the critical aspects just pointed out, is proposed.

3. Methodology for a comprehensive energy management

The methodology framework is shown in Figure 1. The single steps have been discussed in detail by the authors in previous papers (Cesarotti et al., 2007), (Cesarotti et al., 2009), (Di Silvio et al., 2007), (Andreassi et al., 2009). In this chapter the whole methodology and the importance of links and interconnections among the different phases and their role in reducing costs are highlighted.

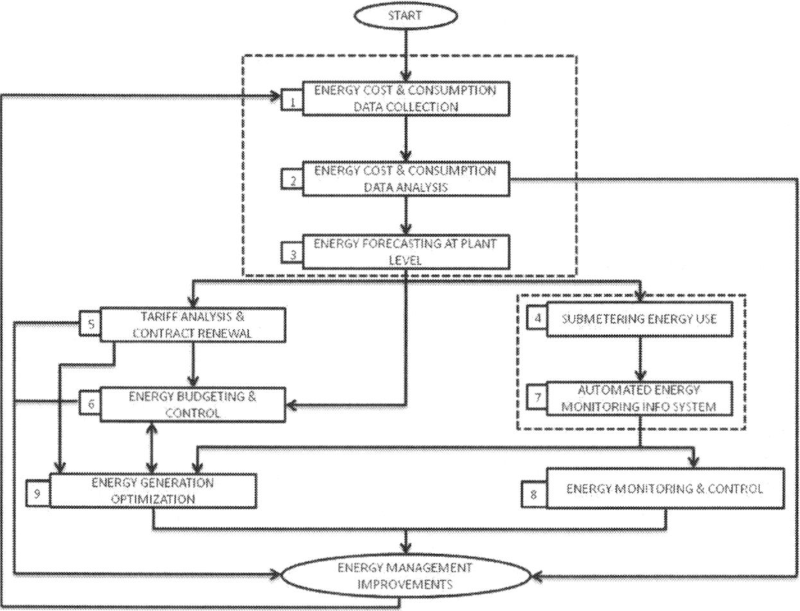

Fig. 1. Framework of the proposed methodology for Energy Management improvements

The main issues of the proposed methodology are: historical data analysis, energy consumption characterization, energy consumption forecasting, energy consumption control, energy budgeting and energy machines management optimization. The methodology supports an industrial plant to:
- identify areas of energy wastage - for example by determining the proportion of energy that does not directly contribute to production and that is often a source of energy savings;
- understand energy consumption of the processes - by establishing a relationship between energy use and production;
- highlight changes to energy consumption patterns - these are either a result of a specific action to improve efficiency or due to an unknown factor which may have a detrimental effect upon efficiency and may lead to process failure or poor quality product;

- identify sporadic faults or events - by alerting operators if excursions from normal, or predicted, production performance are observed;
- reach an optimal condition in terms of supplying, generation, distribution and utilization of energy in a plant by means of a continuous improvement approach based on energy action cost- benefit evaluation.

The single operation described in the methodology steps has its own effectiveness in a context showing an awareness lack about energy management concept. Nevertheless, our intent is to point out the importance of introducing each step in a non-ending loop, granting continuous energy management improvements and a constant reduction of energy consumptions and costs.

Accordingly, in the following sections each step characterizing the proposed methodology will be described in detail. The different phases are:

- energy cost & consumption data collection;
- energy cost & consumption data analysis;
- energy forecasting at plant level;
- sub-metering energy use;
- tariff analysis and contract renewal;
- energy budgeting and control;
- energy monitoring and control;
- power plant management optimization;

Every step is deeply analyzed in the successive paragraph and an application of each of these steps is shown in the case study of the paragraph 6: this working example will support the explanation of the various aspects of the developed methodology.

4. Description of the methodology steps

4.1 Energy cost & consumption data collection

The first step consists in collecting useful data for characterizing the energy consumptions of an industrial plant. We can essentially distinguish four types of variables which can be collected:

- consumption data;
- production data;
- environmental data;
- technical (users) and operational data.

In general there are four stages in data collection: i) using already collected data, without any further modification; ii) modifying the way of collecting data previously employed in the industrial plant; iii) manually collecting further data; iv) establishing an automatic data acquisition system. Most of the core data on production are usually being gathered for other purposes (e.g. cost and production control), and some analyses should already have been done to determine which information is gathered, by whom, how, and why. Sharing this information for energy monitoring purposes may require modifications to enable a more effective energy monitoring. Its impact on other management functions should be considered it may, or may not, be beneficial.

The energy bills are the primary source of information for the consumption data. They are the first point of reference when trying to understand what is being used, as well as how the organization is being measured and charged. In particular:

- for oil and coal the invoices report information on deliveries and consumptions. It is then necessary to take account of stocks. If stocks are not already recorded, it is important to guarantee somehow the suitability of the data. At the same time, a system for recording the stock before delivery has to be introduced.
- for gas and electricity, the information that appears on the bill depends on the tariff type.

The production information can be divided into three types:

- information on production that relates to amount as weight, volume, number of items, area (waiting time and productive hours fall into this category, as the climate measurement in heating or cooling degree days);
- information on production that does not relate to amount as temperature, density, water content, ratios of constituents (e.g. fat to solid ratios in fried food);
- ancillary information as, for example, breakdown causes, occasional notes and comments.

The first one of these is distinguishable from the other three because items of information are additive; in other terms, information for a week can be obtained by adding daily information. Information of the second type is not additive. In some cases monitoring and targeting can achieve adequate resolution only if information of this second type is utilized. Information which is not additive is difficult to summarize and this is often reflected in the way it is handled in organizations. It is more likely to be hand-written, with few checks on its accuracy, and archived without being processed (Carbon Trust, Practical guide 112).

About the environmental data, they usually can be:

- the sunlight variation for electrical energy for lighting; for these data we could refer to meteorology web sites or databases;
- heating and cooling degree day for consumption of energy for heating and cooling, respectively; we could refer to past data or data recorded by sensors in the plant.

For the last point the realization of energy audits becomes fundamental in addition to the collection of documental information and measurements with opportune campaigns, it allows the recording of useful technical data about the plant energy consumptions.

In particular the audit phase consists of inspection in the analyzed plant, interviews with the internal responsibles, measurements and registrations of the machineries performances.

These data are an integration of the other documental information, in particular for analysing the production area, the use of machineries, the unsatisfied needs of maintenance. Besides, the energy audit constitutes a fundamental step for the checks of an energy management system (Carbon Trust, Good practice Guide 200), for verifying the effective results of the integrate management structure.

The most powerful energy audit instruments available in literature are the check lists and the decisional matrices (Carbon Trust, CTV 023). In particular, these instruments have been adapted to our particular procedures and integrated in this described sequence of steps.

The decisional matrices have essentially three functions:

- assessing the system energy performance;
- planning the necessary action, identifying the priorities;
- monitoring the effects of energy management systems.

Concretely they are tables characterized by three levels of detail. They allows the evaluations of distinct characteristics of a system assessing a score (from 0 to 4). (Carbon Trust, Good Practice Guide 306).

The first level (Top-Level, Energy Performance Matrix) groups the results of the other matrices and allows an overview of the organization.

TOP LEVEL						
PERFORMANCE MATRIX						
LEVEL	**1**	**2**	**3**	**4**	**5**	**6**
ENERGY MANAGEMENT						
FINANCIAL MANAGEMENT						
AWARENESS AND INFORMATION						
TECHNICAL						

Table 2. Top Level Matrix

The second level consists of four tables whose results are reported in the Top Level: Energy Management Matrix, Financial Management Matrix, Awareness and Information Matrix, Technical Matrix.

These tables allows to assess a score for the different aspects of these energy management issues. In particular the technical aspects are more deeply investigated in the third level matrices, which analyze the working and performance characteristics of the different plant end users (cooling system, heating system, HVAC system, compressed air, building characteristics, boilers, lighting system, monitoring and control system, Building Energy Management System (BEMS), etc.).

These last matrices are the most powerful instruments for the audit phase because may be used as a guide for analyzing the users performance.

In Table 3 an example matrix (for the compressed air) originally developed on the basis of the other found in the literary review is reported.

Therefore other instruments developed for helping in energy auditing are the check lists. Those divided every user in Generation, Distribution and Use and, for these sectors, make an analysis which is divided in four sections:

- evaluation: a series of questions to focalize the performance and qualities of the main parameters and assessing a score on their evaluation;
- solution – improvement: a list of possible activities to improve energy performance.
- detailed analysis: different detailed aspects which have to be analyzed and possible activities.
- technical – operational parameters: a guide for collect all the necessary technical and operational parameters of the users.

These check lists are less general than the decisional matrices but they present the advantage of characterizing in a more technical and detailed way all the most common service plant as well as the air handling, cooling system , boiler, HVAC system, etc.

In Table 4 an example of the Technical – operational parameters part is reported for the air handling system.

	III LEVEL - AIR HANDLING			
SCORE	COMPRESSORS	PIPING SYSTEM	ENERGY SAVING DEVICES	MONITORING AND MAINTENANCE
4	Multi-stage compressors, well dimensioned, with additional compressor used on demand. Electronic controls for modulating required power.	Pipe and valves well maintained; losses of 10%. An inspection every 6 months. Max difference of pressure of 0,5 bar. Ring piping system. Where possible the welding is preferable.	Sensors for range pressure individuation. Valves for interruptions compressed air require if not necessary. Avoiding of inaccurate uses.	Operational procedures for monitoring and maintenance are defined. Constant control on the humidity and temperature of the inlet air. Pressure gauges near the filters for their substitution. Periodic controls of the cooling water treatment system.
3	Multi-stage compressors, sufficiently dimensioned, with additional compressor used on demand. Electronic control for modulating required power.	Pipe and valves well maintained; losses of 20-25%. An inspection every 6 months. Max difference of pressure of 0,5 bar. Excessive pressure with loss of efficiency.	Sensors for range pressure individuation. Avoiding of inaccurate uses.	Theoretic procedures for monitoring and maintenance are defined. Constant control on the humidity and temperature of the inlet air.
2	Single-stage compressors, well dimensioned for demand peaks. Absence of electronic controls for modulating required power.	Pipe, valves and flanges with high losses. Annual inspections. High differences of pressure. Use of zone insulation valves with regulation functions.	Time control sensors. Valves for interruptions compressed air require if not necessary	Absence of preventive maintenance. Sensors for the registration of air consumptions for different floors; absence of procedures for using these information.
1	Single-stage compressors, sufficiently dimensioned. Absence of electronic controls. Loss of efficiency due to a bad regulation (compressor often works outside the limit value of pressure).	Losses of 30-40%. Inspection when required. High differences of pressure. Use of zone insulation valves.	Time control sensors.	Ad hoc maintenance; incomplete data about the air supplies. Absence of data about the air losses.
0	Single-stage compressors over dimensioned. Absence of electronic controls.	Losses of 40-50% with frequent open/close of the security valves; presence of dead legs.	Centralized control for the on/off of the system	Ad hoc maintenance; absence of data about the air supplies and the air losses.

Table 3. Third Level Matrix: air handling system

Air handling system						
Technical-operational parameters						
GENERATION		Reciprocating compressor (rotary)		Screw compressor (rotary)		
Compressor:	Roots blower compressor (rotary) single stage	Single/ two stage	Multi stage	Single stage	Two stage	Centrifugal compressor
Capacity (m³/h)						
Pressure (bar)						
T outlet (°C)						
Full-load consumption (kWh)						
Partial-load consumption (kWh)						
Cooling water temperature (°C)						
DISTRIBUTION						
Pipes diameter (mm)						
Pipes length (m)						
Piping distribution scheme						
Pipes material						
USE Compressed air users						
Operating pressure (bar)						

Table 4. Check Lists: Technical – operational parameters of the air handling system

4.2 Energy cost & consumption data analysis

In the first step a group of information enabling energy usage to be managed more effectively within an industrial site has to be collected. Most of the needed data are available from existing meter readings, energy bills and production-related data. The aim of this step is to analyze and to give an an interpretation that allows transforming data into useful information for energy management purposes. At this step a standard spread sheet is adequate for many applications. The main analyses concern the following aspects.

Primary energy sources comparison. Once the data are collected, it is necessary to determine the amount of energy spent in the whole business, whatever is the consumed energy source.

Therefore all energy sources must be expressed in the same unit (i.e. MJ, kWh or TEP) and the proportionate use and cost of each different energy source when compared to the total energy consumption should be determined. This allows highlighting amount, cost and fluctuations throughout the year of each kind of primary energy, thus identifying an upper limit on the amount that can be saved and a benchmark to assess energy saving after improvement measures have been performed. Furthermore, these data could be compared with relevant available benchmarks for the same industrial sector. This information can also help in determining if current energy use is higher or lower than usual, or if any outside factors have an impact on how much is being used. It is possible to establish:

- if too much energy is being used;
- how current energy use compares with past figures;
- how the business compares with the industry average (where benchmarks are available);
- whether any other factors are temporarily affecting the figures; these might include cold weather, extended working hours or increased production. Understanding how these drivers are affecting energy data will give a better picture of site consumption.

In particular a "driver" is any factor that influences energy consumption, as weather is the main driver for most buildings and production is the primary driver for most industrial processes. Drivers are sometimes referred to as variables or influencing factors. There are two main types:

- activity drivers: feature of the organization activity that influences energy consumption. Examples include operating hours, produced tons, number of guests and opening hours.
- Condition drivers: where the influence is not determined by the organization activity but by prevailing conditions. Examples include weather, condition of the raw material and hours of darkness.

Specific Energy Consumption (SEC). This relevant parameter is defined as the ratio between energy consumption and an appropriate production measure (driver). It can be calculated for any fixed time period, or by batch. SECs need to be treated with care because their variability may be caused by several factors beyond energy efficiency, such as economies of scale or production problems not closely related to energy management. There are many process benchmarking schemes based on SEC and their easiness of use makes them attractive to many companies.

Current and past comparisons. This approach, suitable for buildings and industrial plants, is usually performed in a graphical form where a bar or column chart is used to compare the data from the current period with a similar previous. A tabular form of this comparison can also be used with a quantity or percentage figure for the difference. It is useful for monitoring year-on-year changes and cyclical patterns, and can also be used for daily and weekly profiles. This technique can be applied to energy data and its drivers.

Time series analysis. Also this approach is suitable for buildings and industry. Most energy managers are interested in the underlying trend of consumption or cost and trend lines are a graphical way of showing this. Typically, the trend line will be the trend of the data series over time. At its simplest, it is a line graph of the data for each period. A more refined application of the technique is to use moving annual totals or averages. This approach is useful since it reduces seasonal influence and allows highlighting other influencing factors.

Since a trend line can be produced from time-related energy data alone, it is a common technique to use at the early stages of investigating energy consumption.

Energy absorption. It is possible to estimate the energy absorption of different plant areas by measuring actual energy requirements and evaluating utilization rates.

Contour map. It offers a more pictorial use of profile information. Here, half-hourly data, typically for a month, is displayed as a multi-colored contour chart. This provides a very easy way of viewing 1 400 data points (30 days x 48 half-hours).

4.3 Energy forecasting at plant level

The core of the methodology is the definition of a consumption forecasting model that allows identifying the specific consumptions of different manufacturing lines in order to formulate the budget (step 6) and identifying the optimal energy rate in the contract renewal phase. Moreover it provides the reference for real-time energy consumption control (i.e. identifying sporadic faults or events).

The expected energy demand is calculated on the basis of mathematical models describing the influence of relevant factors (energy drivers) on the energy consumption by regression analysis (i.e. production volume is an important energy driver at the plant level). The energy consumption C, in delta time, can be defined as:

$$C\,(\Delta t) = E_0(\Delta t) + \alpha_1 V_1(\Delta t) + \alpha_2 V_2(\Delta t) + \ldots \alpha_m V_m(\Delta t) \tag{1}$$

where:

- E_0 is the constant portion of the consumption regardless of production volumes [kWh];
- V_i is the production volume [unit] of the i-th product;
- α_i is the consumption sensitivity coefficient with respect to the production volume [kWh/unit] of the i-th product.

Equation (1) can be calculated by a multiple regression between production volumes and consumptions. In general the production volumes of the different products are sufficient to create a consumption model but in some cases the use of other variables (such as, temperature, degree days, sunlight variations or other operational variables) is required. The $\alpha_1, \alpha_2, \ldots, \alpha_n$ coefficients have to be assessed with statistical analysis on the historical data previously collected. The model has to be statistically validated.

Multiple linear regression model as statistical model does not mean only mathematical expression but also assumptions supplying the optimal estimation of coefficients α_i. These assumptions are usually connected with random error: the random error has normal distribution, it is equal to zero (on the average), supporting elements have equal variances.

Once a regression model has been constructed, it may be important to confirm the model capability of representing the actual behaviour of the industrial plant (in other terms the model capabilities of well fitting real data) and the statistical significance of the estimated parameters. Commonly used checks of goodness of fit include the R-squared, analyses of the pattern of residuals and hypothesis testing. Statistical significance can be checked by an F-test of the overall fit, followed by t-tests of individual parameters.

Moreover, the validity of the multiple regression analysis is related to the validity of the following hypotheses (Levïne et al., 2005):

- *Homoscedasticity.* The variance of the dependent variable is the same for all the data. Homoscedasticity facilitates the analysis because most methods are based on the assumption of equal variance;

- *Autocorrelation.* Independence and normality of error distribution. Autocorrelation is a mathematical tool for finding repeating patterns, such as the presence of a periodic signal which has been buried under noise, or identifying the missing fundamental frequency in a signal implied by its harmonic frequencies. It is frequently used in signal processing for analysing functions or series of values, such as time domain signals. In other terms, it is the similarity between observations as a function of the time separation between them. More precisely, it is the cross-correlation of a signal with itself.

- *Multicollinearity,* which refers to a situation of collinearity of independent variables, often involving more than two independent variables, or more than one pair of collinear variables. Multicollinearity means redundancy in the set of variables. This can render ineffective the numerical methods used to solve regression equations, typically resulting in a "multicollinearity" error when regression software is used. A practical solution to this problem is to remove some variables from the model. The results are shown both as an individual R^2 value (distinct from the overall R^2 of the model) and a Variance Inflation Factor (VIF). When R^2 and VIF values are high for any of the X variables, the fit is affected by multicollinearity.

4.4 Sub-metering energy use

Metering the total energy consumption at a certain site is important, but it does not show how energy consumption is distributed across operational areas or for different applications. After the first three steps, therefore, it can be hard to understand why and where energy performance is poor and how to improve it. Installing sub-metering to measure selected areas of energy consumption could give a considerably better understanding of where energy is used and where there may be scope to make savings. Sub-metering is a viable option for primary metering where it is not possible or advisable to interfere with the existing fiscal meter. For this purpose, a sub-meter can be fitted on the customer side of the fiscal meter so as to record the total energy entering the site.

When considering a sub-metering strategy, the site have to be broken down into the different end users of energy. This might be by area (for example, floor, zone, building, tenancy or department), by system (heating, cooling, lighting or industrial process) or both. Sub-metering of specific areas also provides more accurate energy billing to tenants, if it is required. The sub-metering strategy should also identify individuals responsible for the energy consumption in specific areas and ensure that the capability to monitor the consumption which falls under their management responsibilities. Additionally, it may be worth separately metering large industrial machines.

By this way, it is possible to optimize the location of meters and minimize the total amounts, after energy absorption analysis, following the sub metering methods that are (Carbon Trust, CTV 027):

- *Direct metering* is always the preferred option, giving the most accurate data. However, it may not be cost-effective or practical to directly meter every energy end-use on a site. For a correct evaluation the cost of the meter plus the resource to run and monitor it has to be weighed against the impact the equipment has on energy use and the value of the data that direct sub-metering will yield.

- *Hours-run metering* (also known as *constant load metering*) that can be used on items of equipment that operate under a constant, known load (for example, a fan or a motor). This type of meter records the time that the equipment operates which can then be multiplied by the known load (in kW) and the load factor to estimate the actual consumption (in kWh). Where possible, it measures the true power of the equipment, rather than relying on the value displayed on the rating plate.

- *Indirect metering*, which means combining the information from a direct meter with other physical measurements to estimate energy consumption. Its most common application is in measuring hot water energy consumption, which is usually known as a heat meter. A direct water meter, for example, is used to measure the amount of cold water going into a hot water heater. This measurement, combined with details of the cold water temperature, the hot water temperature, the heater efficiency and the specific heat of water, enables the hot water energy consumption to be calculated.
- *By difference metering* when two direct meters are used to estimate the energy consumption of a third end-use. For example, if direct meters are used to measure the total gas consumption and the catering gas consumption in an office building, the difference between the two measurements would be an evaluation of the energy consumption associated with space heating and hot water. This form of metering should not be used where either of the original meter readings is estimated, since this could lead to large errors. Also, this form of metering should not be used where a very small consumption is subtracted from a large consumption, because the accuracy margin of the large meter may exceed the consumption of the smaller meter.
- Where none of the above methods can be used, it may be possible to use estimates of small power to predict the energy consumption associated with items such as office equipment (by assessing the power rating of equipment and its usage). This method is very inaccurate and should be supported by spot checks of actual consumption wherever possible.

Generally speaking, the introduction of a monitoring system in a plant is fundamental for an effective energy management approach and it can bring the organization to the creation of a real Energy Information Systems. An EIS can be defined as a system for collecting, analyzing and reporting data related to energy performance. It may be stand-alone, part of an integrated system or a combination of several different systems. Besides meters and computers, an EIS also includes all the organizational procedures and methods that allow it to operate and it may draw on external and internal sources of data.

Energy Information Systems can be used to measure electricity, gas and water supplies. They have been successfully used by energy intensive users for many years to drive down costs and, in general, technology cost has reduced significantly over recent years. Then the approach now offers a good return on investment for less energy intensive businesses in terms of managing energy and water usage. Despite an attractive return on investment, it is not being taken up at the rate one would expect given its benefits. All the previous experience indicates that an Energy Information System, if properly used as a demand management tool, guarantees an energy consumption (and costs) reduction between 10% and 15% (Carbon Trust, Practical guide 231). In addition, effective energy and carbon management (i.e. actively managing risks and opportunities associated with climate change and carbon emissions) relies on the availability of appropriate management information. Therefore metering of energy consumption and flows within companies is an intrinsic element of continuing good energy management and carbon emission reduction. There is also a case for using an Energy Information System to reduce the amount of energy needed to guarantee meeting a given electricity demand. By knowing energy consumption profiles and the opportunities to reduce demand through better energy management, energy suppliers may choose to use demand side management as a tool to more effectively match supply and demand and thus reduce the requirement for additional generating capacity.

For realizing an EIS a useful number of smart meters have to be installed (Carbon Trust, CTV 027). Smart meters can provide reliable and timely consumption data readily usable in an energy management program. Such meters can also eliminate problems associated with estimated bills and the potential consequences of not being able to correctly forecast and manage energy budgets. They also can be used to show the energy consumption profile of the site, which can help an energy manager identify wastage quickly. There is no universal definition for smart metering, although a smart metering system generally includes some of the following features:

- recording of half-hourly consumption;
- real-time information on energy consumption that is immediately available or via some forms of download to either or both energy suppliers and consumers;
- two-way communication between energy suppliers and the meter to facilitate services such as tariff switching;
- an internal memory to store consumption information and patterns;
- an easy to understand, prominent display unit which includes:
  - energy costs;
  - indicator of low/medium/high use;
  - comparison with historic/average consumption patterns;
  - compatibility with PCs/mobile phones;
  - export metering for micro-generators.

The essential features of smart metering are those which relate to consumption data storage, retrieval and display. Smart metering can be achieved by installing a fiscal meter which is capable of these essential tasks. Alternative metering solutions are available to bypass replacement of the fiscal meter with a smart meter. These include the use of sub-metering, for instance, a bolt-on data reader which is capable of storing and transmitting half-hourly consumption data. Other automated solutions, which are sometimes conflated with the term 'smart meters' are AMR (Automated Meter Reading) and AMM (Automated Meter Management):

- AMR: is a term that refers to systems with a one-way communication from the meter to the data collector/supplier. It can apply to electricity or gas, although gas systems require batteries to operate, which adds to the cost. AMR bolt-on solutions are available and appropriate for gas meters that have a pulse output. Remote, automatic reading is beneficial in that impractical manual reads are not necessary, and bills can always be based on actual reads, not estimates. How often a read is taken will depend on the supplier, although customers may request regular reads. However, even with AMR, the data will not be available necessarily, unless they are requested or have been initiated by the customer.
- AMM: they are systems similar to AMR arrangements, except that they allow a two-way communication between the meter and the data collector/supplier. As well as having all the benefits listed above, AMM allows for remote manipulation by the supplier. The advantage to the customer is that there is potential to display real-time tariff data, energy use, and efficiency at the meter. AMM is mostly available for electricity with some safety issues affecting AMM for gas.

The available technology for the transfer of consumption data from metering ranges from GPRS or GSM modems sending data bundles to a receiver, through low power radio technology to ethernet/internet interfaces. When installing a metering system which makes use of remote meter reading, it may be considered which communication option is the most appropriate for each particular application. The system appropriateness depends on practical factors such as:

- meters number (including sub-meters);
- size of site(s);
- location of meters;
- power supply;
- proximity to phone line or mobile/radio network coverage.

In addition to these factors, the communication options employed will depend on the site-specific needs as well as the expertise of the metering company being employed. Therefore, it is advisable to ask the meter provider to offer the most reliable and lowest-cost solution, taking into account all of these factors.

4.5 Tariff analysis and contract renewal

The objectives of this step are to choose the less expensive solution relating to own forecasted energy load profile and to evaluate the impact of the different contractual options on the unit energy cost.

Energy bills are usually very complicated, as they consist of several components that often confuse the customer. For example energy use charges, transmission charges, demand charges, fuel adjustment charges, minimum charges and ratchet clauses are the more common components of electrical rate structures. Their knowledge and their control are the first step toward energy cost minimization. In particular below the electrical tariff is described with a lot of details because electricity is always present in industrial consumptions and it represents the most meaningful example (the electrical costs is made up of a large number of different terms). The structural changes that industries have to take into account in order to save electrical cost concern:

- *Electrical rate structure.* The electrical rate based on kWh bands overcame the flat tariff. This entails the proliferation of different proposals which are difficult to be compared, since they are not homogeneous in their formulation. Electrical energy rate could be influenced by total consumption, power furniture, voltage, time bands (tb), customer forecasting capability, and fuel price. The most common rate schedule in use is the day-time schedule. This rate structure eliminates the flat rate pricing of electricity, replacing it with a pricing schedule that varies with the time of the day, the day of the week and the season of the year. They were developed by utilities as a way to reduce the need for peaking stations. What makes this rate structure particularly effective is the variation in rates among bands. The time bands have a strong impact on the effectiveness of energy conservation measures. Under time of day rates, energy conservation efforts must address both the energy use and the demand portion of the bill. While any reduction in kWh use, regardless of when the reduction takes place, will result in lower energy costs, this rate structure increases the measure cost effectiveness that impact energy use during on peak hours while decreasing the measures cost effectiveness that impact off-peak use. This impact on peak energy use is further increased by savings in demand charges. On the other hand different proposals may not be homogeneous and comparisons could be not easy to perform for industries
- *Electrical bill components.* A careful examination of the own electrical bill is necessary to gain the best tariff option. The main components could be: kWh charges, demand

charges, electrical demand ratchet clauses, power factor charges, fuel adjustment. Indeed price contract proposals could vary as fixed price or combustible-linked variable price.

- *Electrical energy sector organization.* An industrial customer could purchase energy through contracts with wholesale suppliers or from producers on the basis of physical bilateral contracts. Therefore industries, aware of their own historical data on electricity consumption, have to be ready to face contractors. The knowledge of the market and sector organization gives the opportunity to compete on energy unit costs;
- *Power plant optimization or design* as it will be described in paragraph 4.8.

More details about tariff analysis are given in (Cesarotti et al., 2007). Briefly, the proposed methodology follows three steps. First of all it is necessary to understand the historical consumptions in the industrial process. Using the procedure defined in the paragraph 5.3 a mathematical model of the plant consumptions can be obtained. The next step is to use the consumption model to forecast the consumption for the next periods. This requires forecasts of energy drivers included in the model. Different sources could be used for this purpose. For instance, in order to identify:

- production: we could refer to companies production plan or demand forecast;
- sunlight variation: we could refer to meteorology web sites or databases;
- degree day for electrical energy for heating or cooling: we could refer to a mean value obtained by the past years.

Besides the forecasted consumption has to be split among time bands according to the trend of consumption of the previous year. The last step is the tariff analysis: analysis of energy process allows minimization of costs in contract renewal for meeting the forecasted energy load profile. Various factors differ among offers ($f_1, f_2,..., f_m$) and have to be considered during contract renewal to determine the best one f_{opt} minimizing the cost applied to energy consumption forecast, $C(\alpha_i)$ as shown in the following equation:

$$f_{opt}(t) = \min_{j \in \{1,...,m\}} | f_j(t) \cdot C(\alpha_i) | \tag{2}$$

The average kWh cost (total cost divided by forecast consumption) helps point out the less expensive tariff. It is recommended a sensitivity analysis to evaluate how much the results are affected by the different hypothesis (future price of energy, future products demand, etc.). However, for the formulation of the final price it is necessary to consider other factors that affect energy tariff and are different among contractors such as formulation of price methods, costumer forecasting capability that influence the price, penalty about reactive energy, etc. Moreover, price contract proposals could vary (i.e. fixed price or variable price combustible-linked). For the final choice other qualitative factors included in the contract have to be considered, such as bonus relating to customer forecasting capability or natural gas contract with the same supplier.

4.6 Energy budgeting and control

Another important feature of energy management and of the presented methodology is planning for future energy demand. Energy budgeting is an estimate of future energy demand in terms of fuel quantity, cost and environmental impacts (pollutants) caused by the energy related activities.

This step allows formulating an accurate energy budget and monitoring the difference between budget and actual costs. This is performed by means of indicators able to distinguish the effect of a different specific consumption from the effect of different operational conditions, e.g. different prices, volumes, etc.

First of all the energy budget has to be estimated by considering both the outputs of the energy consumption forecasting model (providing specific consumptions) and the industrial plant production plans (providing global volumes). Once energy budgeting of electrical consumptions and costs has been performed, it is possible to setup an "on-line" control.

In (Cesarotti et al., 2009) the authors propose energy budgeting and control methods that have been implemented within a set of first and second level metrics. The first level indicators allow identifying the effect of an increase of specific consumption beyond the predicted. The second level indicators allow to identify the effect of variations of price, volume, mix or load bands from the predicted.

In (Cesarotti et al., 2009), the consumption of electrical energy C (kWh) is defined with the expression in (3):

$$C=E_0+\alpha_1 \cdot V_1+\alpha_2 \cdot V_2+...+\alpha_m \cdot V_m \tag{3}$$

where E_0 is the constant portion of the electrical consumption regardless of production volumes (kWh); V_1, V_2, ... , V_m are the production volumes (unit); α_1, α_2, ... , α_m, are the sensitivity coefficients of the electrical consumption with respect to the production volume (kWh/unit).

The expression in (3) could be calculated by a multiple regression between production volumes and consumptions. The α_1, α_2, ... , α_m coefficients have to be assessed with statistical analysis. The model has to be statistically validated through indicators as p-value, r^2 and analysis of variances.

In order to calculate the specific consumptions it is necessary to split the contribution of the fixed amount E_0 among the different productions. This can be done proportionally to production volumes if:

- data relating to the total production time of different products is not available;
- the different production processes are comparable in terms of electrical absorptions.

From (4) one can calculate the specific consumption SC_j (kWh/unit) of j-the manufacturing line, and therefore of j-th product, as in (4):

$$SC_j=\alpha_j+\frac{E_0}{V_{tot}} \tag{4}$$

where V_{tot} are the total production volumes (unit).

After having characterized energy consumption at a plant level, it is possible to formulate the energy budget. Therefore, we have to consider:

- energy characterization, as in the previous paragraph, that gives us the specific consumptions for each type of products as in (4);
- electrical energy prices as expected by the contract; if prices are linked to combustible (btz, brent) prices then a short-term forecasting of these indicators is requested (Cesarotti et al., 2007);
- forecasted production plans and, if the energy price varies by the TOD, also a short-term demand forecast, in order to match the tariff plan, and determine the budgeted cost

As the tariff could vary by TOD, the budget cost of k-th month, BC_k (€), can be computed from the expected price for each tariff period of the day and the relative production volume as follows:

$$BC_k = \sum_{j=1}^{m} \sum_{i=1}^{n} p_{ijk}^{P} \cdot V_{ijk}^{P} \cdot SC_{ijk}^{P} = \text{sum of all elements}\left[(p^P \, V^P \, SC^P)_{ijk}\right] \qquad (5)$$

where $\left[(p^P \, V^P \, SC^P)_{ijk}\right]$ is a matrix whose ij-th elements are given by the product $p_{ijk}^{P} \cdot V_{ijk}^{P} \cdot SC_{ijk}^{P}$; i denotes the time period of the day referring to the tariff; n is the number of time period; j denotes the product type; m is the number of product type; p^P is the planned price (€/kWh); V^P is the planned production volume (unit); SC^P is the specific consumption (kWh/unit) as calculated with (4).

After energy budgeting of electrical consumptions and costs for the industrial plant, it is possible to setup a "on-line" control. In this step we will look for variations in costs and consumptions and we will have to discern if increases in costs and consumptions have to be linked to:

- an increase of energy consumptions of a product family: in this case we have to investigate on the reason of the modification of energy consumption;
- a variation of production volumes or an increase of electrical energy prices: in this case we have to re-plan the budget.

The authors present a series of indicators for controlling the differences between BC and actual cost. These indicators have been derived from the earned value technique, usually used in project management cost/time control.

The following variables have been defined:

- Estimated Cost EC_k (€): it is the estimated energy cost of k-th month calculated considering the actual production volumes and actual tariff:

$$EC_k = \sum_{j=1}^{m} \sum_{i=1}^{n} p_{ijk}^{a} \cdot V_{ijk}^{a} \cdot SC_{ijk}^{P} = \text{sum of all elements}\left[(p^a \, V^a \, SC^P)_{ijk}\right] \qquad (6)$$

where $\left[(p^a \, V^a \, SC^P)_{ijk}\right]$ is a matrix whose ij-th elements are given by the product $p_{ijk}^{a} V_{ijk}^{a} \cdot SC_{ijk}^{P}$; i denotes the time period of the day referring to the tariff; n is the number of time period; j denotes the product type; m is the number of product type; p^a is the actual price (€/kWh); V^a is the actual production volume (unit); SC^P is the specific consumption (kWh/unit) as calculated with (4);

- Actual Cost AC_k (€): it is the actual energy cost of k-th month really sustained by the company related to the actual production volumes:

$$AC_k = \sum_{j=1}^{m} \sum_{i=1}^{n} p_{ijk}^{a} \cdot V_{ijk}^{a} \cdot SC_{ijk}^{a} = \text{sum of all elements}\left[(p^a \, V^a \, SC^a)_{ijk}\right] \qquad (7)$$

Where $\left[(p^a \, V^a \, SC^a)_{ijk}\right]$ is a matrix whose ij-th elements are given by the product $p_{ijk}^{a} \cdot V_{ijk}^{a} \cdot SC_{ijk}^{a}$; i denotes the time period of the day referring to the tariff; n is the number of time period; j denotes the product type; m is the number of product type; p^a is the actual price (€/kWh); V^a is the actual production volume (unit); SC^a is the specific consumption (kWh/unit).

Details about the calculation of parameters in the (5, 6, 7) are reported below.

Summarizing, the three variables are function of energy price, production volume and, specific consumption planned or actual as shown in the Table 5.

Basing the study on the previous formulation, it is possible to investigate the energy consumption behavior of the company related to the selected production volumes. So the following indicators have been formulated.

	BC	EC	AC
Electrical energy price (P)	Plan	Real	Real
Production volume (V)	Plan	Real	Real
Specific consumption (SC)	Plan	Plan	Real

Table 5. Variables

First of all we have to deal with the difference between AC_k and BC_k at k-th month. The first index is the percentage shift of the actual budget and the planned one as in (8):

$$I_{1k}=\frac{AC_k\text{-}BC_k}{BC_k} \tag{8}$$

In particular, the following situations could arise:
- $I_{1k} > 0$ – a positive value of index in (8) means that the company has spent more than predicted at k-th month.
- $I_{1k} = 0$ – a value of index in (8) equal to zero means that the actual cost complies with the budget at k-th month.
- $I_{1k} < 0$ – a negative value of index in (8) means that the company has spent less than predicted at k-th month.

At the same time, the difference between AC_k and BC_k could depend on a difference between the actual tariff and the planned one or by a difference between actual and planned production (for quantities or mix) or a higher specific consumption. In order to distinguish these cases, separating the contribution due to inefficiency of consumption and due to different energy drivers scheduling, we have to introduce the following indicators:

$$I_{2k}=\frac{AC_k\text{-}EC_k}{BC_k} \tag{9}$$

$$I_{3k}=\frac{EC_k\text{-}BC_k}{BC_k} \tag{10}$$

$$I_{1k}=I_{2k}+I_{3k} \tag{11}$$

A positive value of I_{2k} means a higher specific consumption for unit production for the same amount of production volumes. In this case it is important to analyze the energy behavior in terms of AC_k and EC_k for each production department. Then it is necessary to enquire about the cause of deviation with problem solving tools. There are many approaches to problem solving, depending on the nature of the problem and the process or system involved in the problem.

A positive value of I_{2k} highlights a variation in prices or energy drivers, assuming the consumption model obtained from regression completely reliable; the difference between the actual and scheduled values of energy drivers could depend upon:
- energy price: it could have changed during time, e.g. for electrical energy tariff if linked to combustible basket;
- production volume or mix: they could have changed during time due to for example a difference in production plan or availability of the production system;
- electrical loading in time bands: it could have changed during time due to for example a difference in production plan.

The second level indicators have been introduced in order to investigate in the difference $(EC_k - BC_k)$. The difference could be linked to the following effects that have to be investigated:

- price effect: due to a variation in energy price;
- volume effect: due to a variation in production volume;
- loading effect: due to a variation in production loading;
- mix effect: due to a variation in production mix;
- interaction effect: is the differing effect of one independent variable on the dependent variable, depending on the particular level of another independent variable.

An interaction is the failure of one factor to produce the same effect at different levels of another factor. An interaction effect refers to the role of a variable in an estimated model, and its effect on the dependent variable. A variable that has an interaction effect will have a different effect on the dependent variable, depending on the level of some third variable. In our case, for example, a contemporaneous variation of different factors (volume, mix, load, price) involves a greater consumption (Montgomery, 2005).

In order to distinguish the previous effects the following nomenclature has been adopted:

- $\Delta^P{}_{1k}$ (percent) is the percentage of the j-th production volume V (unit) planned at the i-th time band at k-th month on the total of the j-th production volume planned V (unit) at k-th month as in (12); so it represents the coefficient of electrical load of production volume planned in the different time bands:

$$\Delta^P_{1ijk} = \frac{V^p_{ijk}}{\sum_{i=1}^{n} V^p_{ijk}} \tag{12}$$

- $\Delta^P{}_{2k}$ (percent) is the percentage of the j-th production volume V (unit) planned at k-th month on the total production volume planned V (unit) at k-th month as in (13); so it represents the coefficient of mix of production volume planned for production:

$$\Delta^P_{2ijk} = \frac{\sum_{i=1}^{n} V^p_{ijk}}{\sum_{j=1}^{m} \sum_{i=1}^{n} V^p_{ijk}} \tag{13}$$

where $V^p_{jk} = \sum_{i=1}^{n} V^p_{ijk}$ and $V^p_k = \sum_{j=1}^{m} \sum_{i=1}^{n} V^p_{ijk}$

- $\Delta^\alpha{}_{1k}$ (percent) is the percentage of the j-th production volume V (unit) realized at the i-th time band at k-th month on the total of the j-th production volume realized V (unit) at k-th month as in (14); so it represents the coefficient of load of production realized in the different time bands:

$$\Delta^\alpha_{1ijk} = \frac{V^\alpha_{ijk}}{\sum_{i=1}^{n} V^\alpha_{ijk}} \tag{14}$$

where $V^\alpha_{jk} = \sum_{i=1}^{n} V^\alpha_{ijk}$

- $\Delta^\alpha{}_{2k}$ (percent) is the percentage of the j-th production volume V (unit) realized at k-th month on the total production volume realized V (unit) at k-th month as in (15); so it represents the coefficient of mix of production volume realized for production:

$$\Delta^{\alpha}_{2ijk} = \frac{\sum_{i=1}^{n} V^{\alpha}_{ijk}}{\sum_{j=1}^{m} \sum_{i=1}^{n} V^{\alpha}_{ijk}}$$ (15)

where $V^{\alpha}_{jk} = \sum_{i=1}^{n} V^{\alpha}_{ijk}$ and $V^{\alpha}_{k} = \sum_{j=1}^{m} \sum_{i=1}^{n} V^{\alpha}_{ijk}$

Price effect calculation.

It could contribute in the difference between estimated and planned energy cost (EC_k - BC_k). In order to investigate in the price effect, it is necessary to calculate the change (€) in the electrical costs at k-th month due to a variation of price. This has been calculated as in (16):

$$\text{Change for price}_k = \text{Sum of all elements} \left[(P^{\alpha}-P^{P}) \cdot V^{P} \cdot SC^{P}\right]_{ijk}$$ (16)

where $\left[(P^{\alpha}-P^{P}) \cdot V^{P} \cdot SC^{P}\right]_{ijk}$ is a matrix whose ij-th elements are given by the product $(p^{\alpha}_{ijk}-p^{P}_{ijk}) \cdot V^{P}_{ijk} \cdot SC^{P}_{ijk}$.

Therefore, the price effect (per cent) has been calculated as the ratio between the terms in (14) and the difference between (EC_k - BC_k) as in (17):

$$\text{price effect}_k (\text{percent}) = \frac{\text{Change for price}_k}{(EC_k - BC_k)}$$ (17)

Volume effect calculation. It is a candidate contributor to the difference between estimated and planned energy cost (EC_k - BC_k). Production volume could change over time due to, for example, a different production plan or a variation of availability of the production system.

In order to investigate in the volume effect, it is necessary to calculate the change (€) in the electrical costs at k-th month due to a variation of the production volume in terms of planned and actual one. While there percentage mix and time bands load have not been modified. This has been calculated as in (18):

$$\text{Change for volume}_k = \text{Sum of all elements} \left[P^{P} \cdot \Delta^{P}_{1} \cdot \Delta^{P}_{2} \cdot (V^{\alpha}-V^{P}) \cdot SC^{P}\right]_{ijk}$$ (18)

Where $\left[P^{P} \cdot \Delta^{P}_{1} \cdot \Delta^{P}_{2} \cdot (V^{\alpha}-V^{P}) \cdot SC^{P}\right]_{ijk}$ is a matrix whose ij-th elements are given by the product $P^{P}_{ijk} \cdot \Delta^{P}_{1ijk} \cdot \Delta^{P}_{2ijk} \cdot (V^{\alpha}_{ijk} -V^{P}_{ijk}) \cdot SC^{P}_{ijk}$.

Therefore, volume effect (per cent) has been calculated as the ratio between the term in (16) and the difference between (EC_k - BC_k) as in (19):

$$\text{volume effect}_k (\text{percent}) = \frac{\text{Change for volume}_k}{(EC_k - BC_k)}$$ (19)

Mix effect calculation. It is another potential contributor to the difference between estimated and planned energy cost (EC_k - BC_k). Production mix could have changed over time due to, for example, a difference in production plan or a variation of availability of the production system.

In order to investigate the mix effect, it is necessary to calculate the change (€) in the electrical costs at k-th month due to a variation of production mix. The difference in the mix coefficient, as in (13) and in (15), has been introduced to calculate the changed cost.
While the production volumes and the percentage of time band load have not been modified.
The difference in energy costs, due to a variation of production mix, has been calculated as in (20):

$$\text{Change for mix}_k = \text{Sum of all elements} \left[P^P \cdot \Delta_1^P \cdot (\Delta_2^\alpha - \Delta_2^P) \cdot V^P \cdot SC^P \right]_{ijk} \tag{20}$$

Where $\left[P^P \cdot \Delta_1^P \cdot (\Delta_2^\alpha - \Delta_2^P) \cdot V^P \cdot SC^P \right]_{ijk}$ is a matrix whose ij-th elements are given by the product $P^P_{ijk} \cdot \Delta^P_{1ijk} \cdot (\Delta^\alpha_{2ijk} - \Delta^P_{2ijk}) \cdot V^P_{ijk} \cdot SC^P_{ijk}$.

Therefore, the mix effect has been calculated as the ratio between the term in (18) and the difference between $(EC_k - BC_k)$ as in (21):

$$\text{mix effect}_k (\text{percent}) = \frac{\text{Change for mix}_k}{(EC_k - BC_k)} \tag{21}$$

Loading effect calculation. Finally, also it could be potential contributor to the difference estimated and planned energy cost $(EC_k - BC_k)$. Production loading could be changed during the time due to, for example, a variation of the production plan. In order to investigate the load effect, it is necessary to calculate the change (€) in the difference in the costs at k-th month due to a variation of the production load. The difference in the loading coefficient, as in (12) and in (14), has been introduced to calculate the changed cost. Whilst production volume and percentage mix have not been modified. The difference in energy costs, due to a different loading production than planned in the budget, has been calculated as in (22):

$$\text{Change for load}_k = \text{Sum of all elements} \left[P^P \cdot (\Delta_1^\alpha - \Delta_1^P) \cdot \Delta_2^P \cdot SC^P \right]_{ijk} \tag{22}$$

where: $\left[P^P \cdot (\Delta_1^\alpha - \Delta_1^P) \cdot \Delta_2^P \cdot SC^P \right]_{ijk}$ is a matrix whose ij-th elements are given by the product $P^P_{ijk} \cdot \Delta^P_{2ijk} \cdot (\Delta^\alpha_{1ijk} - \Delta^P_{1ijk}) \cdot SC^P_{ijk}$

Therefore, the load effect has been calculated as the ratio between the term in (20) and the difference between $(EC_k - BC_k)$ as in (23):

$$\text{load effect}_k (\text{percent}) = \frac{\text{Change for load}_k}{(EC_k - BC_k)} \tag{23}$$

Moreover, it is necessary to consider an interaction effect due to contemporaneous variation of different factors as discussed before. It is possible to calculate the contribution of interaction effect as in (24):

$$\text{contribution effect } (\%) = 100\% - (\text{load effect}(\%) + \text{volume effect}(\%) +$$

$$\text{mix effect}(\%) + + \text{price effect}(\%)) \tag{24}$$

4.7 Energy monitoring and control

The aims of this step are:

- to distinguish between "justified" variability due to different setting of energy drivers (i.e. summer or winter for cooling) and "unjustified" variability that implies necessity to inspect equipment in order to evaluate the need of corrective action;
- to distinguish if variability is random due to common causes or it is due to assignable causes.

The authors propose a methodology for real time decision strategies based on statistical techniques of process control as CuSum (Cumulative sum of differences) control charts that differentiate variability thanks to their high sensitivity.

The point in the CuSum chart at time t is defined as:

$$\text{Cusum value } (\Delta t) = Cp(\Delta t) - Ca\,(\Delta t) \tag{25}$$

where:

- $Cp(\Delta t)$ is the planned consumption calculated by the forecasting consumption model;
- $Ca(\Delta t)$ is the actual consumption.

This technique is relatively simple, but very effective to identify energy savings (downward trending line) or higher rates of consumption (upward trending line). If the energy performance of a building or of an industrial process is consistent, its actual consumption will be roughly equal to the expected values (however calculated). In some periods actual consumption will exceed expected one and in others it will be less, but in the long term the positive and negative variances cancel out and their cumulative sum ('CuSum') will remain roughly constant. If, however, a problem occurs that causes persistent energy waste, even if the problem is minor, positive weekly variances will outweigh the negative and their cumulative sum will increase. The CuSum chart would switch from the baseline to a rising trend (Elovitz, 1995) and (Cesarotti et al.,2010).

4.8 Power plant management optimization

The main target of this step is to define the power plant component (thermal/cogenerative engines, boilers, chillers, etc.) set points satisfying the energy load of a buildings/industrial plant, pursuing a specified optimization criteria (i.e. system efficiency, costs, pollutant emissions). An optimal (accurate and appropriate) management of the energy system may lead to substantial energy (and costs) savings and/or environmental benefits without any improvement on the power plant components.

In general the equipments that can be investigated with this approach are:

- gas engines;
- gas steam boilers;
- hot water boilers;
- mechanical chillers;
- absorption chillers.

Being understood that any power plant may be treated by the proposed method. All the integrated equipments are considered as energy converters. They are characterized by inputs and outputs and are modeled as black-boxes. The outputs depend on the component load. It is worth of noting that, although the output could be more than one, as in the case of a gas engine cogenerator (electricity and hot water for example), each equipment is usually defined by only one input (fuel or electric energy).

Conservation equations are considered to solve each subsystem with a quasi-steady approach (i.e. the variables are considered constant between two time-steps) (Weron, 2008), (Farla & Blok, 2000).

The input variables involved in the mathematical representation are subdivided into two main classes, as proposed in (Barbiroli, 1996): controllable and non-controllable variables. The non-controllable inputs are those related to the energy requirements (i.e. dependent on plant production plan or the building operation), as, at each time-step, the power plant has to supply the "non-controllable" energy demand.

The energetic non-controllable inputs are the cooling demand ($\dot{Q}_{CD}$), the low temperature heat demand ($\dot{Q}_{HwD}$), the high temperature heat demand (steam) ($\dot{Q}_{SD}$) and the electricity demand (P_{EID}). The economic non-controllable inputs are the fuel cost (c_f) and the electricity cost. Considering that electricity can be purchased by or sold to the public network, as the power plant electricity output may be higher or lower than the electric demand, the energy costs in sale (S_{EI}) and in purchase (c_{EI}) are considered. The controllable inputs are the power plant component set points varying from 0 (representing switching off) to 1 (representing maximum load). The total cost (TC), the electricity cost and consumption (ElC, P_{ElBal}), the fuel cost and consumption (FC, $\dot{m}_{Tf}$) are the model outputs. The optimization procedure is performed on one or a combination of the above outputs.

Simulations are performed pursuing the goal of optimizing the equipment operation, in order to satisfy specified criterion. Currently, three "optimization criteria" have been implemented:

1. minimum cost of operation;
2. minimum fuel consumption;
3. minimum pollutant emissions (CO, NO_x, SO_x, soot, CO_2).

For the last strategy different weights of the different pollutant emissions may be applied. In the present work, we have assumed that they are proportionally weighted with the Italian legislation maximum limits, as reported below. A back-tracking algorithm is used for the optimal solution identification. The numerical representation of every subsystem is summarized in Table 1. Each equation is representative of the energy transformations taking place into the correspondent equipment between input and output. Efficiency forming equations are set point dependent, according to the manufacturer specifications. The efficiency (h) of each equipment (x) is represented by a k-th order polynomial function as it follows:

$$\eta = \sum E^k \cdot SP_x \tag{26}$$

where E is the primary input energy and SP_x the equipment set point at every time-step.

As an example, a cogenerator can be represented as a black-box where fuel is converted, through an efficiency function like (26), in electricity, thermal energy (both low and high temperature) and cooling energy, as shown in Figure 2. The energy model can be divided into two main submodels: the electricity balance and the thermal balance.

5. Case study

The proposed methodology has been applied to an industrial plant that does not adopt any particular energy management strategy. The company is involved in the production of household ovens and cooking planes for kitchens.

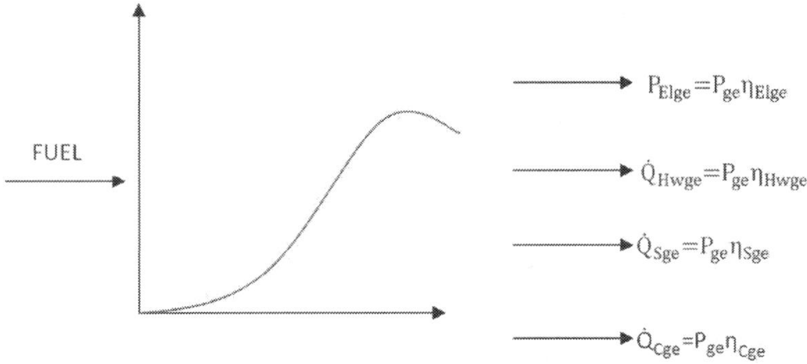

Fig. 2. Representative model of a trigenerator

Equipment	Electrical power	Chemical power	Hot water power	Steam power	Cold power
Gas engine	$P_{Elge}=P_{ge}\eta_{Elge}$	$P_{ge}=\dot{m}_{fge}H_iSP_{ge}$	$\dot{Q}_{Hwge}=P_{ge}\eta_{Hwge}$	$\dot{Q}_{Sge}=P_{ge}\eta_{Sge}$	$\dot{Q}_{Cge}=P_{ge}\eta_{Cge}$
Mechanical chiller					$\dot{Q}_{Cmc}=P_{mc}COP_{mc}S$
Absorption chiller					$\dot{Q}_{Cac}=\dot{Q}_{Hwac}COP_{ac}$
Hot water boiler			$\dot{Q}_{Hwb}=\dot{m}_{fHwb}H_i\eta_{Hw}$		
Steam boiler			$\dot{Q}_{Sb}=\dot{m}_{fSb}H_i\eta_{Sb}$		

Table 6. Subsystem characterization

The production volume has been grouped in 5 product families representing the entire production: household oven n°1 (HO1), household oven n°2 (HO2), household oven n°3 (HO3), cooking plane n°4 (CP4) and, cooking plane n°5 (CP5). The plant produced several different products classified by shape and size, and identified by a specific tag.

The production process is made up of the following working cycle: sheet metal forming by means hydraulic presses; welding and folding; glazing; quality control; sticking; assembling; final quality control.

Fig. 3. ASME process description

The aim of the project is reducing specific energy costs through the application of the methodology previously illustrated.

The steps 1 and 2 of the proposed methodology have allowed identifying overall plant energy costs and consumptions due to electrical energy and gas, and to evaluate their distribution among the different production areas. Electrical energy consumption was 10 GWh/year which took to a cost of about 1.1 million €. Regarding the gas, a consumption of about 3 MSm³/year took to a cost of just little more of 1 million €. The amount of global cost was about 2 100 000 €/year with an incident on final product cost of about 3 €/u.

The primary energy consumption (TEP) distribution is 48% electricity (1MWh = 0.23 TEP) and 52% natural gas (1000 Sm^3= 0.82 TEP), while energy cost is distributed 52% for electricity and 48% for natural gas due to the higher unit price (€/TEP) of electricity as shown in Figure 4.

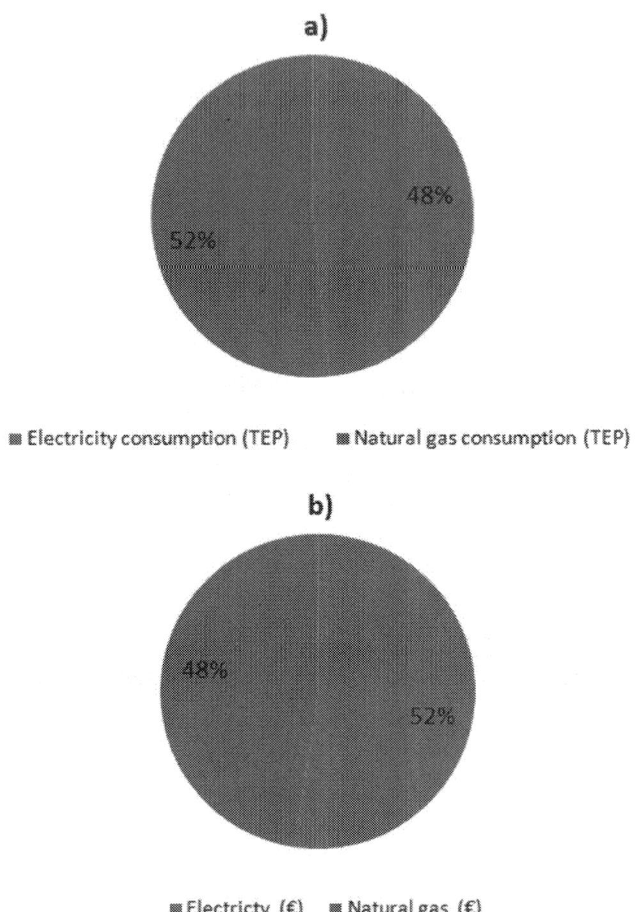

Fig. 4. Distribution of energy consumption (a) and cost (b)

The industrial plant features only one electrical meter on the main electrical transformer booth and only one gas meter on the main panel. Therefore, in order to identify the energy consumption distribution, an assessment has been carried out and a measure campaign of absorbed active electrical power for zone has been performed. As a result the compressors, hydraulic presses and welding machines represented 60% of the whole electrical consumption as shown in Fig. 5a.

A measure campaign of consumed natural gas in each zone of the industrial plant revealed that 72% of the whole gas consumption is consumed for the glazing area (41% for furnace 1 and 31% for furnace 2) and, 28% in the boilers employed for heating and hot water production for grease removal as shown in Fig. 5b.

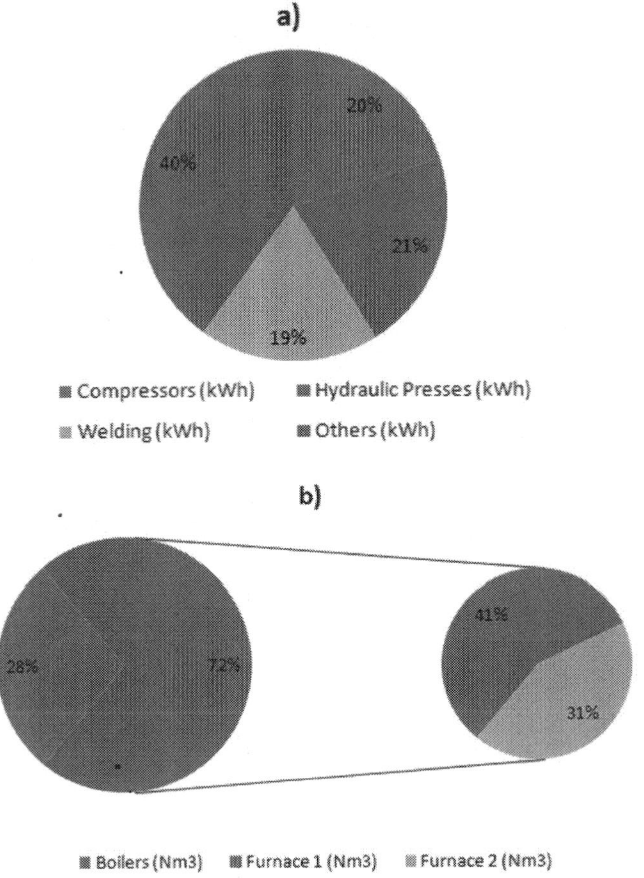

Fig. 5. Distribution of electrical consumption (a) and natural gas (b)

The hourly data recorded by the central electrical meter allow the characterization of the consumption in terms of main load profile of the entire plant and the realization of useful graphical representations as well as the contour map. Two examples of these further analysis are reported in Figure 6 and Figure 7.

A detailed analysis of the energy bills for the previous thirty-six months revealed a mean electricity cost of 0.11 €/kWh, which is sufficiently high to highlight a saving opportunity by changing the electrical contract. On the contrary, the gas costs did not appeal for saving.

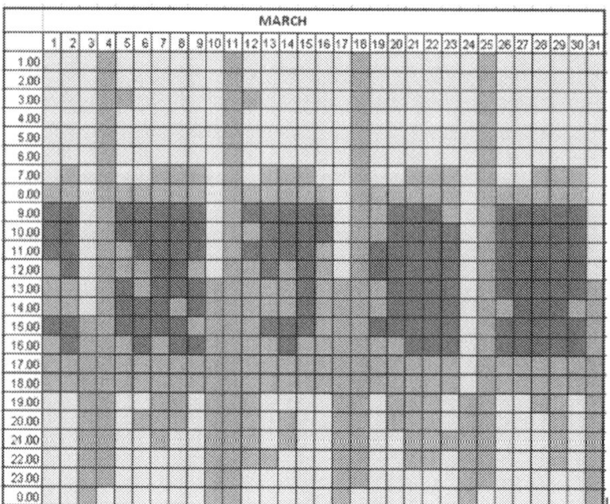

Fig. 6. Contour map: electrical consumptions in March

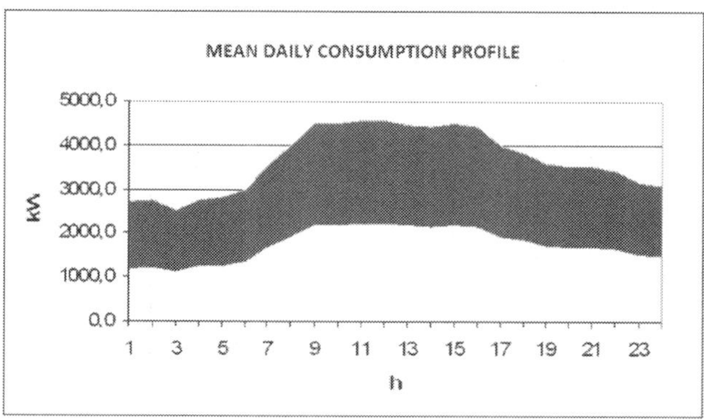

Fig. 7. Mean load profile for a weekday

Therefore after this preliminary step, following the methodology step 3, the energy drivers for the characterization of electrical consumption of the main and unique electrical meter and gas meter have been identified. As already remarked, the production volume has been grouped into 5 product families, assumed to be the energy drivers for electricity consumption.

Data related to the electrical consumption of the years 2005, 2006, 2007 with monthly time resolution and the production volume per each product family have been considered and the statistical software MINITAB has been used . The output of the regression model is:

$$C\left(\frac{kWh}{month}\right) = 460\,602\left(\frac{kWh}{month}\right) + 4.21\left(\frac{kWh}{unit}\right)\cdot V1\left(\frac{u}{month}\right) + 1.55\left(\frac{kWh}{unit}\right)\cdot V2\left(\frac{u}{month}\right) +$$

$$+4.42\left(\frac{kWh}{unit}\right)\cdot V3\left(\frac{u}{month}\right) + 5.51\left(\frac{kWh}{unit}\right)\cdot V4\left(\frac{u}{month}\right) + 10.5\left(\frac{kWh}{unit}\right)\cdot V5\left(\frac{u}{month}\right) \quad (27)$$

```
Predictor      Coef   SE Coef        T        P
Constant     460602     34472    13,36    0,000
H01           4,212     1,018     4,14    0,000
H02           1,547     1,044     1,48    0,149
H03           4,423     1,067     4,14    0,000
CP4          5,5098    0,9181     6,00    0,000
CP5          10,527     1,147     9,18    0,000


S = 3304,60    R-Sq = 87,6%    R-Sq(adj) = 85,6%


Analysis of Variance

Source            DF            SS           MS        F        P
Regression         5    2323097489    464619498    42,55    0,000
Residual Error    30     327611160     10920372
Total             35    2650708649
```

Fig. 8. Statistical parameter for model validation

The statistical validation of the regression model, shown in Figure 8, is characterized by a squared regression coefficient, R^2, of about 87.6%, thus denoting a strong correlation. The p-values in the table show the reliability of the regression coefficients and the analysis of variance shows a controlled residual error. The model is consistent for the analysis: after the analysis of statistical measures and their positive results we can discuss the characterization model and in particular we pointed out that 460 602 kWh/month was a consumption independent of production volumes.

As the energy consumption model is now available, the forecast of energy consumption in the year 2008 can be performed on the basis of the predicted production volume of each product family for the same year provided by the company.

By this way it is possible to work with a reliable forecasting consumption for the contract renewal, following the methodology in the step 5.

The original electric energy contract features three time bands, F1, F2 and F3, with unit costs 0.15 €/kWh, 0.09 €/kWh and 0.06 €/kWh, respectively. The consumptions distribution per band was 41% in F1, 53% in F2, 6% in F3.

The methodology application to this plant allowed the contract renewal, enabling the choice of the best tariff among the Italian free energy market. Ten different tariff proposals (both fixed and combustible basket linked) considering 2 (peak – off-peak) and 3 (F1, F2, F3) bands have been considered and compared. The tariff proposals and the resulting energy costs are summarized in Table 7.

The consumption forecast based on the production volumes yields an overall consumption of about 15 GWh/year subdivided into the rate bands as follows:
- 52% in F1, 35% in F2, 13% in F3;
- 81% peak, 19% off-peak.

The predicted consumption per band allows calculating the unit energy cost, in order to identify the optimal electrical tariff. This is the only way to have a clear vision of the electricity unitary price and a homogeneous basis to compose the total cost. Following this approach, the best tariff is the bidder 4 (see Table 7), that is a three-time bands characterized by the following unitary prices (F1 = 0.13 €/kWh, F2 = 0.1 €/kWh, F3 = 0.05 €/kWh).

Fixed	Combustible basket linked	F1, F2, F3	P , OP	# of bidder	Average cost with regression model (€/kWh)	Average cost without regression model (€/kWh)
X		X		**Bidder 1**	**0.1173**	0.1128
X			X	Bidder 2	0.1123	0.1099
	X		X	Bidder 3	0.1157	0.1123
X		X		**Bidder 4**	**0.1091**	0.1093
	X	X		Bidder 5	0.1143	0.1109
	X	X		Bidder 6	0.1122	0.1098
X		X		Bidder 7	0.1201	0.1189
	X	X		Bidder 8	0.1134	0.1109
X		X		**Bidder 9**	**0.1121**	0.1087
	X		X	Bidder 10	0.1267	0.1178

Table 7. Characteristic and calculation of optimal tariff

After contract renewal, the company aimed to understand the evolution of energy cost and consumption and defined a reliable budget, following the methodology in the step 6. Therefore, after energy consumption characterization and prediction, the budget has been calculated considering the following information:

- the forecast of production volume for 2008 that has been provided by the company;
- the electrical energy tariff has been fixed equal to (F1 = 0.13 €/kWh, F2 = 0.1 €/kWh, F3 = 0.05 €/kWh);
- the production has been scheduled 52% in F1, 35% in F2, 13% in F3.

The plant has to be operated mostly during peak hours due to the constraint stated by union agreement and to the convenience of factory workers hourly cost during peak time. This component had more influence on the final product cost than the energy cost. In particular the different products were made in different lines operating simultaneously during the production time. So there was no difference in terms of absorption. The 2008 planned budget was 1 636 500. €. It has been evaluated considering a reliable forecasting of consumption, the best tariff renewal and the optimization of the energy machines management.

The effectiveness of the proposed approach is highlighted by the real energy consumption of the industrial plant in 2008.

The optimal tariff led to a mean energy cost of 0.1091 €/kWh, against 0.1173 €/kWh of the original one, thus yielding a whole saving of about 120 000 €. It is worth of underlying that a tariff comparison on a fair basis could be done thanks to the forecasting model (i.e. integrated approach), as the same comparison based on the simple historical data analysis would have led to wrong choices. Only considering the historical data, in fact, the "best" tariff would have been the bidder 9, a 3 time bands with the following unitary costs F1 = 0.14 €/kWh, F2 = 0.09 €/kWh, F3 = 0.06 €/kWh. The application of this tariff would have given an actual energy cost of 11.21 cent€/kWh, about 7% higher than that given by the bidder 4. Choosing bidder 9 in place of bidder 4 would have led to a loss of 45 000 € in 2008. The industrial plant behavior, in fact, may significantly change from year to year, especially in the case of multi-product plants, thus leading to energy drivers modifications. This means that simply employing historical energy consumption data would not take into account these changes, thus leading to wrong conclusions. It is obvious that the more the industrial plant production is variable, the more the integrated approach is effective.

In relation to the energy budgeting, the planned budget error was only of 1% relating to the actual data for energy expense for the 2008. Formulating the energy budget only considering the historical data and the old tariff not renew, we would have obtained a budget of 1 173 000 € with an error of 10% respect the actual energy expense for the 2008 even under hypothesis to increase the forecasting of 30% linked to an increase of the production volume. This error would have entailed not correct allocation of the budget cost with a consequence on the final cost balance of the year.

For the 2008, in order to monitor the energy intensive areas of the plant, the company decided to install both electrical and gas meters in the plant. A measure campaign has been carried out as described above in paragraph 5.4. Accordingly to the previous consumption splitting up, following the methodology step 4, the planned distribution of electrical and gas meters are shown in Figure 9. An energy information system has been implemented in order to analyze energy data and to control real time the consumption following the methodology step 7.

Measuring system installation allowed to implement a real time control of consumption both on compressors and hydraulic presses. The authors show an application on the hydraulic press as an example. First of all the statistical model of electrical consumption has been defined considering as energy driver the strokes of hydraulic press at quarter hour (strokes/15 min).

A linear regression model has been built on the hydraulic press meter, with a quarter hour time resolution, as follows:

$$C(kWh)=6.5 \ (kWh)+ 0.5 \ \left(\frac{kWh}{strokes}\right) \ S(strokes) \tag{28}$$

$$R^2=98\% \tag{29}$$

Then a CuSum control chart has been implemented to monitor deviation to normal consumption. The cumulative sum of difference between actual and predicted value of consumption was automatically plotted on the chart as in Figure 10. The CuSum can be used to monitor consumption process variability and it allowed to distinguish between random variability and variability due to different utilization conditions. Such a situation occurred as energy drivers were included in the predicting model. Hence a deviation in normal consumption is pointed out when the points in the chart exceed a previously defined statistical limit. The CuSum were implemented and automatically upgraded with data registered by electrical meters and sensors.

Figure 10 shows part of the CuSum evolution. In the first part the CuSum has a flat trend and is below the first limit value, thus highlighting a good agreement with the prediction of the consumption model.

Then a significant and progressive increase is observed, due to an unexpected energy consumption rise, which is to say an extra energy consumption not related to the chosen energy drivers.

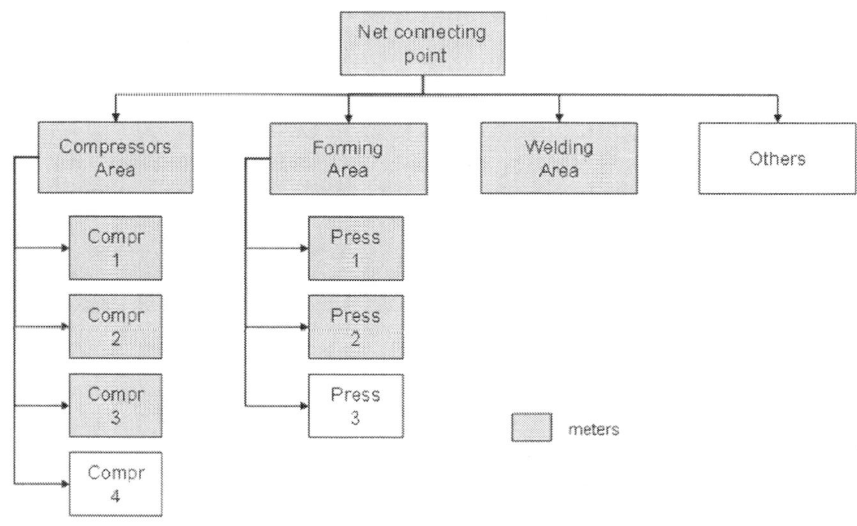

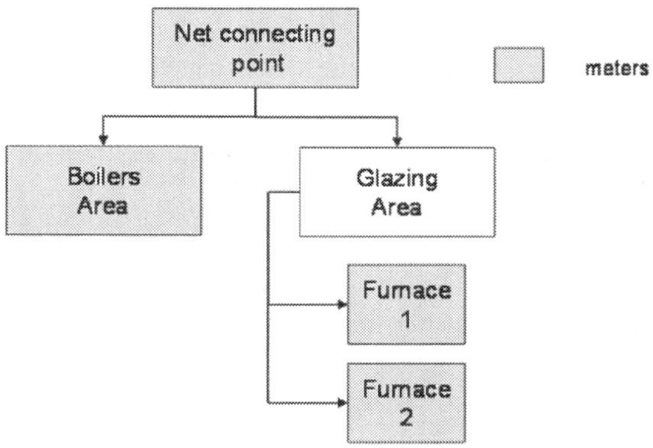

Fig. 9. Distribution of electrical and gas meters

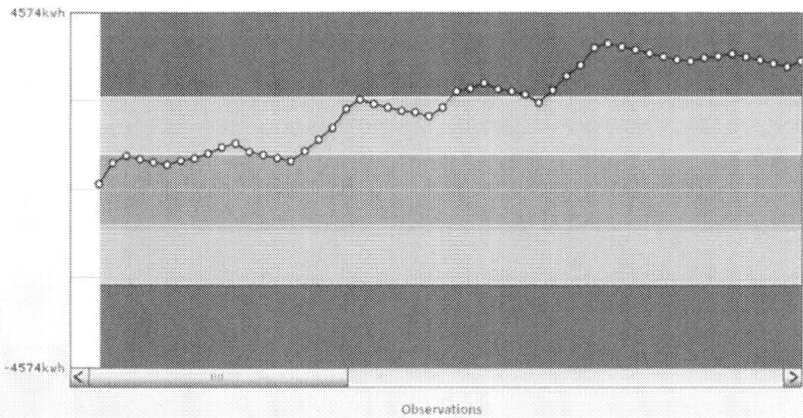

Fig. 10. CuSum of the hydraulic press energy consumption

Due to the modality of CuSum construction a meaningful change in the slope of the curve highlights the presence of energy consumption anomalies. A warning or an alarm for the operator could be set when the CuSum reaches an upper or a lower limit. As proposed in (Cesarotti et al., 2010), the first (warning) limit values are the $\pm3\sigma$ of initial population, the second (alarm) limit values are set evaluating the particular sensitiveness of the monitored users.

Using these limit values, an alert has been given (in October 2008) to point out that energy was being wasted; the emerged problems were essentially linked to bad maintenance procedures and an excessive heating of hydraulic oil.

The improvement in these two topics bring a great change in the press performance, as it's reported in Figure 11; in Table 8 an estimate of the reached saving is also described.

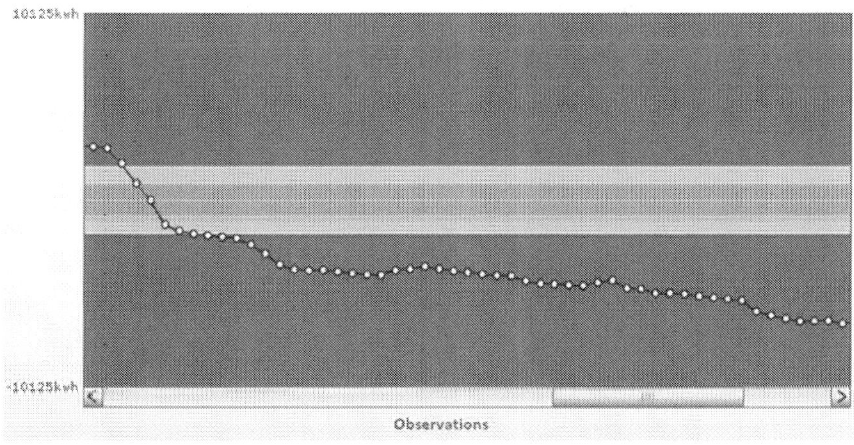

Fig. 11. CuSum of the hydraulic press energy consumption after maintenance

	kWh/year	€/year
2008 assessment	827 030	€ 104 206
2009 assessment	734 518	€ 92 549
Difference	92 513	€ 11 657
% Saving	11%	

Table 8. Savings evaluation

The implemented method allowed a control that it was not a simple monitoring of the actual consumption of the hydraulic press but it was a control based on the comparison with the planned consumption. Indeed the planned consumption was referred to the strokes/min that drive the consumption of the press and statistically reliable. Finally the accurate setting out of the sub-meters in the plant allowed to circumscribe the analysis of deviation.

The use of control chart allowed to find out different behaviors depending on the monitored system as:

- anomalous use of the system (systems or components left on during no operating time);
- physical limit of the system users (i.e. compressor with constant power absorption that does not adapt to variable demand of air of the final user);
- anomalous system operating conditions due to need of maintenance (i.e. inefficient thermal transfers due to calcareous coat, anomalous press consumption due to lack of lubrication, etc.).

Finally the company has been interested, for the strategic future plans, to simulate a power plant to produce energy.

The simulated power plant consisted of a cogenerative gas engine producing part of the plant electrical and thermal energy for hot water and steam. The engine was used to be on during daily time (i.e. 8 a.m. – 18 p.m.) and the other equipments were used to satisfy the company energy loads. No particular strategy was applied to optimize the use of the cogenerative engine. The power system behavior has been translated into a mathematical model, as the one described in (Andreassi et al., 2009), which emulates the energy/mass balances existing between the power plant and the building. The model allows matching the industrial plant energy demands (electricity, hot water, cold, etc.) through an analysis of the system performance characteristics, taking into account the main subsystems integration issues, their operation requirements and their economic viability. All the integrated equipments are considered as energy converters. They are characterized by inputs and outputs and are modeled as black-boxes. Conservation equations are considered to solve each subsystem with a quasi-steady approach (i.e. the variables are considered constant between two time-steps). Simulations are performed pursuing the goal of determining conversion efficiency and energy cost with optimised equipment operation, in order to satisfy specified criterion. In this case the minimum energy cost have been chosen as the optimization criterion (other could be minimum fuel consumption or minimum pollutant emissions). It is worth to underline that this kind of analysis takes into account the possibility of selling excess energy and the different cost of the same fuel as a function of its utilization (i.e. different taxes are applied if the same fuel is used for heat or electricity production).

Beyond the saving obtained through the power plant management optimization, it is important to highlight its strong correlation with the other methodology steps, and in particular the forecasting model and the tariff analysis. The economic and consumption advantages descending from a comprehensive application of the proposed methodology is shown in Table 9. As expected an increasing modules integration maximized the cost saving that was about 220 000 €/year.

Electrical power (cosφ=1)	kW$_e$	1 063
Thermal power	kW$_e$	642
Electrical energy	kWh$_e$	2 750 194
Thermal energy for hot water about 90°C	kWh$_t$	1 694 880
Thermal energy for steam	kWh$_t$	1 502 160
A) Electrical energy costs	€	336 649
B) Hot water energy costs	€	61 721
C) Steam costs	€	54 703
D) Natural gas costs	€	233 601
Saving	€	219 472

Table 9. Economic plan of the investment

6. Conclusions

A methodology pursuing the energy management improvements is presented. Each step constituting the proposed process is illustrated, underlying the main operational aspects and the distinctive characteristics. The relations between the methodology steps and some significant results emphasizing the main aspects are reported.

In particular the importance of establishing a complete monitoring system is underlined and the methodological instruments for controlling the energy performance of an organization are described. The proposed methodology helps the organizations to establish an effective energy management system which can:

- develop and understand of how and where energy is used in the facility;
- develop and implement a measurement method to provide feedback that will measure performance;
- benchmark energy use against other comparable facilities to determine how energy efficient an organization is;
- identify and survey the energy using equipment;
- identify energy conservation options and prioritize their implementation into an energy management plan;
- review the progress on an ongoing basis to determine the program's effectiveness.

The application of this methodology to a case study highlights the effective convenience of this approach. The data collection and analysis allowed the characterization of the energy profile of the organization, in terms of consumption, costs and future trends. Useful instruments (as the contour map and the mean profiles) have been applied. A forecasting model has been calculated for studying the future consumption and make possible correct budget consideration: in particular a 10% saving has been obtained with a contract renewal and the final error in budget allocation is about 1%. The case study also demonstrated the effectiveness of an energy monitoring system in order to identify in short time inefficiencies of the energy users; it allows a rapid alarm and the possibility to plan the necessary actions to reduce energy costs. In this case the organization cost reduction was 11%, eliminating inefficiencies in the hydraulic press.

7. References

Andreassi, L., Ciminelli, M.V., Feola, M., Ubertini, S., (2009). Innovative method for energy management: Modeling and optimal operation of energy systems, *Energy and Buildings Volume*, vol.41, pp. 436-444

Arivalgan, A., Raghavendra, B.G., Rao, A.R.K., (2000). Integrated energy optimization model for a cogeneration in Brazil: two case studies, *Applied Energy*, vol.67, pp. 245-263

Barbiroli, G., (1996). New indicators for measuring the manifold aspects of technical and economic efficiency of production processes and technologies, *Technovation*, vol. 16, No.7, pp. 341-374

Brandemuehl, M.J., Braun, J.E., (1999). *The impact of demand-controlled and economizer ventilation strategies on energy use in buildings*, ASHRAE Trans 105 (Part 2), pp. 39-50.

Cape, H. T., (1997). *Guide to Energy Management*, Fairmont press inc.

Carbon Trust, (1996). Good Practice Guide 200, A strategic approach to energy and environmental management

Carbon Trust, (2001). Good Practice Guide 306, Energy management priorities: a self assessment tool

Carbon Trust, (2007). CTV 023, Management overview-Practical energy management

Carbon Trust, (2007). CTV 027, Metering. Introducing the techniques and technology for energy data management

Carbon Trust, (2007). CTV 027, Technology Overview, Metering. Introducing the techniques and technology for energy data management

Carbon Trust, (2007). Practical guide 112, Monitoring and Targeting in a large companies

Carbon Trust, (2007). Practical guide 231, Metering. Introducing information systems for energy management

Cesarotti, V, Di Silvio, B., Introna, V., (2007). Evaluation of electricity rates through characterization and forecasting of energy consumption: A case study of an Italian industrial eligible customer, *International Journal of Energy Sector Management*, vol.1, No.4, pp. 390-412

Cesarotti, V, Di Silvio, B., Introna, V., (2009). Energy budgeting and control: a new approach for an industrial plant, *International Journal of Energy Sector Management*, vol.3, No.2, pp. 131-156

Cesarotti, V., Deli Orazi S., Introna, V., (2010). Improve Energy Efficiency in Manufacturing Plants through Consumption Forecasting and Real Time Control: Case Study from Pharmaceutical Sector, *APMS 2010 International Conference Advances in Production Management Systems*

Demirbas, A, (2001). Energy balance, energy sources, energy policy, future developments and energy investments in Turkey, *Energy Conversion Manage*, vol.42, pp. 1239–1258

Di Silvio, B., Introna, V., Cesarotti, V., Barile, F., (2007). Condition based maintenance of industrial cooling system through energy monitoring and control, *9th International Conference on The Modern Information Technology in the Innovation process of the Industrial Enterprises (MITIP 2007) Proceedings*, pp. 327-332

Elovitz, D.M., (1995). *Minimum outside air control method for VAV systems*, ASHRAE Trans 101 (Part 1), pp. 613–618

Farla, J.C.M., Blok, K., (2000). The use of physical indicators for the monitoring of energy intensity developments in the Netherlands, 1980–1995, *Energy*, vol.25, pp.609–638

Frangopoulos, C.A., Lygeros, A.L., Markou, C.T., Kaloritis, P., (1996). Thermoeconomic operation optimization of the Hellenic Aspropyrgos. Refinery combined cycle cogeneration system, *Applied Thermal Eng.*, vol.16, pp. 949-958

Kannan, R., Boie, W., (2003). Energy management practices in SME––case study of a bakery in Germany, *Energy Conversion and Management*, vol.44, pp. 945-959

Krakow, K.I., Zhao, F., Muhsin, A.E, (2000). *Economizer control*, ASHRAE Trans 106 (Part 2), pp. 13–25

Levine, D. M., Krehbiel, T. C., Berenson, M. L., (2005). *Basic Business Statistics: Concepts and Applications*, Prentice Hall, NJ, USA

Montgomery, D.C., (2005). *Design and Analysis of Experiment*, Wiley, New York, NY

Petrecca, G., (1992). *Industrial Energy Management*, Springer NY

Piper, J., (2000). *Operations and Maintenance Manual for Energy Management*, Sharpe Inc.,NY

Puttgen, H.B., MacGregor, P.R., (1996). Optimum scheduling procedure for cogenerating small power producing facilities, *IEEE Trans Power Systems*, vol.4, pp. 957-964

Sarimveis, H. K., Angelou, A. S., Retsina, T. R., Rutherford, S. R., Bafas, G. V., (2003). Optimal energy management in pulp and paper mills, *Energy Conversion and Management*, vol.44, No.10, pp. 1707-1718

Skantze, P., Gubina, A., Ilic, M., (2000). *Bid-based Stochastic Model for Electricity Prices: The Impact of Fundamental Drivers on Market Dynamics*, Energy Laboratory Publications, Massachusetts Institute of Technology, Cambridge, MIT EL 00–004

Temir, G., Bilge, D., (2004). Thermoeconomic analysis of a trigeneration system, *Applied Thermal Energy*, vol.24, pp. 2689-2699

Tstsaronis, G., Pisa, J., (1994). Exoergonomic evaluation and optimization of energy systems – application to the CGAM problem, *Energy*, vol.19, pp. 287-321

Tstsaronis, G., Winhold, M., (1985). Exoergonomic analysis and evaluation of energy conversion plants. I: A new methodology. II: Analysis of a coal-fired steam power plant, *Energy*, vol.10, pp.81-84

Von Spakovsky, M.R:, Curtil, V., Batato, M., (1995). Performance optimization of a gas turbine cogeneration/heat pump facility with thermal storage, *Journal of Engineering of Gas Turbines and Power*, vol.117, pp. 2-9

Weron, R., (2008). Market price of risk implied by Asian-style electricity options and futures, *Energy Economics*, vol.30, pp. 1098–1115

Worrell, E., Price, L., Martin, N., Farla, J.C.M, Schaeffer, R., (1997). Energy intensity in the iron and steel industry: a comparison of physical and economic indicators, *Energy Policy*, vol.25, pp. 727–744

Energy Optimization:
a Strategic Key Factor for Firms

Stefano De Falco
School of Sciences and Technologies
University of Naples
Italy

1. Introduction

This chapter will discuss aspects related to the variables of firm governance from the viewpoint of energy optimization. This is an important aspect because, in the current highly competitive market, in addition to competing on the characteristics of the products produced or services rendered, become strategic factors also important parameters of production efficiency, which often force companies to relocate in remote areas where energy costs of production are lower. Instead, another possible solution is to increase the efficiency of its industrial system for an enterprise of production or reduce consumption of any enterprise in the field of services to avoid such delocalization.

The recent debate on the energy has seen a plurality of views and actions initiate a broader discussion of what does not happen just a few years ago. The combination of environmental effects is clearly measurable emissions generated by anthropogenic climate, and the crisis in prices energy produced with the explosive global demand, has produced a transformation of the importance of that perspective as to the terms of a violent debate acceleration, such as to require all players to such sensitive issues to rethink their positions.

The weight of energy production from renewable sources of total production, continues to be dramatically lower, and not aligned to the objectives of reduction of emissions. This is compounded by the fact that, in the price system of fossil of today, the cost of Kilowattora product with the most economic renewables now available (large wind blades in windy areas) is more that three times that produced by traditional methods, such as from coal. This heavy gap making it unacceptable to think that the solution to the problem could come from the side of improvement in the production of energy, shows that the greatest gains can be reached quickly and with more low investment costs are on the energy savings. This is essentially to rethink the development model, especially for urban development and settlement, identifying ways in which to reach the lowest levels of energy consumption while maintaining sustainable economic growth rates, breaking the existing link between economic growth and energy consumption.

2. Some emerging issues

A. Currently, energy policies are all related to buildings existing and / or new construction, (supported by a large number of cultural projects), and the rules are all finalized to the improving of the climate, with a capacity of incision extremely limited if proportionate to the complexity of the topic. The above actions will inevitably occur with a frequency much time slow (30-40 years). This makes this process slow, in fact, the response generated by interventions are not commensurate, neither predictable in terms of quantity, with the development of environmental problems and the availability of sources fossil energy occurring currently underway at both local and global. The awareness of this situation requires a different and wider strategy approach to the problem. Evidences of the endemic slow, causes of the fragmentation in the standards and the establishment of initiatives for energy policies, are the lack of rules for the approval and the low implementation of facilities for the production of renewable energy. Because of this gap, the regulatory framework, characterized by a highly fragmented, leads to a different approach from region to region, often hostile towards the projects.

B. It is now given irrefutable that the heart of the problem of climate emissions is physically concentrated in medium and big cities, in which the temperature is higher than at least two degrees compared to less densely urbanized area. Hence the choice in European headquarters, to identify as the seventh thematic strategy of the urban environment, complex and multi-space within which it manifests the need for mandatory affirmation of the principle of integration of environmental policies on the "other" policies. In a large number of activities now under way around the energy issues, the environment fails to a systematic approach, which sees the re-location of different actions and initiatives. The theme of this strategy, which refers to the concept of integration, limits to the urban environment to its scope. From the perspective of the territorial structure is precisely this point today debate. More and more forms of settlement are abandoning the traditional partition between city and countryside, while the settlement process more violent and more consumption of soil invest today the wide margins of regional transport infrastructure road, with the inevitable growth in demand for private mobility by road, adding unsustainable land (waterproofing, concrete) unsustainable environmental (pollution, release of CO_2) and unsustainable energy. The model of environmental thought to determine the benefits of a program reordering settlement should first assess the savings resulting from the indicators such as:
- demolition of buildings that spend Energy
- reconstruction of buildings zero emissions and implementation of integrated systems for urban production and distribution of energy (central heating, cogeneration, tri-generation, biomass, etc.).
- reduction of land (increased density)
- reduce the heat to a local scale (less surfaces paved / cemented to the highest density)
- a reduction in private mobility mass (less commuting to distant destinations, less commuting to the exchange with the iron)
- reduction of congestion (traffic flowing more)
- increased pedestrian generated by the deployment of new centrality around Iron stations;

C. The spatial diffusion of contemporary forms of renewable energy production (wind, solar active and passive generation of biogas, etc..) is now changing the historical characteristics of the national electricity grids. Where once his role was to distribute energy produced in the territory in a few centralized energy policy, the spread of those new ways of sustainable production and the liberalization of electrical output measures is relying increasingly on the network collection of role of energy. No longer a one-way, but a network of integration / interdependence. In turn, the infrastructure of a national scale is not most describe as the backbone infrastructure in charge of bringing the energy from one end to another country, but becomes the infrastructure for interconnection of territories production consumption characterized by its energy self. This conceptual transformation but it is not happening in terms of reality. These new features needed in relations between the network and the area also produced a new conceptualization of both the network that the territory itself. The territory in terms of energy changes, becoming space liabilities through a field by the interconnected through active infrastructure, and each system has territorial identifiability thus allowing the Construction of a specific energy balance and sustainability assessments energy - environment, even in view of the allocation of certificates to the white under the Kyoto Protocol. The provision requires the independent choice closest between the "collection" of renewable energy technology and their use, namely the orientation the potential for further investigation on natural land.

D. Finally, the liberalization initiatives in the field of municipal is producing, in different contexts, groups of companies in multi-communal area forming a system of spatial mesh already made substantially

corresponding to the spatial mosaic of local energy markets over recalled. There is an optimum growth Multiutilities beyond which the costs the complexity of risk management overhang the benefits from synergies. It is not can identify the optimum size, but expected to read the current processes aggregation is coming to set up poles "regional".

But what now takes on greater significance is the only partial liberalization of markets. This still remains the problem of fragmentation of supply in too many units productive.

3. Energy optimization in industrial farms

In industrial field, one of the most advantages, derived from the industrial automation process implementation, is possibility to regulate process control parameters. This possibility allows to determine an optimal configuration of control parameters, useful to reduce the energy consume and at the same time, to guarantee the same quality level of the production.

The importance of energy usage escalates rapidly due to the international task of reducing global emissions of carbon dioxide. According to a recent research report from Cambridge, significant changes are needed in order to make the industrial system sustainable. Therefore energy becomes an increasingly important issue, especially for the process industries that normally use a relatively large amount of energy. Even though some process industries are not that dependent on external supply of energy, since energy often becomes a by-product when the incoming raw materials are transformed in the main production, effective and profitable use of energy is still an important and strategic issue. In addition, in times of high electricity prices, some process industries are forced to reduce, or even stop, their production, further highlighting the strategic dimension of effective energy planning.

From a general perspective, process industries include firms that deal with powders, liquids, or gases that become discrete during packaging. They include the pipeline industries such as refining, chemical processing, food processing, textiles, and metals.

Process manufacturing is defined as: *Production which adds value by mixing, separating forming, and/or chemical reactions. It may be done in either batch or continuous mode.* Process industries make up a high proportion of the manufacturing operations in the early stages of the overall production cycle of converting raw materials into finished products. Most process industries can be classified as either basic producers or converters, and sometimes a combination of the two. A basic producer is a manufacturer that produces materials from natural resources to be used by other manufacturers, whereas a converter changes these products into a variety of industrial and/or consumer products. As such, process manufacturers would be positioned in the lower right hand corner of the product-process matrix, typically producing commodities in high volume/limited variety.

Whereas fabricators and assemblers can be labor intensive, process industries rather have a high cost of capital invested in facilities and in many cases also a high cost of energy usage. In many process industries the cost of energy can be between 10-20 % of the total cost of goods sold, in other words similar to the cost of direct labor in many labor intensive companies. For process industries with a high cost related to the supply of energy, it is imperative to establish an energy management system and to analyze its effect on productivity and efficiency. The supply of energy also plays a central role for the profitability of the company, in terms of e.g. the relationships linking the value of energy to its influence on product prices. Furthermore, many process industries have the possibility to extract an energy surplus from the by-products, thereby offering the possibility to sell electricity, heating, etc., to the surrounding. Hence, there are many areas to improve and optimize, and effective energy planning plays a central part in overall operations management for many process industries.

In this section an innovative methodology for the productive processes qualification based on quality characteristics improvement and on their simultaneous evaluation cost, is proposed, and an industrial farm (Leghe Leggere spa) application of the proposed technique is discussed.

3.1 The proposed approach

The proposed approach uses statistical tools in original way obtaining an innovative qualification activity in term of measurement, diagnostic and optimisation of industrial systems.

The proposed methodology, is based on the following five steps:

I) Definition of a P-Diagram as reported in Figure 1.

In the diagram of Figure 1 the system performances $\underline{y}^T = (y_1,...,y_v)$ (quality characteristics) are linked to the input signals $\underline{m}^T = (m_1,...,m_q)$ through a certain function. The system desired performances are obtained through opportune control parameters $\underline{x}^T = (x_1,...,x_n)$ that are all system parameters able to change deterministically performances, while $\underline{u}^T = (u_1,...,u_k)$ are the noise factors, whose effects on the performances variations are not controlled by desired deterministic regulations.

For the evaluation of the control parameters and noise factors are used cause-effect diagrams in which all variations of the quality characteristics values, according to existing models or to experimental dates, are attributed to all possible sources.

In the design of system a general function between control parameters Xi (i=1..n) and the quality characteristic selected Y is:

$$y = f\left(x_1, x_2, ..., x_n\right) \tag{1}$$

Control parameters Xi are considered random variables with an evaluated mean $\tilde{x}_i$ and evaluated variance $s^2(x_i)$, than the quality characteristic is

$$\tilde{y} = f\left(\tilde{x}_1, \tilde{x}_2, ..., \tilde{x}_n\right) \tag{2}$$

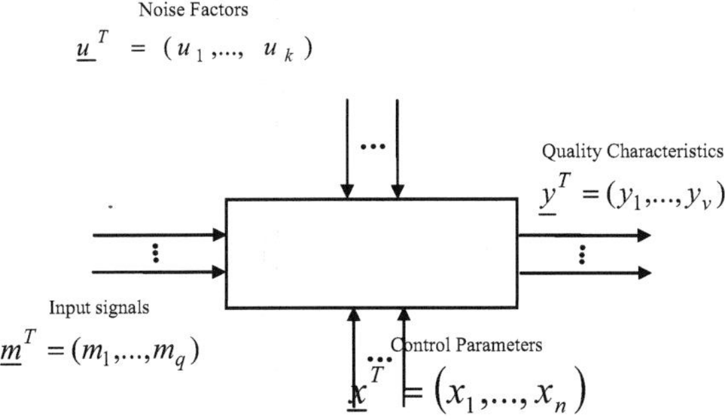

$$\underline{u}^T = (u_1,..., u_k)$$

$$\underline{y}^T = (y_1,...,y_v)$$

$$\underline{m}^T = (m_1,...,m_q)$$

$$\underline{x}^T = (x_1,...,x_n)$$

Fig. 1. P-Diagram.

III) Then, an ANOVA analysis is performed to determine the effective effects of the control parameters selected in the past step on the quality characteristic to be optimized.

Through this technique total variation SST (Total Sum of Square) of the monitored quality characteristic can be divided in more components according to the number of the control parameters.

$$SS_T = [\sum_{i=1}^{N} \eta_i^2] - \frac{T^2}{N} \tag{3}$$

where N is the total number of the experiments, η_i are the objective function values, used to represent the quality characteristic, in the different experiments, and T is the sum of η_i.

Variation of every parameter is estimated from the (4):

$$SS_V = \frac{V_1^2}{n_{V_1}} + \frac{V_2^2}{n_{V_2}} + \frac{V_3^2}{n_{V_3}} +..+ \frac{V_i^2}{n_{Vi}} - \frac{T^2}{N} \tag{4}$$

where V_i is the value of the parameter considered and n_v is the number of times in which it's in 1-level.

Calculus of variations of each parameter allows to know the effective incidence of itself on the quality characteristic. At this aim, variations are normalized in variances through the division of themselves for the degree of freedom (DoF).

Finally a Fisher test is conducted and each parameter variance is compared with error variance and the result compared with the statistical F.

$$F_V = \frac{\sigma_V^2}{\sigma_e^2} \tag{5}$$

With

$$\sigma_V = \frac{SS_V}{g_V}, \quad \sigma_e = \frac{SS_e}{g_e} \tag{6}$$

IV) Then, an experimental design is defined and is performed to reduce the experimental test points. Each control parameter selected needs not less three-variation levels to allow measure its curvature. If system is characterized by too many control parameters is possible use orthogonal matrices, to reduce the experimental plan.

Finally, to optimise the quality characteristic an objective-function is selected. The structure of the objective-function is dependable from the specific case dealt, and the main ones are reported in Literature.

V) In the last step, the costs, related to the improvement activities on the quality characteristic selected, are evaluated through a Quality Loss Function:

$$L(y) = \frac{A_o}{\Delta_o^2}(y-m)^2 \qquad (7)$$

in which $\frac{A_o}{\Delta_o^2}$, generally indicated with "K", is constant defined as quality cost coefficient,

whose determination is conducted fixing the tolerance limit behind output product is reworked and evaluating the relative cost through a complex analysis of all economic impact factors (people, energy use, devaluation).

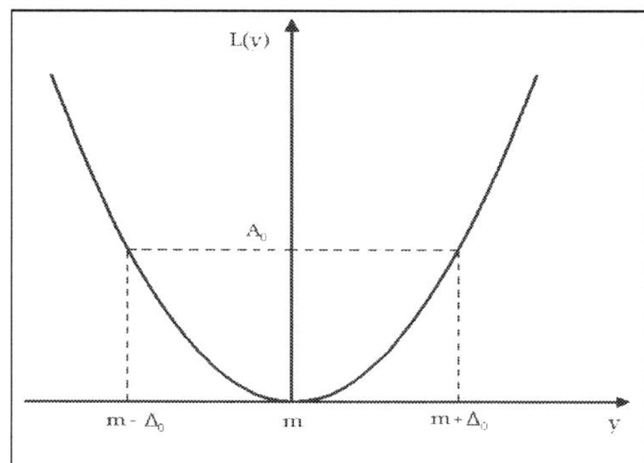

Fig. 2. Quality Loss Function

The expression (7) has to be applied in the two operative conditions, pre and post experimentation, to verify the presence of an increment of cost function. In fact, certainly the quantity (y-m)2 is reduced after the experimentation, as imposed by the objective-function, but the coefficient K value should be incremented according to new distribution of the economic impact factors.

Proposed methodology application allows the industrial processes quality characteristic optimisation, through the choice of the control parameters opportune parametrical combination that make system insensible to noise factors, and through the analysis of the cost function.

3.2 Case study

Here the results of the application of the proposed methodology to industrial farm (Leghe Leggere spa) are reported.

Quality characteristic selected is superficial hardness of the aluminium bar produced in the farm analyzed.

First step is verify of process normality. For this purpose a χ^2 test is been conducted and results are reported in the following.

χ^2 TEST

$$\chi^2_{k-p-1} = \sum \frac{(F_0 - F_t)^2}{F_t}$$ (8)

where:
F_0 = observed frequencies
F_t = theoretical frequencies
k = classes number
p = parameters number
In the dealt case: k=5;p=3 than: k – p – 1 =1
from (8) results:

$$\chi^2_1 = 5.66$$

From the tables, for one DoF and 0.01 significant level, results:

$$\chi^2_{1,0.01} = 6.63$$

It's possible to accept normality Hypothesis with significant level of 0,01 %, id est, a confidence interval of 99,99%.

In table 1 control parameter and their levels are reported.

Levels	$\dfrac{Mg}{Si}$	t_a (h)	T (C°)	T_p (h)
1	$\dfrac{Mg}{Si} = 0,90$ Mg=0,38 % Si=0,42 %	16	175	8
2	$\dfrac{Mg}{Si} = 1,09$ Mg=0,48% Si=S0,Si=0,44%	14	185	6
3	$\dfrac{Mg}{Si} = 1,21$ with Mg=0,58 % Si=0,48 %	12	200	4

Table 1. Levels

In the dealt case, we have a limited nominal value (70 Brinnel) of the quality characteristic, superficial hardness, to be included in the interval 60-80 Brinell, so we have used a signed-target objective function:

$$\eta = -10 \log \sigma^2$$ (9)

From calculus of σ results:

$$\sigma_1^2 = \frac{1}{3}\left(6^2 + 3^2 + 6^2\right) = 27,00 \tag{10}$$

$$\sigma_2^2 = \frac{1}{3}\left(3^2 + 3^2 + 5^2\right) = 14,33$$

$$\sigma_3^2 = \frac{1}{3}\left[(-2)^2 + 0 + 2^2\right] = 2,67$$

$$\sigma_4^2 = \frac{1}{3}\left(15^2 + 13^2 + 8^2\right) = 152,67$$

$$\sigma_5^2 = \frac{1}{3}\left(3^2 + 10^2 + 10^2\right) = 69,67$$

$$\sigma_6^2 = \frac{1}{3}\left[(-1)^2 + (-8)^2 + (-6)^2\right] = 33,67$$

$$\sigma_7^2 = \frac{1}{3}\left(14^2 + 8^2 + 8^2\right) = 108,00$$

$$\sigma_8^2 = \frac{1}{3}\left(10^2 + 15^2 + 10^2\right) = 141,67$$

$$\sigma_9^2 = \frac{1}{3}\left(15^2 + 15^2 + 10^2\right) = 183.33$$

than for the objective function selected results:

Experimental number	$\dfrac{Mg}{Si}$ [%]	Oven remaining time [hours]	Oven temperature [°C]	Oven waiting time [hours]	η [decibel]
1	0,90	8	175	16	-14,31
2	0,90	6	185	14	-11,56
3	0,90	4	200	12	-4,27
4	1,1	8	185	12	-21,82
5	1,1	6	200	16	-18,43
6	1,1	4	175	14	-15,27
7	1,2	8	200	14	-20,33
8	1,2	6	175	12	-21,51
9	1,2	4	185	16	-22,63

Table 2. Experimental results of orthogonal matrix

For the hypothesis of independence of control parameters results:

$$\overline{A_1} = \frac{-14,31 - 11,56 - 4,27}{3} = -10,05 \tag{11}$$

$$\overline{A_2} = \frac{-21,82 - 18,43 - 15,27}{3} = -18,51$$

$$\overline{A_3} = \frac{-20,33 - 21,51 - 22,63}{3} = -21,49$$

$$\overline{B_1} = \frac{-14,31 - 21,82 - 20,33}{3} = -18,82$$

$$\overline{B_2} = \frac{-11,56 - 18,43 - 21,51}{3} = -17,17$$

$$\overline{B_3} = \frac{-4,27 - 15,27 - 22,63}{3} = -14,06$$

$$\overline{C_1} = \frac{-14,31 - 15,27 - 21,51}{3} = -17,03$$

$$\overline{C_2} = \frac{-11,56 - 21,82 - 22,63}{3} = -18,67$$

$$\overline{C_3} = \frac{-4,27 - 18,43 - 20,33}{3} = -14,34$$

$$\overline{D_1} = \frac{-14,31 - 18,43 - 22,63}{3} = -18,46$$

$$\overline{D_2} = \frac{-11,56 - 15,27 - 20,33}{3} = -15,72$$

$$\overline{D_3} = \frac{-4,27 - 21,83 - 21,51}{3} = -15,77$$

Than is possible to reach the optimum configuration of control parameters to maximize the objective function.

Variable	Parameter	Optimum level
A	$\dfrac{Mg}{Si}$	0,90
B	Oven remaining time	4 h
C	Oven temperature	200°C
D	Oven waiting time	14 h

Table 3. Parameters levels optimum choice

ANOVA
Total variation SS_t can be divided in its five components:
SS_A variation owned to factor A
SS_B variation owned to factor B
SS_C variation owned to factor C
SS_D variation owned to factor D
SS_e variation owned to error

$$SS_T = SS_A + SS_B + SS_C + SS_D + SS_e \qquad (12)$$

$$SS_T = [\sum_{i=1}^{N} \eta_i^2] - \frac{T^2}{N} \qquad (13)$$

$SS_T = 290,03$

$$SS_V = \frac{V_1^2}{n_{V_1}} + \frac{V_2^2}{n_{V_2}} + \frac{V_3^2}{n_{V_3}} - \frac{T^2}{N} \qquad (14)$$

with V = A, B, C, D
Main effects of control parameters are shown in table 9.

	A	B	C	D
1	-30,14	-56,46	-51,09	-55,37
2	-55,52	-51,50	-56,01	-47,16
3	-64,47	-42,17	-43,03	-47,60
TOTAL	-150,13	-150,13	-150,13	-150,13

Table 4. Main effects.

$SS_A = 211,42$
$SS_B = 35,09$
$SS_C = 28,62$
$SS_D = 14,21$
$SS_e = SS_A + SS_B + SS_C + SS_D - SS_T = 0,68$
Fischer Test
Fischer test results are shown in table 10

Source	Variation (SS)	DoF	Variance	F
A	211,42	2	105,71	459,70
B	35,09	2	17,54	76,26
C	28,63	2	14,32	62,26
D	14,21	2	7,10	30,86
e	0,68	3	0,23	
T	290,03	11		

Table 5. Fischer Results

From the table results that quality characteristic variation is owned to parameter variation and not to the error, and above all it depends from factor A variation.

In the last step, the costs, related to the improvement activities on the quality characteristic selected, are evaluated through Quality Loss Function:

$$L(y) = \frac{A_o}{\Delta_o^2}(y-m)^2 \tag{15}$$

$$k = \frac{A_0}{\Delta_0^2} = \frac{0,71}{10^2} = 0,0071 \tag{16}$$

Optimum combination of control parameters that maximize the objective function η is $A_1\, B_3\, C_3\, D_2$.

Than, in these conditions, mean μ results:

$$\mu = \overline{A_1} + \overline{B_3} + \overline{C_3} + \overline{D_2} - 3\overline{T} = -4,13 \tag{17}$$

and:

$$-10\log\sigma^2 = -4,13 \Rightarrow \sigma^2 = 2,59$$

So results:

$$L_d(y) = \frac{A_0}{\Delta_0^2}(y-m)^2 = k\sigma^2 = 0,02\left[\frac{\text{€}}{kg}\right] \tag{18}$$

Pre –experimentation control parameter combination was A_1, B_2, C_2 e D_2, characterized by μ :

$$\mu = \overline{A_1} + \overline{B_2} + \overline{C_2} + \overline{D_2} - 3\overline{T} = -11,57 \tag{19}$$

and:

$$-10\log\sigma^2 = -11,57 \Rightarrow \sigma^2 = 15,35$$

than the pre-experimentation value of quality Loss was:

$$L_p(y) = \frac{A_0}{\Delta_0^2}(y-m)^2 = k\sigma^2 = 0,11\left[\frac{\text{€}}{kg}\right] \tag{20}$$

from (10) and (12), after experimentation results a economic improvement of:

$$L_p(y) - L_d(y) = 0,11 - 0,02 = 0,08\left[\frac{\text{€}}{kg}\right] \tag{21}$$

In this case, the proposed technique has produced a strong improvement of the quality characteristic selected (superficial hardness of aluminium bar) and contemporary has produced a reduction of associate productive unitary cost through the preliminary check of the critical productive phases in term of energy use.

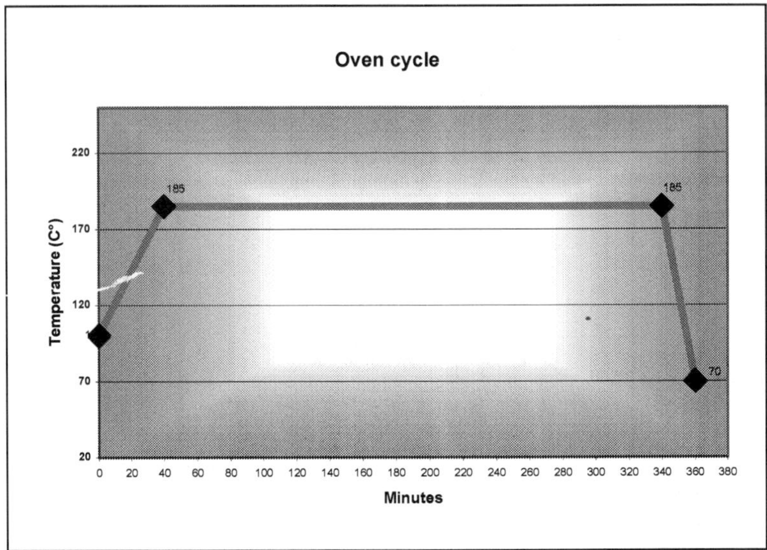

Fig. 3. Oven cycle

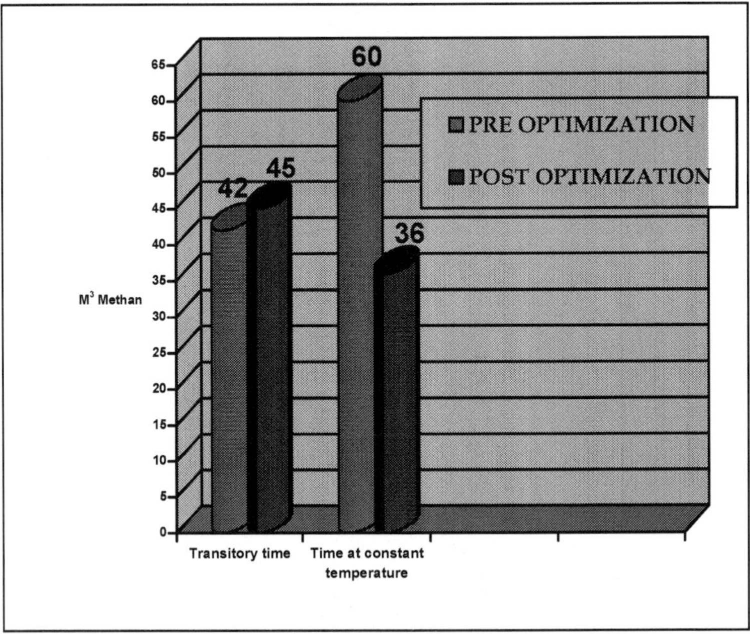

Fig. 4. Methane consume

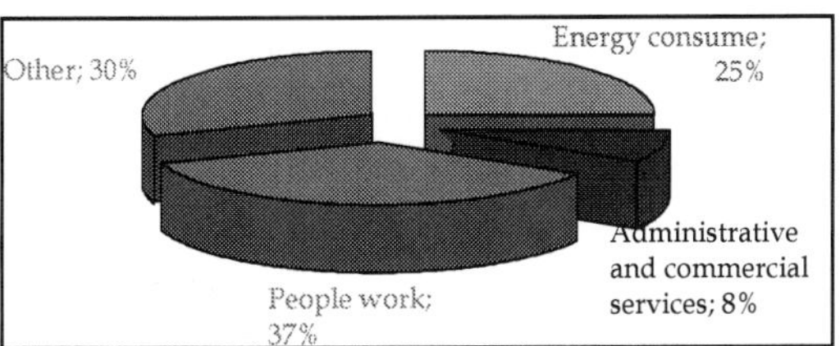

Fig. 5. Distribution of impact factors for k calculus

4. Energy optimization in service farms

The present section gives a quality measurement methodology based on a complex analysis of internal and external indicators and of the links existing between the two ones, oriented to the energy optimization in service farms.. The result of the proposed methodology application is to dispose of an operative tool to apply appropriate corrective actions to get the quality characteristic monitored on the nominal value. In this section an application of the proposed methodology to a water supply company is proposed. The starting assumption is based on the belief that delivering a service through a control quality system is a condition that ensures an energy optimization in the work processes of the company.

Nowadays we are attending the continuous proliferating of Quality Systems applied in more and more several fields; but differently from some years ago, a recent trend turns the use of these models not only towards the supplying of products but also of services. This development exercises an ever-growing influence on the organization and management of those companies interested to keep step in a competitive environment like the modern one. Moreover this new approach upsets the traditional economical policy, in which, not the efficiency, but the profit is in first place. On the other hand, presence of non-quality, results, on the whole, more onerous than to adopt a Quality Management System. So quality measurements hold an important role, proving certified information about the efficaciousness and the efficiency of a productive process.

The modernizing process is involving also the Public Organizations, as the Utilities Supplying Companies. Care must be addressed, above all, to organizations that supply indispensable public services: electric power, water, gas. In fact, often, a monopolistic management characterizes the distribution of these Utilities; this is due, in most cases, to high production costs that would make difficult the rising of a more competitive environment of small and medium enterprises. Absence of an alternative choice for the consumer could take off any stimulus at continuous improvement that instead is a typical result of the competition presence. Therefore the Quality Measurements can assure an objective valuation of the offered service quality, and the characterization of right quality indexes can assume an essential function in the definition of those criteria, that are basic in the modern process of optimization for services production and management.

4.1 The proposed methodology

Diffusion of new approach has certainly contributed to give a more managerial feature to the Public Organization, pursuing as a target the efficaciousness but also the efficiency of the service's delivery. This line of action left the hierarchic structure that put at the top the object of service and believed less important the management aspects of its delivery. The starting point is represented by a new vision of the service delivery, where a circular structure get a foothold; so the object of service represents simply a basic service that develops oneself in the delivery of an infrastructural service.

The purpose is to verify the features conformity of any supplied service at the prefixed targets. So it will lead to single out a set of internal and external quality indexes; the former represent a direct measure of the quality for the infrastructural service as regards the internal process, the latter represent a measure of the quality perceived from the user. The developed procedure is characterized by two stages, which allow to make systematically a detailed analysis of the problem and to execute properly the quality measurements. The first one is the planning stage; it constitutes a preliminary step for preparing the measurement process. In this stage it needs to define the measure variables, which describe better the case examined. A useful tool is the 'processes approach', that is to single out the component processes of the internal and external activities accomplished by the Organization, with the respective responsibilities. Later on it's opportune to define the quality indexes to monitor and their ranges with the enclosed corrective actions. The last step foresees the choice of the internal and external indexes for every process with the reciprocal relations of dependence; the last allow executing the efficacious corrective actions. A particular care must be addressed also for the choice of the informative system.

The second stage is relative to the execution of the measurement process, from the data collection to their interpretation. It consists also to realize a control panel for verifying the conformity of data to the fixed ranges, and if necessary for adopting the corrective actions. A positive aspect of the proposed methodology is its possible application on any Organization or Management System. Now below we present its validation on a concrete case: a Water Supply Company of a big City.

4.2 Validation field: a water supply company

The process of water supply for the considered Organization can be schematized by a block, where two interfaces are present, on one hand there is the Company, on the other hand the final service user. By the 'processes approach', it's possible to recognize three structural processes that are representative of the Organization Core Business: a) management of the installations and water network; b) management of the relation with the consumers; c) monitoring of water quality. The total output of these processes forms altogether the service delivered to the consumers, moreover by a careful analysis it's to observe a reciprocal influence among the structural processes, such interaction is schematized by other four infrastructural processes for internal services: a) management of the provisions; b) management of the staff professional training; c) process of internal communication; d) management of measure equipment.

This approach results propaedeutic to define a set of opportune quality indexes of the several processes, in order to value the conformity of the delivered service at the fixed ranges. The indexes singled out are classified as internal ones for checking the internal service efficiency, and external ones for valuing the service efficaciousness and the customer's satisfaction. The former are the warning lights of a complex control panel, that is able to indicate possible out-control situations. The tools used for the preliminary analysis are graphic instruments, as the graphs of the index trend in comparison with the average level and control limits, histograms and radar charts. Besides by cause-effect diagram it has been possible to proceed with the Decision Making Analysis (DMA), in order to search for correlations among the indexes.

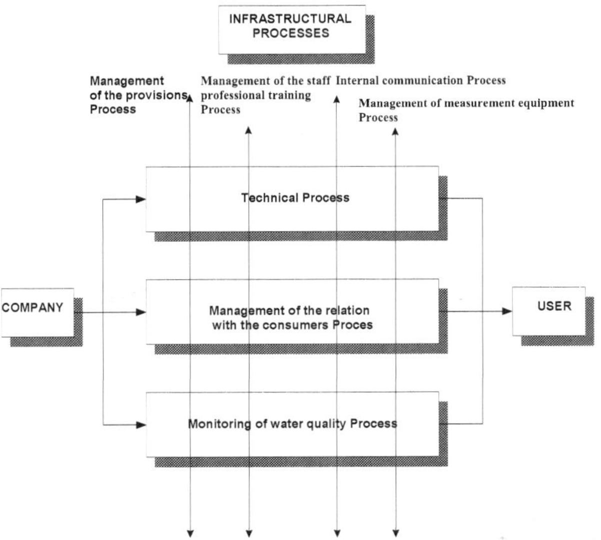

Fig. 6. Block-diagrama

Internal Quality Measures

In table 6 quality internal indexes for a water supply company are reported in term of effective measured value and its standard value.

In figure 7, as an example, a point to point trend of estimation time is reported; in the graph is also reported the standard value line, values measured average and trend line.

In figure 8, a histogram of estimation time is reported; in the graph is also reported the standard value line.

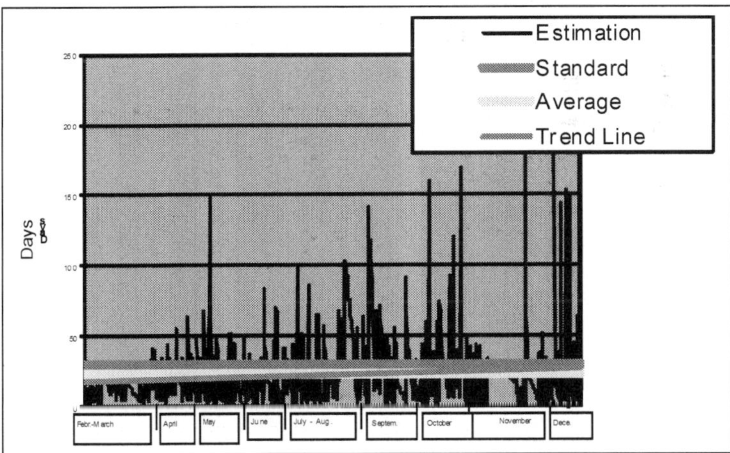

Fig. 7. Estimation Time – Point to point trend

QUALITY INDEX	Standard	Average Values		Absolute Values	
		Average	Average Position vs STD (%)	N° Int.	Out of STD
Estimation Time	30 days	22,64	75,47	1020	20%
Works Execution Time	60 days	12,25	20,42	1745	1%
Connection Time	10 days	6,93	69,30	1275	11%
Contract Cessation Time	30 days	17,92	59,74	2263	7%
First Intervention Time	8 hours	1,50	23,00	1865	1%
Service Restoration Time	24 hours	15,57	66,48	1378	19%
Service Reactivation Time after Payment	1 days	1,02	101,89	53	21%
Check Time of Water-meter	30 days	12,58	41,94	24	0%
Notification Time of Water-meter Operation	30 days	15,05	50,16	21	9%
Response Time for Complaint	30 days	17,86	59,53	85	0%

Table 6. Example of monitoring

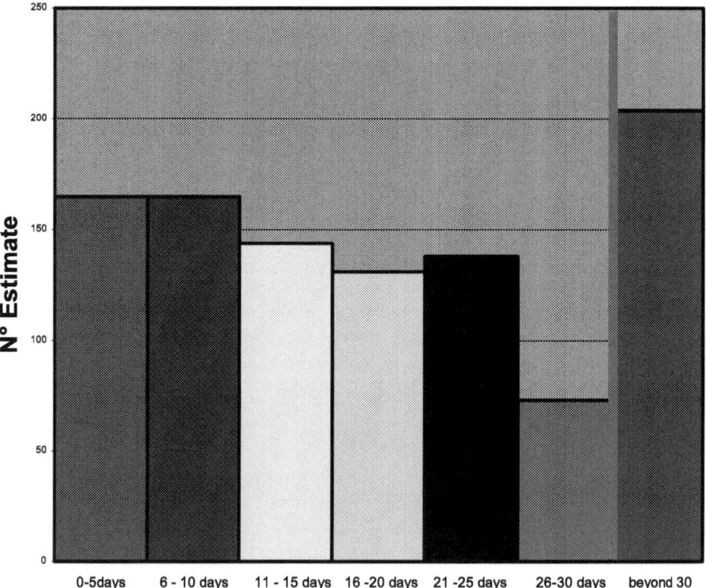

Fig. 8. Estimation Time – Frequency Histogram

The figure 9 shows radar chart evaluated on all internal indicators.

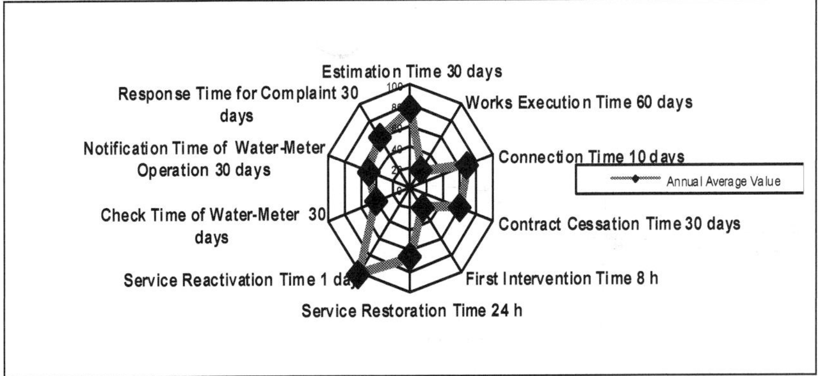

Fig. 9. Radar Chart: Quality Factors

The figure 10 shows monthly average trend of each quality internal index.

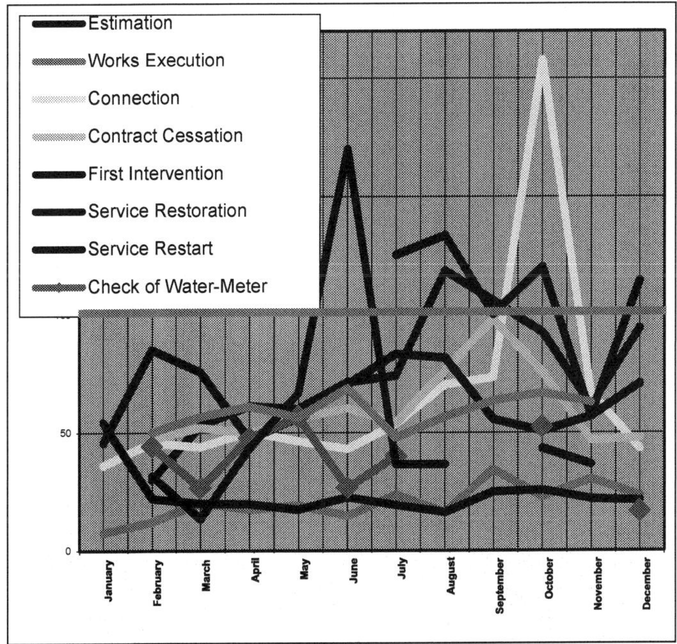

Fig. 10. Monthly Average Trend

Results of decision making analysis, in which, as described in the proposed methodology, internal indicators have to be associated with the interested process (Technical Process, User-Management Process and Measurement Systems Management Process) are reported, respectively, in tables 7, 8 and 9.

Stages of Technical Process	Efficacy	Efficiency
Leak Search RS[1]	N° Recognized Leak/km of inspected network km of inspected network/ km of total network	Working hours/Km inspected network km of inspected network/year Inspection cost km network/year
Emergency and Damage RS/RA	Service Restoration Time Time of service cessation for emergency Check of Water-Meter Operation N° users involved by service cessation First Intervention Time	N° emergency interventions/year Working hours/N° emergency interventions
User request for intervention RS/RA	Connection Time Average Time of Water-Meter replacement Works Execution Time N° installed Water-Meter Time of on the spot investigation N° projects of ampliation network/year	N° interventions/N° workers N° realized projects/N° total projects
Network Management RS/RA	Average pressure of network Interruption Time of intervention	Cost of network maintenance/year Km of network in maintenance/year

[1] Where:RS: Underground network; RA: Aerial network.

Table 7. Indexes of Technical Process

External Quality Measures
Servqual Method allows the measure of the external quality, id est, the quality perceived from service users. It consists in the data analysis through a questionnaire proposed at a statistical significative sample of customers. The Servqual index is a measure of the customer satisfaction, in terms of the measured gap between perception and expectation. The user expresses his estimate in a scale 1 up 10, subsequently by the valuation of average Servqual indexes; the zones of force and improvement are got. A vision of these zones allows recognizing the processes, which need corrective actions.

In table 10 are reported external indexes chosen for each service parameter proposed by Servqual method.

Stages of User-Management Process	Efficacy	Efficiency
Survey of consumptions	N° annual measures/total users	N° measures/N° workers
Invoice/ Management of Payment and Default	N° errors of invoice/N° issued invoices Time of invoice rectification N° defaulting users/total users N° users non-defaulting/total defaulting users Notification of service suspension for default	Average time among measure and bill Average time among bill and consignment Average time among consignments and takings Volumes of invoiced water Takings/turnover
Contracts	Time of estimate N° new contracts N° notices of cessation/N° new contracts Notice of water-meter control N° contractual modifications	Time of contractual cessation
Informations and claims	Wait Time at counter window N° information requests for bill/N° total information request N° reached complaints/year Response time to complaints Response time to written requests. N° information requests/year Opening hours of counter windows/week	N° workers of information service/total workers N° workers of counter windows/total workers

Table 8. Indexes of User-Management Process.

	Efficacy	Efficiency
Measurement Systems Management Process	N° controlled measurement systems per annum Average Time of internal calibration Average Time for replacement of measurement system under calibration	Annual Cost of calibration operation Average Time among the forwarding of instrument to Metrological Institute and its return

Table 9. Infrastructural Indexes of Measurement Systems Management Process.

The control of the provided service quality requires, as above, from a side to verify the customer satisfaction and from the other one a valid control panel monitoring the process indexes. The proposed methodology represents an integrated system of measure, where the data of efficaciousness and efficiency influence each other themselves producing the improvement corrective actions according to the Standard UNI EN ISO 9001:2000. Moreover it represents a valid solution in cases where the complexity of the measurement process or data entity is considerable.

Servqual Parameter	Indexes
Accessibility	Timetables opening front office shops Billing Facility to obtain information
Professionality	Competences Personal availability
Efficaciousness	Errors in bill Costs/quality Interrupt information Time billing
Safety	Confidence in the drink water Emergency rapidity
Tangible Aspects	Clarity of the invoice Water quality Continuity of distribution without pressure decrease

Table 10. Servqual Parameters

In figure 11 are shown the results of the data acquired by questionnaires.

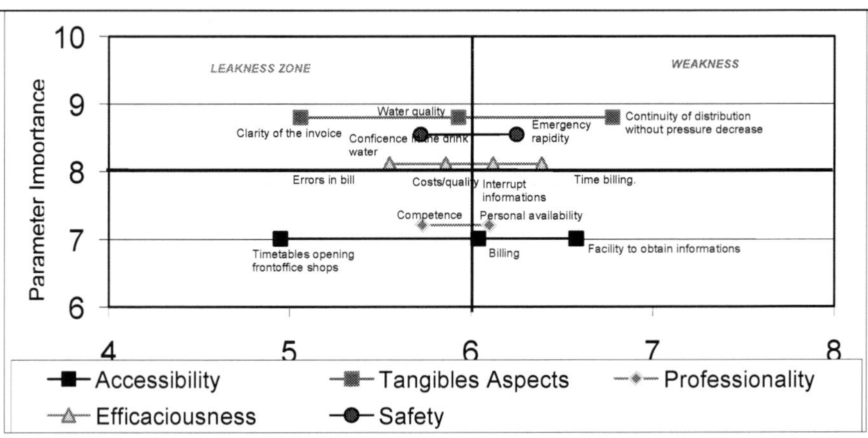

Fig. 11. Weakness and leak ness zones

5. Other aspects of energy optimization: ergonomics in maintenance for energy sustainable management

Also with reference to energy, the sustainability concept recalls, implicitly, the human factor concept, since the energetic sustainability has two key components: one related to production (renewable sources exploitation), the second one associated to the consumption and, then, to energy efficiency and saving. In this field building maintenance could represent a key strategy. This section shows potentialities of the ergonomic approach, particularly referring to building maintenance for a sustainable management of energy. It highlights the central role played by final users and identifying all conditions contributing to maintenance efficiency, able to assure, in the same time, energetic resources optimization and environmental comfort.

5.1 Human factors and sustainability

As the Brundtland Commission has defined, "sustainable development is development that meets the needs of the present without compromising the ability of future generations to meet their own needs", therefore, it is a process where resources exploitation, investment strategies, technology development trends and institution innovations are all harmonized, increasing present an future potentialities for human needs and wishes fulfillment. The concept of sustainability provides a values set having a cross role among the single sciences and disciplines, bringing a substantial and paradigmatic change in scientific approach, thanks to the integration of fields of knowledge traditionally distant.

This new course involves science, culture, ethics, religion, entrepreneurship on the basis of the will to establish an equilibrated connection between used resources in human activities. In this sense, sustainability is meant as a methodological reference able to affirm universal values such as wellbeing, equity, ethics fully respecting roles of all stakeholders but, also, diversities in cultures and contexts, involving psychological, social, economical and cultural needs, of which the environmental dimension is the framework synthesizing technical, socio-economical and cultural components.

This perspective highlights the necessity to understand needs of all various peoples involved in a process, in their specific environmental context, in order to configure operational scenarios which are sustainable thanks to their ability to meet their expectancies, and, finally, to their capacity to be easily accepted and promoted.

Also with reference to energy, the sustainability concept recalls –implicitly- the human factor concept, since the energetic sustainability has two key components: one related to production (renewable sources exploitation), the second one associated to the consumption and, then, to energy efficiency and saving.

Strategies oriented to eco-sustainability in energy related issues, involve the focus on systems and technologies able to increase the retention of produced energy and its saving rather than a general improvement at the production stage, emphasizing the role of efficiency in the final uses stages.

Energy consumption is, then, linked to a general problem of adequacy, involving specificities of both systems and technologies, which are asked to become more and more effective and functional, as well as more conscious procedures for their use.

Against the pressing necessity of lifestyles and approaches compatible with the optimal resources consumption, the issue of pervasiveness of appropriate individual and collective behaviours is now emerging. This perspective enhance the role of energy end-users, which are asked to fit their needs to the conscious usage of resources, also by mean of tools and devices controlling more and more sophisticated functions.

The consideration of human and behavioural variables in requirements design affects effectiveness and efficiency of systems, even relating to energy retention. In fact, especially for what concerns building and construction field, it has been demonstrated that comfort and eco-compatibility goals are more coherent rather than competitive.

In this framework it can be particularly helpful the availability of methodological and operational tools, able to analyze human activities, observe and understand needs and expectancies coming form users in order to produce interfaces compatible with them. It is matter of understanding and assessing ways of human-system interaction, as well as designing devices and procedures able to improve their efficiency assuring all stakeholders satisfaction in a balanced relation with environment.

5.2 Maintenance activities and energy efficiency in buildings

Main scope of maintenance is the continuity in keeping of building estate capacity to perform required functions, so that its utilization by users is complete and uninterrupted. Explicit goals of maintenance are of productive type, focused on the maintenance object and aimed to keep systems in efficiency; on the opposite, implicit maintenance goals come form a mainly constructive perspective, in a process view of maintenance, mostly focused on interactions among various actors, instruments, agencies and organizations. In the domain of sustainable energy management, building maintenance can represent the strategic tool and also a great opportunity. In fact, maintenance operations scopes imply a combination of heterogeneous activities to be done in order to limit functional decay, performance increasing and resources optimization. Up-keeping of constant efficiency levels prevents the risk of broken-downs or performances declining in technical elements, so that one of maintenance outcomes can be the energy consumption reduction and, in general, the environmental resources optimization.

In detail, maintenance actions allow to precede/assure that buildings systems, components and plants:

- have adequate tightness performances, avoiding problems coming from water introduction and accumulation;
- minimize impacts on environment and inhabitants
- guarantee correct indoor ventilation;
- guarantee thermal comfort, allowing the HVAC plants optimal management ;
- guarantee lighting comfort, allowing the illumination plants optimal management;
- make inhabitants able to personally control environmental conditions, in order to adjust excess and insufficiencies in HVAC plants performances on the basis of their activities.

Under the energetic point of view, building efficiency produces effects on both energy retention and environmental impacts reductions, with reference to:

- wastes limitation, in terms of energy keeping, resources exploitation reduction and management costs decreasing;
- pollution reduction, with concerns to good repair of plants, reduction of contaminants sources and, consequently, decreasing of global environmental costs;
- increasing of comfort for inhabitants, thanks to the increased adjusting features, with personnel costs reduction and increased perceived quality and satisfaction by end users.

5.3 The role of users

Latest trends in standardization, oriented to Total Productive Management show that maintenance is meant as the whole of actions aimed not only to broken-downs repair, rather then to prevention, continuous improvement and handover of simplest maintenance actions to building tenants and occupants. Compliantly to quality assurance principles, a key aspect is the mounting involvement in the maintenance process of operational figures which were before extraneous to. In this view, it is foreseeable that building user can act without intermediaries, doing autonomously simplest maintenance tasks.

Users consciousness and participation introduce in maintenance strategies a practice able to overcome the strict task assignment of functional competences, focusing on the fact that only a small number of maintenance activities actually requires advanced skills, while many of them can be successfully carried out with any particular competence or instrument. System state monitoring allows to grasp any indication, as far as feeble, about their decay conditions and gives the possibility to adjust their working to assure comfort and well-being in a wide range of use conditions, so that final users and/or tenants play a key role for many maintenance operations. They act in advance and opportunely, with clear positive effects on resources optimization, waste and financial costs reduction. For instance, for what concerns thermal comfort, the possibility to adjust indoor temperature at any time or to program time-frame of heating reduction or switching-off brings to a twofold result: wellbeing improvement and fuel consumption reduction. Similar benefits come from the possibility to regulate autonomously the quantity of incoming solar radiation with mobile shielding devices, or, also, the air flows tuning thanks to movable grids and mouthpieces in windows. But, if users involvement becomes more and more relevant, adequacy of operational contexts is a crucial matter, for both aspects technical and organizational, in order to assure the effectiveness of not-specialized operations.

5.4 The relevance of maintainability

Given the complexity of maintenance system due to its human, technical and organizational variables, building energetic efficiency depends on how systems are prearranged -that is designed- to be maintained. Maintainability is the primary parameter by which it is possible to assure efficiency of maintenance system. It can be defined as the success probability of a specific maintenance action on a given element in a specific timeframe by a determined skill/professional profile, with specific tools and procedures.

Then, maintainability comes from the combination of technical specifications of maintained systems with all other factors in the maintenance context, from human ones to procedural, infrastructural, financial ones. This pragmatic approach is aimed to shape architectural and plants design in order to guarantee high efficiency levels of building life-cycle, through the definition of technical specifications centered on issues such the possibility to reach elements to be maintained, the management of transport and use of spare parts and tools where they have to be used, the possibility to easily and accurately execute maintenance actions, in an "supportive" operational context. For these reasons, heterogeneous aspects such as: physical accessibility, components visibility, problems detection, sub-systems isolation, logic and physical simplicity of parts to be disassembled and re-assembled, availability and exchangeability of spare parts, availability of tools and instruments and their adequacy to the maintenance tasks, comprehensibility of operational directions, technical sheets and information, postural comfort in task execution, immediacy in errors detection and their remedies, the use of human resources in the mid-range of their abilities have to be analyzed in an integrated way. Therefore, maintainability design has to be oriented to qualify and quantify all factors of the maintenance system, by way of an approach like the ergonomic one, allowing to highlight the various interactions involving human in organized contexts.

Ergonomics, or human factors, is the discipline concerning human-system interactions applying theories, principles, data and methods for design and assessment of tasks, activities, environments and devices, in order to make them compliant with needs, abilities and limitations of peoples. The scope is the optimization at the same time, human wellbeing, also meant as safety and satisfaction for done activities, and the whole system efficiency, achieving established goals caring of resources and costs.

In the case of building and urban maintenance, human-system-environment interaction is more complex. In fact, the scope of maintenance actions is to accomplish tasks for problem fixing or functional up-grading, doing controls, substitutions and adjustment operations. Tools and devices are then instrumental entities, and can be considered as intermediary elements between human and maintained systems. This mediation has a significant role in maintenance, since effectiveness of maintenance actions on buildings or their component comes, mostly, from effectiveness of tool usage doing those actions. In this view, many-sided interactions among peoples, maintained objects and tools, set-up: people-object direct interactions (users-building) or tool-mediated interactions (operators-tools-components), people-tool interactions, tools-component interactions, which all should be adequately investigated in order to understand how to arrange conditions for an operating maintenance.

Finally, human-human interaction hasn't to be neglected, being it concerned with interpersonal relationships and communication process establishing among the several maintenance figures as well as users-technicians, often triggering accidental interactions, with consequent potential conflicts and annoyance situations.

Maintainability conditions

Maintainability conditions are the whole of context features allowing to perform maintenance activities effectively and efficiently. They concern the way in which actions are carried out, the specificity of places in which those activities are performed, appliances features used to maintain systems and skills and motivations of various people involved in maintenance tasks.

In order to guarantee effective maintenance activities, operators and final users must clearly detect the different elements of maintenance context, access and reach them with any effort, easily understand functioning modalities of work tools and easily follow the maintenance procedures in a safe and comfortable way supporting, with documentary evidence, actions they have to do and actions they have done.

Moving from the premised that maintainability conditions may be seen as the whole of necessary situations to assure adequate quality levels of maintenance activities, we can detail them referring to each element of the context :

- systems to be maintained;
- instruments and tools supporting maintenance;
- operational space where maintenance activities are executed;
- maintenance tasks to be carried out.

Each element has to be featured by a set of requirements complying maintenance needs. A group of qualitative end quantitative criteria can detail each requirement, in order to identify specifications to design and control building maintenance conditions.

Detectability

It is a condition of the building/technical element/system component allowing and favoring its identification to gain maintenance goals. It can be specified in terms of capacity to easy localize each element within a system (Localization), to execute actions for troubleshooting nd/or for controlling and inspections (Testability and Checkability).

Accessibility
It is a condition of the building/technical element/system component allowing and favouring to operators arrive and access to execute maintenance activities. It can be specified in terms of capacity to point out to human senses (Visibility), and to offer reachable and dimensionally adequate access points (Reachability).

Comprehensible
It is a condition of the building/technical element/system component allowing and favouring to simply understand their operational using tasks. It can be specified in terms of capacity to quick recognize systems functioning modalities (Self-explainability) also giving, where needed all information about their use (Contextual information). We can refer this condition also to maintenance actions, in order to effectively transmit task guidance to users and operators according to their skills, experience and attitude (Clearness and effective of communication).

Operability
It is a condition of the building/technical element/system component allowing and favouring the comfortable execution of maintenance activities. It can be specified in terms of capacity to realize adequate postural conditions for operators acting on maintenance systems (Postural adequacy); to present function independence and standard units (Modularity and Standardization) easy to be replaced (Replaceable) and exchanged with a slightly different part (Interchangeable); to permit that maintenance actions can be easily exerted by hands, avoiding any accidental switch-on (Resistance to accidental activation); to be easily cleaned (Cleanable); repaired (Repairability), disassembled and removed (Subassembly end removal); error tolerant; easy to adjust (Adjustable). The condition of easy to operation can be helpfully referred also to maintenance tasks in terms of operators physical and mental effort control and activities practical execution easiness.

Documentability
It is a condition of the building/technical element/system component allowing and favouring maintenance data collection and updating. It can be specified in terms of capacity to easy identify input and output technical data about maintenance task (Information finding) and easily detect and update actors and steps of maintenance processes (Information traceability).

Users interfaces usability
User interfaces for systems and installation controlling are crucial elements for building energetic efficiency, because they are devices by which depend quality actually perceived by users, in terms of effectiveness of functions and environmental comfort. We can define interfaces as the places where human and system communications are exchanged. It is made of system parts presenting information to user about its functions and internal state, and receiving information from user about how to change system state. The user activates system functions operating on it by senses and motions to achieve his goals. User and system then can be seen as subjects of a dialog realized by inputs that user send to activate some functions and consequent outputs that system gives back by answers confirming or denying users requests.

A supportive maintenance context is crucial to guarantee buildings energetic efficiency, in order to assure resources optimization and inhabitants wellbeing and satisfaction. Consideration of human factors perspectives in requirements design, able to plan and execute maintenance activities together with physical interface realization, can contribute to improve energetic performances in the whole. In fact a better usability of devices and tools increases systems efficacy but also brings to increase autonomous and conscious usage by final users. This issue is widely considered as strategic to encourage environment friendly, an then sustainable, behaviours.

6. The importance of energy monitoring system in the sustainable energy management

To remain in a healthy and competitive market, companies are using the identification of the production process to help and guide the procedures for using these facilities to achieve economic gains in the most efficient way possible. In this case, and with this type of control on electric power systems, both business users and the companies producing electricity can draw considerable benefits and advantages. So far the systems are being used and that most companies are implementing are designed for internal monitoring of the establishment, for the award of costs, management of the loads and the collection of information that can be used to highlight and identify the equipment problems of operation. Furthermore, these systems are useful for significantly reducing the capital invested to increase the power undisciplined and expansion of power conversion. A general control over the management to ensure a single company under the cost-efficiency, is capable of streamlining the systems and operations maintenance, trying to act, according to the timetables, resource stock, the availability of personnel, etc.. to operate on non-viable subsystems and not in production. The rationality of the system is that it should use renewable energy sources such as primary energy sources and those from fossil fuels as energy sources subsidiary to keep available during peak hours. Such a system will be feasible when the energy produced from renewable sources has become a major against the world production of macro or continental areas. There is the need for a global monitoring system because as the energy from renewable sources although inexhaustible, independent, at least in general from sites where they are produced, and environmentally friendly, at least at first sight, by their nature and defect , are not always consistently available and therefore are extremely unpredictable, as is the unpredictable amount of energy produced. As mentioned earlier, you should create a network of general supervision and control that can handle data from the central production of energy from renewable sources, power production from non-renewable sources, and finally to make an assessment of consumption by end-users in order to establish a balance between the needs of production and consumption. The development strategy consists of a network of programmable logic controllers (PLC) that manage the local control of each plant. These local controllers are linked together through an appropriate system interface and communication, through a master / slave network that makes it available and accessible operational monitoring of each installation. A supervisor of a network so it would be a structured system Supervisory Control And Data Acquisition (SCADA), decentralized management to enable a user friendly in the world. The focus on the rational use of energy has been great interest so prevalent after the oil crisis of 1970, in fact from that year onwards, was tried to minimize wastage and improve the efficient equipment of any kind. Logically consumption are by no means diminished, quite the contrary, there was a request for several reasons, such as consumer lifestyles, achieving a better family welfare, the increase of the transport sector. Based on these criteria of growth in general an initial savings, with consequent pollution of the electricity network for the introduction of harmonics, it has had with the use of inverter technology that enabled a reduction in demands for electricity for all types of users, from the Sunday and therefore low power to large industrial powers. The purpose of the described approach, however, is not substantially save energy as such, but a rational use of energy and an intelligent distribution of requirements among the various sources of production.

7. Conclusion

Today, energy issues as such often origin in the strive for sustainability and a strong version of the ISO 14001 Environmental Management System (EMS) triggers companies to appropriately find, measure and manage their environmental obligations and risks. The integrated EMS has been argued to achieve both environmental and financial benefits, but environmental impact can also be decreased in terms of effective and efficient modes of operations. Policy-makers are under pressure to formulate and adopt energy policies aimed at different sectors of the economy due to the use of fossil fuels and its result in global warming. The manufacturing industry accounts for a yearly consumption of about 75 % of the global use of coal, 44 % of the global use of natural gas, and 20 % of the global use of oil, and in addition around 42 % of all the electricity.

In many Countries, the energy intensive industry accounts for some 70 % of the aggregated industrial energy use. Energy efficiency in the industry sector hence plays a central role in terms of environmental impact.

Some researchers early identified the possibility to reduce emissions trough regional energy supply cooperation. Later on, regional cooperation between different companies regarding energy issues has been shown to be both financially and environmentally beneficial with extensive potential in reducing CO_2 emissions.

In summary, one can conclude that effective energy management has notable environmental impact, but energy issues are seldom a top priority, even in energy intensive organizations.

After an overview regarding the general approach to the problem of sustainable energy, in the chapter have been analyzed some different aspect of the same theme of energy optimization.

An innovative methodology for the productive processes qualification based on quality characteristics improvement and on their simultaneous evaluation cost, is proposed.

The proposed approach is applicable to every type of productive processes after giving them a p-diagram structure as described. Its innovative idea to optimize quality characteristics through an objective function and contemporary minimize the related cost function, related to the energy consume, allows compliance industrial farm strategy improvements.

Energy optimization is also a problem with an high impact on service. The starting assumption is based on the belief that delivering a service through a control quality system is a condition that ensures an energy optimization in the work processes of the company. In the chapter this aspect is discussed and a real case regarding a water supply company is reported.

The control of the provided service quality requires, from a side to verify the customer satisfaction and from the other one a valid control panel monitoring the process indexes. The proposed methodology represents an integrated system of measure, where the data of efficaciousness and efficiency influence each other themselves producing the improvement corrective actions according to the Standard UNI EN ISO 9001:2000. Moreover it represents a valid solution in cases where the complexity of the measurement process or data entity is considerable.

To consider the energy optimization process as key factor for firms it's necessary to consider also the rule and the relevance of maintenance activities and the importance of human factor.

A supportive maintenance context is crucial to guarantee buildings energetic efficiency, in order to assure resources optimization and inhabitants wellbeing and satisfaction. Consideration of human factors perspectives in requirements design, able to plan and execute maintenance activities together with physical interface realization, can contribute to improve energetic performances in the whole. In fact a better usability of devices and tools increases systems efficacy but also brings to increase autonomous and conscious usage by final users. This issue is widely considered as strategic to encourage environment friendly, and then sustainable, behaviours.

8. Appendix: some brief discussion on standards

The future ISO 50001 standard for energy management was recently approved as a Draft International Standard (DIS).

ISO 50001 will establish a framework for industrial plants, commercial facilities or entire organizations to manage energy. Targeting broad applicability across national economic sectors, it is estimated that the standard could influence up to 60% of the world's energy use.

The document is based on the common elements found in all of ISO's management system standards, assuring a high level of compatibility with ISO 9001 (quality management) and ISO 14001 (environmental management). ISO 50001 will provide the following benefits:

- A framework for integrating energy efficiency into management practices;
- Making better use of existing energy-consuming assets;
- Benchmarking, measuring, documenting, and reporting energy intensity improvements and their projected impact on reductions in greenhouse gas (GHG) emissions;
- Transparency and communication on the management of energy resources;
- Energy management best practices and good energy management behaviours;
- Evaluating and prioritizing the implementation of new energy-efficient technologies;
- A framework for promoting energy efficiency throughout the supply chain;
- Energy management improvements in the context of GHG emission reduction projects.

ISO 50001 is being developed by ISO project committee ISO/PC 242, Energy management. The secretariat of ISO/PC 242 is provided by the partnership of the ISO members for the USA (ANSI) and Brazil (ABNT). Forty-two ISO member countries are participating in its development, with another 10 as observers.

Now that ISO 50001 has advanced to the DIS stage, national member bodies of ISO have been invited to vote and comment on the text of the standard during the five-month balloting period.

If the outcome of the DIS voting is positive, the modified document will then be circulated to the ISO members as a Final Draft International Standard (FDIS). If that vote is positive, ISO 50001 is expected to be published as an International Standard by june 2011.

9. References

Arpaia P., (1996). *Experimental Optimisation of Flexible Measurement Systems*, IEEE Proc. (Pt. A: Science, Meas. and Tech.) vol.143, n.2, March, 1996, pp.77-84.

Arpaia P., Daponte P. , Sergeyev Y.D. , Bordeaux (F) (1999). *A-priori statistical parameter design of ADCs by a local tuning algorithm of global optimisation*, Proc. of IV IMEKO International Workshop on ADC Modelling and Testing 9-10 September 1999, pp.236-244

Arpaia P., Grimaldi D. , (2002) "Metrological performance optimisation of a displacement magnetic transducer", Instrumentation and Measurement Technology Conference, 2002 IMTC/2002, Vol. 2, 21-23 May 2002, pp. 1295 – 1300.

Arpaia P., Polese N., (2001). *Uncertainty reduction in measurement systems by statistical parameter design*, 6th IMEKO TC-4 Int. Symp., Lisboa, Sept., 2001

Attaianese, E. (2008), *Progettare la manutenibilità. Il contributo dell'ergonomia alla qualità delle attività manutentive in edilizia*, Liguori, Napoli, 2008

Campbell, J. D. and Reyes-Picknell, J., (2006): *Strategies for Excellence in Maintenance Management, Productivity* Uptime, 2nd Edition Press, 2006

Countryside Agency (2005) *The countryside in and around towns: a vision for connecting town and country in the pursuit of sustainable development*, Countryside Agency, Cheltenham (www.countryside.gov.uk).

DTI (2005) *Micro-generation strategy and low carbon buildings programme – consultation'*, DTI, UK (www.dti.gov.uk).

Evans, S. (ed.) (2009) *Towards a sustainable industrial system*, University of Cambridge, Research Report, UK

Ferrero A. , Gamba R. , Salicone S. (2004). *A method based on random-fuzzy variables for on-line estimation of the measurement uncertainty of DSP-based instruments*, IEEE Trans. Inst. Meas., vol. 53, 2004, pp. 1362-1369

Flemming, S., Hilliard, A. and Jamieson G.A. (2008). *The need of human factors in the sustainability domain* in Proceedings of the Human Factors and Ergonomics Society 52nd Annual Meeting, 2008, 750

Flemming, S.A.C., Hilliard, A., & Jamieson, G. A. (2007). *Considering human factors perspectives on sustainable energy systems*, Poster presented at the International Society for Industrial Ecology Conference, February 6, Toronto, Canada, 2007

HM Government (2003) *Energy white paper: our energy future – creating a low carbon economy* (www.dti.gov.uk/energy/whitepaper)

HM Government (2005) *Securing the future –UK Government sustainable development strategy* (www.sustainable-development.gov.uk).

Hulme M., Jenkins G., Lu X., Turnpenny J., Mitchell D., Jones R. K., Lowe J., Murphy J., Hassell D., Boorman P., McDonald R., Hill S. (2002) *Climate change scenarios for the UK: the UKCI2 scientific report'*, *Tyndall Centre for Climate Change Research*, School of Environmental Sciences, University of East Anglia, Norwich, UK (www.ukcip.org.uk)

Kettinger, W. J., & Lee, C. C. (1994). *Perceived service quality and user satisfaction with the information services function*. Decision Sciences, 25(5-6), 737766.

London Renewables (2004) *Integrating renewable energy into new developments* toolkit for planners, developers and consultants Sustainable and Secure Buildings Act.

Mayor of London (2004) *The London Plan: spatial development strategy for Greater London*, GLA, London (www.london.gov.uk/mayor/strategies)

Mayor of London (2004) *The Mayor's energy strategy: a green light to clean power*, GLA, London (www.london.gov.uk/mayor/strategies)

Nunnally, J. (1978). *psychometric theory* (2nd ed.). New York: McGraw-Hill.

Özdamar, L. and Birbil, S.I. (1999) *A hierarchical planning system for energy intensive production environments*, International Journal of Production Economics, Vol. 58, No. 2, pp.115-129.

Parasuraman, A., Zeithaml, V A., & Berry, L. L. (1988). SERVQUAL: *A multiple-item scale for measuring consumer perceptions of service quality*. Journal of Retailing, 64 (1), 12-40.

Parasuraman, A., Zeithaml, V A., & Berry, L. L. (1994).*Reassessment of expectations as a comparison in measuring service quality: Implications for future research*. Journal of Marketing, 58(1), 111-124.

Parasuraman, A., Zeithaml, V. A., & Berry, L. L. (1985). *A conceptual model of service quality and its implications for future research*. Journal of Marketing, 49, 41-50.

Parasuraman, A., Ziethaml, V. A., & Berry, L. L. (1991). *Refinement and reassessment of the SERVQUAL scale*. Journal of Retailing, 67, 420-450.

Planning Policy Statements: PPS1 (delivering sustainable development) and PPS22 (renewable energy). Others also refer to sustainable development, energy and climate change (see www.odpm.gov.uk)

Taguchi, G. Elsayed, E.A.&Hsiang, T. (1989). *Quality Engineering in production systems*, New York, Mc Graw Hill 1989

TCPA (2004) *Biodiversity by design: a guide for sustainable communities*, TCPA, London

Teas, R. K. (1994). *Expectations as a comparison standard in measuring service quality: An assessment of a reassessment*. Journal of Marketing, 58(1), 132139.

Wise, J.A. (2007) . *Measuring the Occupant Benefits of Green Buildings* in Human Factors and Ergonomics Society Annual Meeting Proceedings, Environmental Design , pp. 495-499(5), 2007

Wise, J.A. (2007). *Human Factors & the Sustainable Design of Built Environments*, in Human Factors and Ergonomics Society Annual Meeting Proceedings, Environmental Design , pp. 808-812(5), 2007

4

Use of Online Energy System Optimization Models

Diego Ruiz and Carlos Ruiz
Soteica Europe, S.L.
Spain

1. Introduction

Modern industrial facilities operate complex and inter-related power systems. They frequently combine internal utilities production with external suppliers, including direct fired boilers, electric power generation with turbo alternators or gas turbines, heat recovery steam generators, have different drivers (i.e., turbines or motors) for pumps or compressors and several types of fuels available to be used. Tighter and increasingly restrictive regulations related to emissions are also imposing constraints and adding complexity to their management. Deregulated electric and fuels markets with varying prices (seasonally or daily), contracted and emissions quotas add even more complexity. Production Department usually has the responsibility for the operation of the facility power system but, although Operators are instructed to minimize energy usage and usually tend to do it, a conflict often is faced as the main goal of Production is to maintain the factory output at the scheduled target. The power and utilities system is seen as a subsidiary provider of the utilities needed to accomplish with the production target, whichever it takes to generate it.

Big and complex industrial facilities like Refineries and Petrochemicals are becoming increasingly aware that power systems need to be optimally managed because any energy reduction that Operations accomplish in the producing Units could eventually be wasted if the overall power system cost is not properly managed. However, process engineers always attempted to develop some kind of tools, many times spreadsheet based, to improve the way utilities systems were operated. The main drawback of the earlier attempts was the lack of data: engineers spent the whole day at phone or visiting the control rooms to gather information from the Distributed Control System (DCS) data historian, process it at the spreadsheet and produce recommendations that, when ready to be applied, were outdated and not any more applicable.

The evolution from plant information scattered through many islands of automation to unified and centralized Plant Information Systems was a clear breakthrough for the process engineering work. The long term, facility wide Plant Information System based historians constitute what is known as an enabling technology, because they became the cornerstone from where to build many other applications. Besides others, advanced process control, optimal production programming, scheduling and real time optimization technologies were built over them and flourished after data was stored for long terms and became easily retrievable.

Process engineers, still working on their original spreadsheets, attempted to improve the early days calculations by linking them to real time data and performing some sort of optimization. They used the internal optimizers or solvers provided with the spreadsheet software. One of the authors went though the same path when he was a young process engineer at a Petrochemical Complex. After many years of exposure to dozens of manufacturing sites worldwide, he found that almost all Process Engineering Departments had in use an internal spreadsheet over which several process engineers worked when on duty at the power house or utilities unit. The spreadsheets usually evolved wildly during many years and became extremely complex as hundreds of tags were added and, very often, they remained completely undocumented.

That kind of "internal tools" are in general fragile: it is very easy to severely harm one of such spreadsheets by simply adding or moving a line or column, up to the extent to become completely unusable. Plant Information Systems provided, from the very beginning, a way to link real time or historical data very easily into a spreadsheet but, as raw data is contaminated with unexpected problems, sometimes hard to identify and filter from errors, the use of such a tools for real time optimization was seldom a real success. We commonly found they were used for a while, usually requiring a lot of effort from the process engineer who developed it but, as soon as the creator was promoted or moved to other position, their use steeply declined and became an unused, legacy application. It was becoming more and more clear that a certain kind of intelligence should be added to those energy management tools in order to produce good results in a consistent way, dealing with real time information potential errors and maintained green and usable for long periods, with minimal engineering effort.

The authors found that, for the Process Industry, the definition of an intelligent system is generic and not very well defined. The industry usually call "intelligent" to any piece of software that helps to automate the decision making process, efficiently control a complex process, is able to predict properties of products or process variables, alerting or preventing hazardous situations or, in last instance, optimize process or business economics. For the practical engineers, the definition of an intelligent system is factual, not methodological. But always there is a computer behind, accessing data and running a piece of software. The above mentioned systems comply, up to certain extent, with one of the classical definitions of intelligence (Wiener, 1948): the intelligent behavior is a consequence of certain feedback mechanisms, based on the acquisition and processing of to accomplish with a certain objective. A coherent engineering environment providing all the needed tools into a single shell, starting with real time data acquisition and information validation, flexible and versatile modeling and simulation environment, robust mixed integer non linear optimization techniques, appropriate reporting tools, results historization and easily interpretation of the site wide optimization solution and constraints was a real need. Once available, energy systems became optimized and operated under an optimal perspective.

During the past 20 years, Visual MESA optimization software evolved from the earlier text based, offline application of the 1980's to an online, real time, graphical user interfaced, highly sophisticated intelligent system considered today as the industry standard Energy Management System (EMS) real time online optimizer (Nelson et al., 2000). Software development time line is presented in Fig.1, showing the main landmarks of the past years. It has been widely implemented in the processing industry and it is applied routinely to reduce the cost of operating the energy systems at power, chemical, petrochemical, and refinery plants worldwide. Several projects where the authors participated are cited in the References list below. A few of them will be also commented.

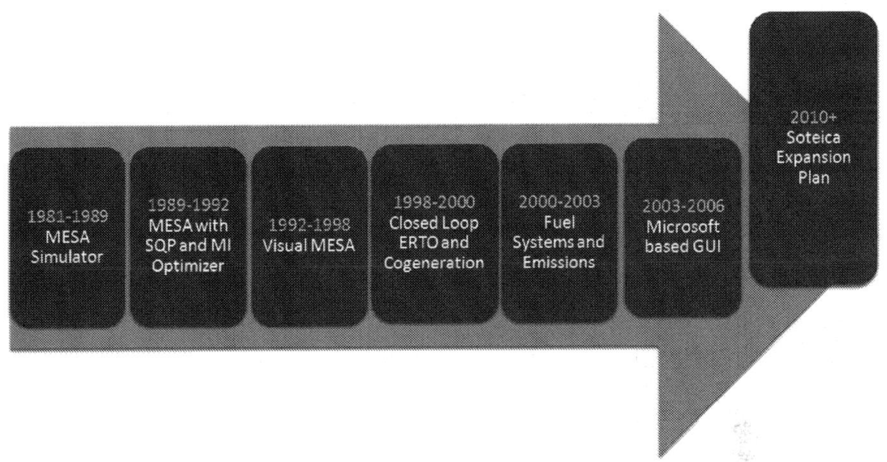

Fig. 1. Visual MESA Energy Management Software Development Time Line

2. Energy management system description

A detailed model of the energy system (fuels, steam, electricity, boiler feed water and condensates network) is built within the EMS environment and it is continuously fed with validated, real time data. It includes all the actual constraints of the site and decision variables for their operation. Optimization is configured to minimize the total energy cost. Continuous performance monitoring is also done, since the model writes back its results to the Real Time Data Base (Plant Information System). It also provides reliable data that helps to audit the energy productions and usages within the site energy system, and in that way wastes can be detected and eliminated. The model is also often used in standalone mode to perform case studies for economical evaluations of potential investments and for planning the operation of the energy system. Greenhouse emissions are also taken into account.

2.1 Manual, open loop versus closed loop operation

Although the Operators still need to close the loop manually and projects have proven to save substantial amounts of money, with very fast pay-back, even more economic benefits can be obtained if some of the manual optimization handles are automated under a closed loop scheme (Uztürk et al., 2006). Closed loop applications are expected to be increasingly popular during the next few years, to become in future the Closed Loop Energy Real Time Optimizers (CLERTO) a new standard.

There are several and important additional advantages of using EMS's for online, real time, closed loop optimization because it increases the benefits already obtained in open loop, especially when fuels and power prices are market driven and highly variable. Several implementations of this kind have already been done (Wellons et al., 1994; Uztürk et al., 2006).

It is important to emphasize the fact that a successful online optimization application is much more than just providing 'a model and an optimizer'. It also requires the project team provides real time online application implementation experience and particular software capabilities that, over the life of the project, prove to be crucial in deploying the online application properly. These software features automate its execution to close the loop, provide the necessary simple and robust operating interface and allow the user to maintain the model and application in the long term (i.e., evergreen model and sustainability of the installation).

2.2 Online capabilities

The online capabilities are a relevant portion of the software structure and key to a successful closed loop implementation. A proper software tool should provide standard features right out of the box. Therefore, it should not require any special task or project activity to enable the software to easily interact and cope with real time online data. The EMS based models are created from scratch acquiring and relying on real time online data. A standard OPC based (OLE for process control) protocol interface has been provided to perform a smooth and easy communication with the appropriate data sources, such as a distributed control system (DCS), a plant information system, a historian or a real time database. Sensor data is linked to the model simulation and optimization blocks by simply dragging and dropping the corresponding icons from the builder's palette and easily configuring the sensor object to protect the model from measurement errors and bad values through the extensive set of validation features provided. Fig. 2 shows an example of the configuration options in case of sensor data validation failure.

Properly designed software need to provide all the main features to implement online and closed loop optimization including:

- Sensor data easily tied to the model (drag and drop).
- Data validation, including advanced features such as disabling optimizers or constraints depending on the status of given critical variables.
- Steady state detection capabilities, based on a procedure using key variables' fast Fourier transform (FFT) based technique to identify main process variables steadiness.
- Online model tuning and adaptation, including the estimation of the current imbalances and maintaining them constant during the optimization stage.
- Control system interfaces for closed loop, online optimization, sending the decision variables set points back to the DCS via OPC.
- Closed loop model and control system reliability and feasibility checks (i.e., communications watchdog capabilities), to ensure the proper communication between the optimizer and DCS, via OPC.

Fig. 3 shows typical installation architecture for closed loop real time optimization, including the proper network security layers and devices, for example firewalls and demilitarized zones (DMZ) domains.

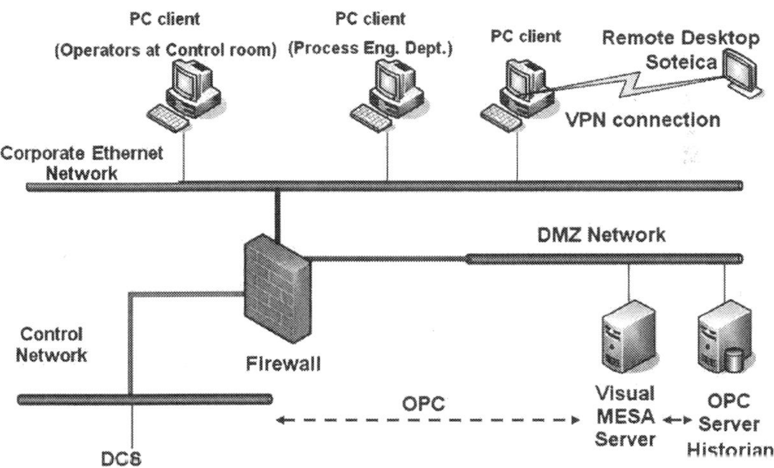

Fig. 2. Sensor Configuration Options

Fig. 3. Installation Architecture for Closed Loop Implementation

2.3 Optimization variables and constraints configuration for closed loop optimization

Building a model that realistically represents the utilities and energy system topology, includes all the optimization variables and constraints and, at the same time, includes all the system economic details, especially the fuels and electricity contractual complexity.

Such a complex optimization problem can be represented and solved in a straightforward manner when using a proper software tool, even when the model is to be executed as a closed loop, real time application.

During the model and optimization building, the following set of variables must be identified and properly configured:

Optimization variables are those where some freedom exists regarding what value might be. For example, the steam production rate at which a particular boiler operates is a free choice as long as the total steam production is satisfied, thus the most efficient boiler's production can be maximized.

There are two main kinds of optimization variables that must be handled by an online energy management system optimizer:

- Continuous variables, such as steam production from a fired boiler, gas turbine supplemental firing and/or steam flow through a steam-driven turbo generator. Those variables can be automatically manipulated by the optimizer writing back over the proper DCS set points.

- Discrete variables, where the optimizer has to decide if a particular piece of equipment will operate or not. The most common occurrence of this kind of optimization is in refinery steam systems were spared pump optimization is available, one of the drivers being a steam turbine and the other an electric motor. Those variables cannot be automatically manipulated. They need the operator's manual action to be implemented.

Constrained variables are those variables that cannot be freely chosen by the optimiser but must be limited for practical operation.

There are two kinds of constraints to be handled:

- Direct equipment constraints. An example of a direct equipment constraint is a gas turbine generator power output. In a gas turbine generator, the fuel gas can be optimized within specified flow limits or equipment control devices constraints (for example, inlet guide vanes maximum opening angle). Also, the maximum power production will be constrained by the ambient temperature. Another example of a direct equipment constraint is a turbo generator power output. In a turbo generator you may optimize the steam flows through the generator within specified flow limits but there will also be a maximum power production limit.

- Abstract constraints. An abstract constraint is one where the variable is not directly measured in the system or a constraint that is not a function of a single piece of equipment. An example of this type of constraints is the scheduled electric power exported to the grid at a given time of the day. Economic penalties can be applied if an excess or a defect. Another example of this type of constraint is steam cushion (or excess steam production capacity). Steam cushion is a measure of the excess capacity in the system. If this kind of constraint were not utilized then an optimizer would recommend that the absolute minimum number of steam producers be operated. This is unsafe because the failure of one of the units could shutdown the entire facility.

3. Project activities

An Energy Management System (EMS) Implementation project is executed in 9 to 12 months. The main steps are presented in Fig. 4. and discussed below.

3.1 Required information

After the Purchase Order is issued, a document would be submitted to the Site with all the informational requirements for the EMS project sent it to the project owner. By project owner we understand a Site engineer who, acting as a single interface, will provide the needed information and coordinate all the project steps. The EMS server machine would need to be configured with the required software, including the OPC connectivity server and made available prior to the Kick-Off Meeting.

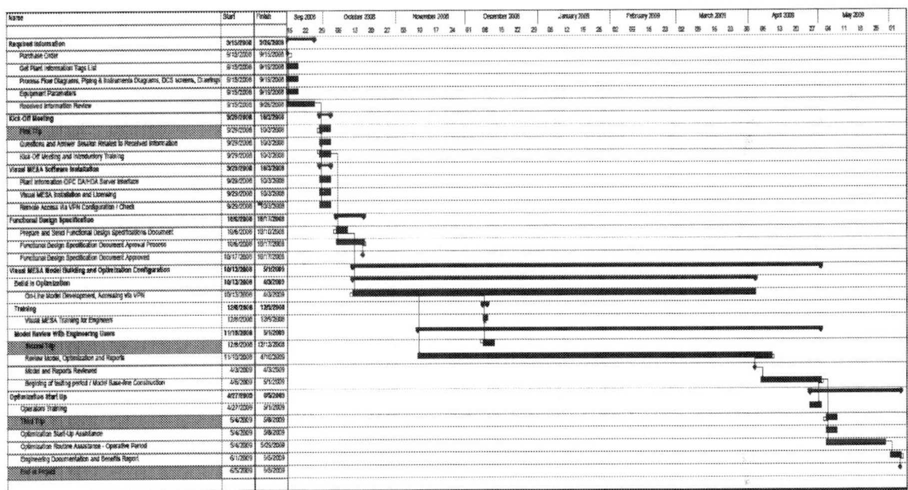

Fig. 4. Typical Energy Management System Implementation Project Schedule

3.2 Kick-off meeting

Prior to the Kick-Off Meeting, the provided information will be reviewed to have a better understanding of the Site facilities and process. Additional questions or clarifications would be sent to the Site regarding particular issues, as required. During the week of the on-site Kick-Off Meeting, all information would be reviewed with the Site staff, and additional information required for building the model would be requested, as needed. At that time, the optimization strategy would also be discussed. During the same trip, an introduction to the EMS will be given to the project owner in order for him to have a better understanding of the scope, information requirements and EMS modelling. The EMS software would be installed at this time.

3.3 EMS software installation

The software is then configured and licensed on the EMS server PC. It would also be connected to the OPC server. Remote access to the model would also need to be made available at this time and would need to be available throughout the rest of the project.

3.4 Functional design specification

With the information provided during the Kick-Off meeting, a Functional Design Specification document would be prepared, revised by both parties in concert, and then approved by the Site. In this document, a clearly defined scope of the model and optimization is provided and will be the basis for the rest of the project work.

3.5 Visual mesa model building and optimization configuration

During this stage, the model and the report are built working remotely on the EMS server. The model grows with access to online real time data. Every time a new piece of equipment or tag is added, it can instantly begin to gather information from the Plant Information System via the OPC interface. Periodic questions and answers regarding the equipment, optimization variables, and constraints may be asked to the Site. The second trip to the facility would occur during this stage and would be used for mid-term review of the model and optimization. Also, an EMS training course for engineers is given at that time. Continuing forward, the model is continually reviewed by both parties and any improvements are made, as required. After reviewing the model and confirming that it meets the requirements of the Functional Design Specification, the Site would give its approval of the model.

Upon model approval, a month-long testing period would commence, the results of which would form the model "burn-in". During the "burn-in" period, the EMS would run routinely, but optimization recommendations would still not be implemented by the operations staff. A base line could be obtained based on the cost reduction predicted by the optimizer during this period, in order to compare with the full implementation of the suggestions at the end of the project. The project owner would review the optimization recommendations with the project developing staff. Minor modifications would be made to the model, as needed.

3.6 Optimization startup

Site engineers would then train the operations staff to use Visual MESA and to implement the recommendations. The trainers could use the provided training material as a basis for their training if they preferred. Continuing in this period, operations staff would begin implementation of the optimization recommendations. Project developing staff would return to the Site facility a third time to review implementation of the optimization recommendations and make any final adjustments to the model, as required. Throughout this stage, the model would be improved and adjusted according to feedback from Site staff. Lastly, engineering documentation specific to the Site implementation would be provided and a benefits report would be submitted, comparing the predicted savings before and after the optimum movements are applied on the utilities system.

4. Key Performance Indicators (KPIs)

Besides the real time online optimization, during the EMS project appropriate energy performance metrics can also be identified and performance targets could be set. Also, within the EMS model calculation and reporting infrastructure, corrective actions in the event of deviations from target performance could be recommended.

Those metrics are usually known as Key Performance Indicators (KPI's) and can be related to:

- High level KPI's that monitor site performance and geared toward use by site and corporate management. For example: Total cost or the utilities system, predicted benefits, main steam headers imbalances, emissions, etc.
- Unit level KPI's that monitor individual unit performance and are geared toward use by unit management and technical specialists. For example: plant or area costs, boilers and heaters efficiencies, etc.
- Energy Influencing Variables (EIV's) that are geared towards use by operators. For example: Equipment specific operation parameters, like reflux rate, transfer line temperatures, cooling water temperature, etc.

The metrics are intended for use in a Site Monitoring and Targeting program where actual performance is tracked against targets in a timely manner, with deviations being prompting a corrective response that results in savings. They are calculated in the EMS and written back to the Plant Information System.

5. Project examples

The first two examples correspond to open loop implementations. The third one corresponds to a closed loop implementation. Finally, the last two examples correspond to very recent implementations.

5.1 Example one

In a French refinery a set of manual operating recommendations given by the optimizer during an operational Shift have been (Ruiz et al., 2007):

- Perform a few turbine/motors pump swaps.
- Change the fuels to the boilers (i.e., Fuel Gas and Fuel Oil).

As a result of the manual actions, the control system reacted and finally the following process variables:

- Steam production at boilers.
- Letdown and vent rates.

Figures 5, 6, 7 and 8 show the impact of the manually-applied optimization actions on steam production, fuel use and CO_2 emissions reduction.

Obtained benefits can be summarized as follows:

- Almost 1 tons per hour less Fuel Oil consumed.
- Approx 7 tons per hour less high pressure steam produced.
- Approx 2 tons per hour less CO_2 emitted.
- Approx 200 kW more electricity imported (which was the lowest cost energy available).

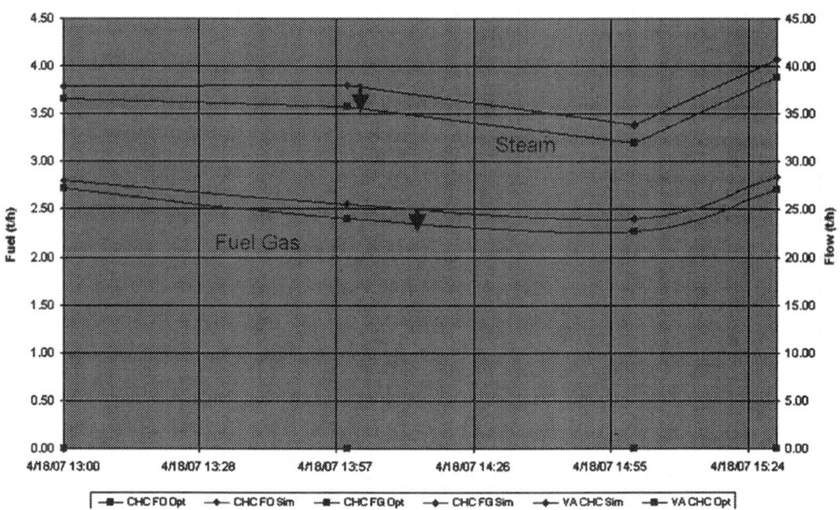

Fig. 5. Boiler C (100% Fuel Gas); 2 tons per hour less of steam

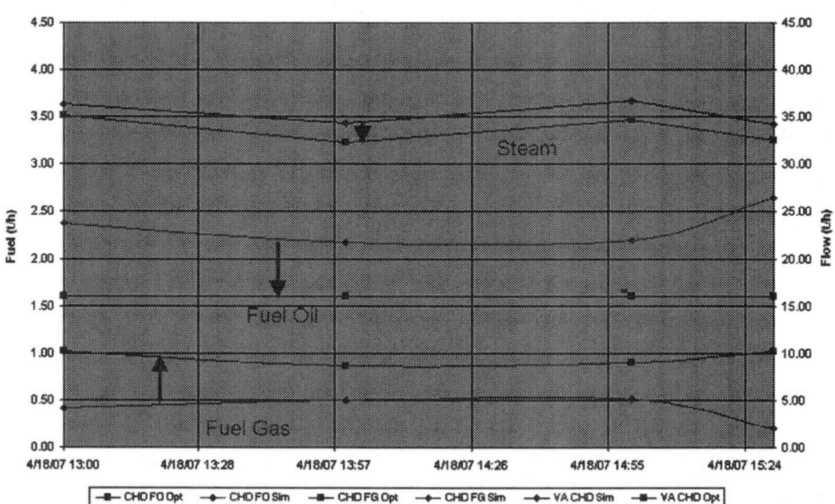

Fig. 6. Boiler D (Fuel Oil and Fuel Gas); 2 tons per hour less of steam and Fuel Oil sent to the minimum

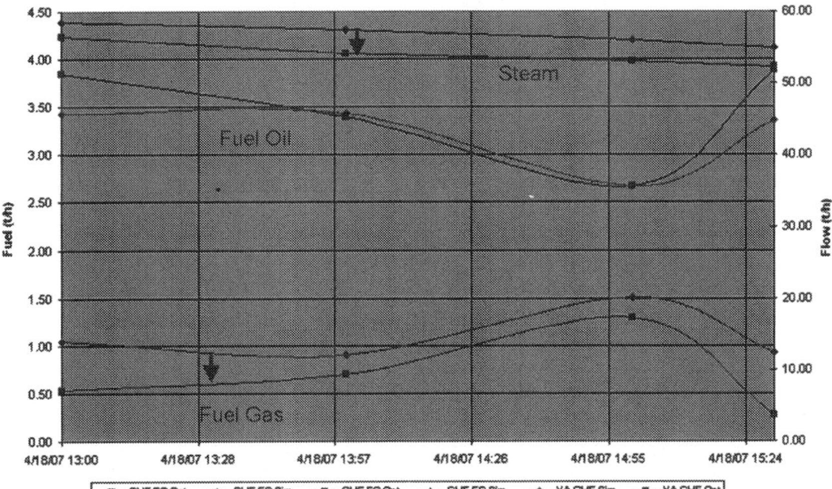

Fig. 7. Boiler F (Fuel Oil and Fuel Gas); more than 3 tons per hour less of steam

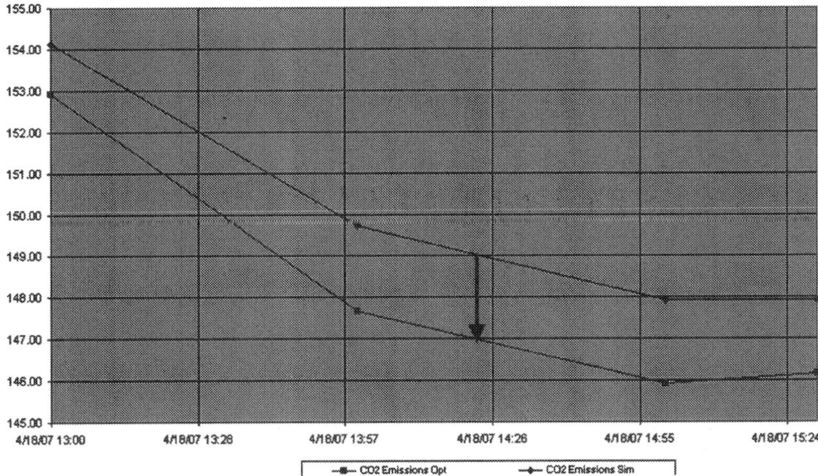

Fig. 8. CO_2 emissions; 2 tons per hour less

5.2 Example two

The second example corresponds to the energy system of a Spanish refinery with an olefins unit (Ruiz et al., 2006). In order to accurately evaluate the economic benefits obtained with the use of this tool, the following real time test has been done:

- First month: Base line, The EMS being executed online, predicting the potential benefits but no optimisation actions are taken.
- Second month: Operators trained and optimization suggestions are gradually implemented.
- Third month: Optimization recommendations are followed on a daily basis.

Fig. 9 shows the results of this test. Over that period, in 2003, 4% of the energy bill of the Site was reduced, with estimated savings of more than 2 million €/year.

5.3 Example three

The third example corresponds to a Dutch refinery where the EMS online optimization runs in closed loop, the so-called energy real time optimizer (Uztürk et al., 2006).

Typical optimisation handles include letdowns, load boilers steam flow, gas turbine generators/steam turbine generators power, natural gas intake, gas turbine heat recovery, steam generators duct firing, extraction of dual outlet turbines, deaerator pressure, motor/turbine switches, etc. Typical constraints are the steam balances at each pressure level, boiler firing capacities, fuel network constraints, refinery emissions (SO2, NOx, etc.) and contract constraints (for both fuel and electric power sell/purchase contracts).

Benefits are reported to come from the load allocation optimisation between boilers, optimised extraction/condensing ratio of the dual outlet turbines, optimised mix of discretionary fuel sales/purchase, optimised gas turbine power as a function of fuel and electricity purchase contract complexities (trade off between fuel contract verses electricity contract penalties).

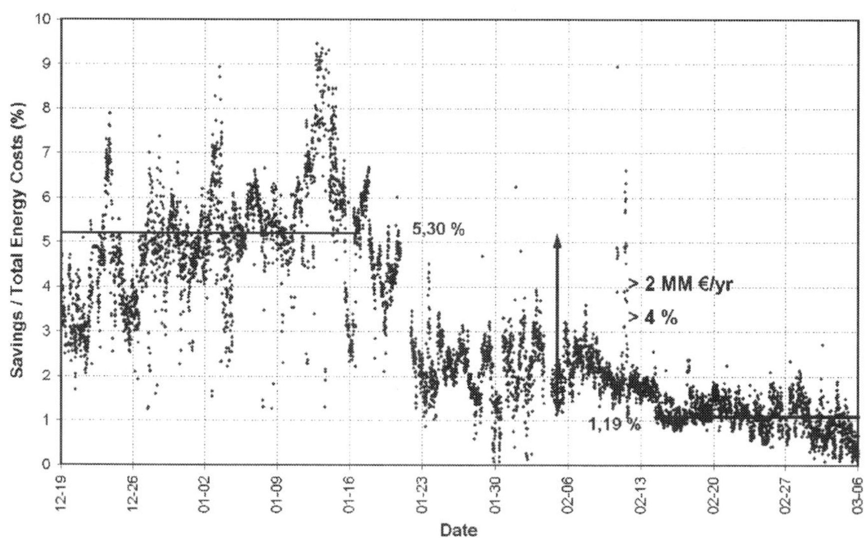

Fig. 9. Energy cost reduction evolution by using an online energy management tool

5.4 Example four

The fourth example corresponds to a French petrochemical complex, where the energy management system helps in emissions management too (Caudron, et al, 2010).

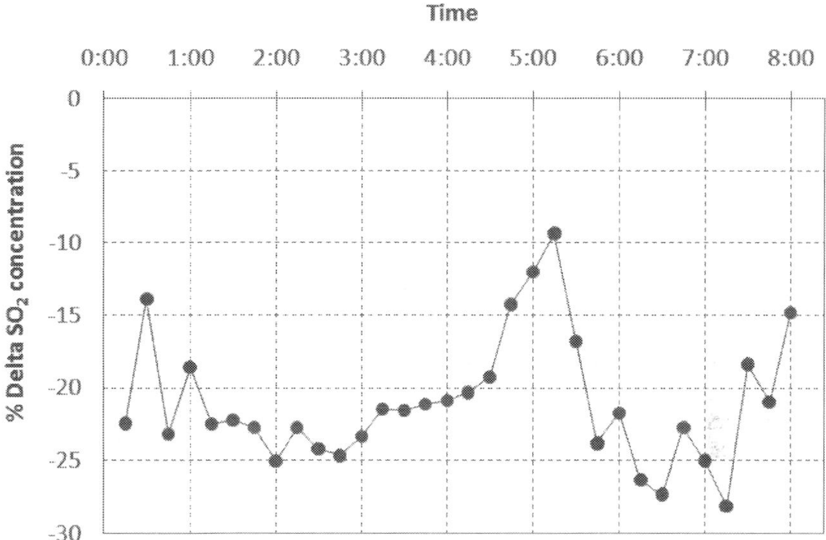

Fig. 10. Identified SO₂ emissions reduction along a shift

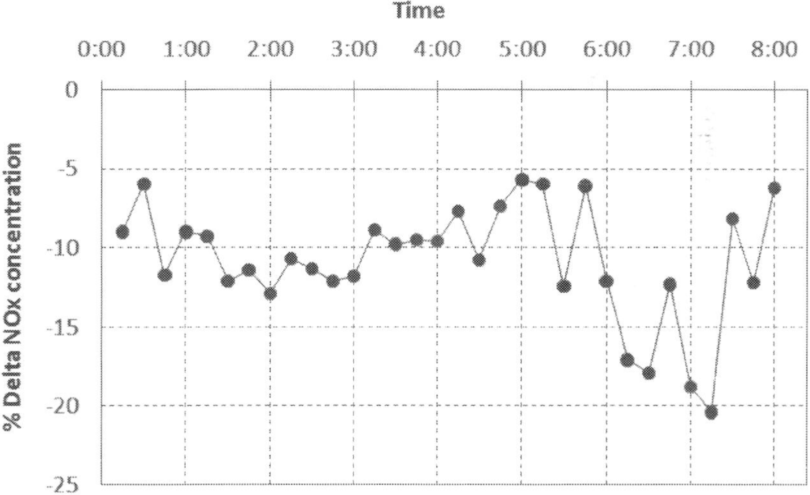

Fig. 11. Identified NOₓ emissions reduction along a shift

While reducing the energy costs, Figures 10 and 11 show respectively an example of the corresponding potential reduction in SO2 and NOx emissions (in terms of concentration with data corresponding to one of the main stacks) found during the same operational shift period applying the optimization recommendations. In this example, due to fuel management, a reduction in SO2 (around 20% less in concentration) and in NOx (around 10% less in concentration) in one of the main stacks has been also obtained.

5.5 Example five

This last example corresponds to the implementation of the energy management system in a Polish refinery (Majchrowicz et al, 2010). Visual MESA historizes important key performance indicators (KPIs). The most important ones are the economic energy operating cost, the optimized one and the predicted savings. Figure 12 shows an example of potential savings reduction due to the application of optimizer recommendations meaning effective energy costs reduction achieved. Each point in the figure corresponds to an automatic Visual MESA run. The variability along some days in the predicted savings can have different reasons, such as the changes in the operating conditions (e.g. weather, changes in producers and consumers). When a set of recommendations are followed by operators on day to day basis based on site wide optimization, the predicted savings are closer to zero.

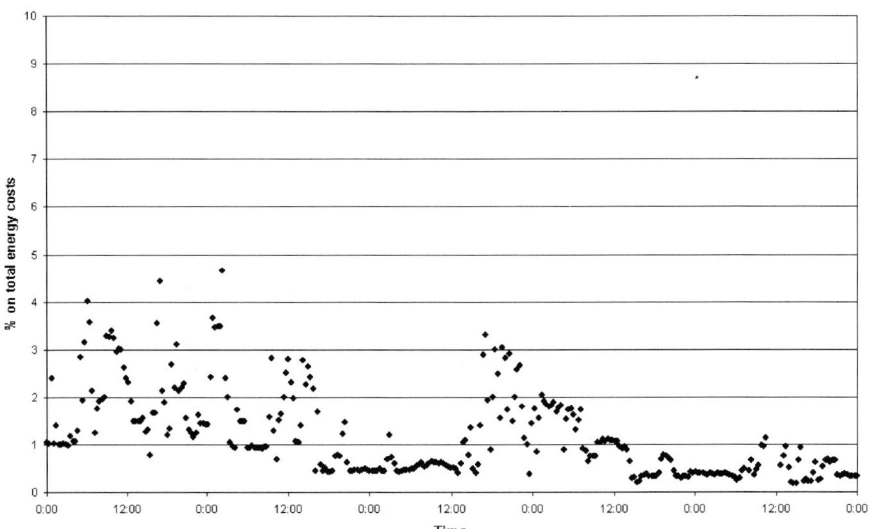

Fig. 12. Example of energy costs reduction follow-up

6. Conclusion

Online energy system optimization models are being used successfully throughout the Processing Industry, helping them to identify and capture significant energy cost savings. Although wide opportunities still exist for a growing number of real time online Energy Management Systems executed in open loop, an increased number of Closed Loop applications are expected in the near future. This evolution will bring additional economic benefits to the existing user base, especially when fuels and power prices are market-driven and highly variable. High frequency optimization opportunities that cannot be practically addressed by manual operating procedures would be captured and materialized. Of critical importance is having a robust and mature solver that reliably converges in a reasonable period of time in order to ensure buy-in from Operations for the continuous use of the system. More focus will be also on key process side operations, when tightly related with the Energy Network. Besides the Refining and Petrochemical industries, who were the early adopters of this kind of technology, other industries will take advantage of the real time energy management. For example, the Alcohol and Pulp & Paper industries, where waste fuel boilers, electric power cogeneration and evaporators systems could be also optimized together and District Heating and Cooling companies (i.e., power houses which are providers of heating and cooling services to cities, towns, campuses, etc.), that produce steam, chilled water and many times include cogeneration of electricity.

7. References

Benedicto, S.; Garrote, B.; Ruiz, D.; Mamprin, J. & Ruiz, C. (2007). Online energy management, *Petroleum Technology Quarterly (PTQ)*, Q1 (January 2007), pp. 131-138.

Caudron, M.; Mathieu, J.; Ruiz, D.; Ruiz, C. & Serralunga, F. (2010). Energy and emissions management at Naphtachimie petrochemical site. *ERTC Energy Efficiency Conference 2010*, Amsterdam, Netherlands.

García Casas, J.; Kihn, M.; Ruiz, D. & Ruiz, C. (2007). The Use of an On-line model for Energy Site-Wide Costs Minimisation. *European Refining Technology Conference (ERTC) Asset Maximisation Conference*, Rome, Italy.

Kihn, M.; Ruiz, D.; Ruiz, C. & García Nogales, A. (2008). Online Energy Costs Optimizer at Petrochemical Plant. *Hydrocarbon Engineering*, Vol. 13, No. 5, (May 2008), pp. 119-123, ISSN 1468-9340

Majchrowicz, J.; Herra M.; Serralunga, F. & Ruiz, D (2010). Online energy management at Grupa LOTOS refinery. *ERTC Annual Meeting 2010*, Istambul, Turkey.

Nelson, D.; Roseme, G. & Delk, S. (2000). Using Visual MESA to Optimize Refinery Steam Systems, *AIChE Spring Meeting*, Session T9013, Georgia, USA

Reid, M.; Harper, C. & Hayes, C. (2008). Finding Benefits by Modeling and Optimizing Steam and Power System. *Industrial Energy Technology Conference (IETC)*, New Orleans, USA

Ruiz, D.; Ruiz, C.; Mamprin, J. & Depto. de Energías y Efluentes Petronor (2005). Auditing and control of energy costs in a large refinery by using an on line tool, *European Refining Technology Conference (ERTC) Asset Maximisation*, Budapest, Hungary

Ruiz, D.; Ruiz, C.; Nelson, D.; Roseme, G.; Lázaro, M. & Sartaguda, M. (2006), Reducing refinery energy costs, *Petroleum Technology Quarterly (PTQ)*, Vol. Q1 (January 2006), pp. 103-105.

Ruiz, D.; Mamprin, J.; Ruiz, C. & Département Procédés - Energie, Logistique, Utilités, TOTAL - Raffinerie de Feyzin (2007). Site-Wide Energy Costs Reduction at TOTAL Feyzin Refinery. *European Refining Technology Conference (ERTC) 12th Annual Meeting*, Barcelona, Spain.

Ruiz, D.; Ruiz, C. & Nelson, D. (2007). Online Energy Management. *Hydrocarbon Engineering*, Vol. 12, No. 9, (September 2007), pp. 60-68, ISSN 1468-9340

Ruiz, D. & Ruiz, C. (2008). A Watchdog System for Energy Efficiency and CO2 Emissions Reduction. *European Refining Technology Conference (ERTC) Sustainable Refining*, Brussels, Belgium

Ruiz, D. & Ruiz,C. (2008). Closed Loop Energy Real Time Optimizers. *European Refining Technology Conference (ERTC) Annual Meeting Energy Workshop*, Vienna, Austria

Uztürk, D.; Franklin, H.; Righi, J. & Georgiou, A. (2006). Energy System Real Time Optimization. *NPRA Plant Automation and Decision Support Conference*, Phoenix, USA.

Wellons, M.; Sapre, A.; Chang, A. & Laird, T. (1994). On-line Power Plant Optimization Improves Texas Refiner's Bottom Line. *Oil & Gas Journal*, Vol.22, No.20, (May 1994)

Wiener N. (1948). *Cybernetics, Control and Communication in the Animal and the Machine* (second edition), MIT Press, Cambridge, Mass.

5

Energy Demand Analysis and Forecast

Wolfgang Schellong
Cologne University of Applied Sciences
Germany

1. Introduction

Sustainable energy systems are necessary to save the natural resources avoiding environmental impacts which would compromise the development of future generations. Delivering sustainable energy will require an increased efficiency of the generation process including the demand side. The architecture of the future energy supply can be characterized by a combination of conventional centralized power plants with an increasing number of distributed energy resources, including cogeneration and renewable energy systems. Thus efficient forecast tools are necessary predicting the energy demand for the operation and planning of power systems. The role of forecasting in deregulated energy markets is essential in key decision making, such as purchasing and generating electric power, load switching, and demand side management.

This chapter describes the energy data analysis and the basics of the mathematical modeling of the energy demand. The forecast problem will be discussed in the context of energy management systems. Because of the large number of influence factors and their uncertainty it is impossible to build up an 'exact' physical model for the energy demand. Therefore the energy demand is calculated on the basis of statistical models describing the influence of climate factors and of operating conditions on the energy consumption. Additionally artificial intelligence tools are used. A large variety of mathematical methods and ideas have been used for energy demand forecasting (see Hahn et al., 2009, or Fischer, 2008). The quality of the demand forecast methods depends significantly on the availability of historical consumption data as well as on the knowledge about the main influence parameters on the energy consumption. These factors also determine the selection of the best suitable forecast tool. Generally there is no 'best' method. Therefore it is very important to proof the available energy data basis and the exact conditions for the application of the tool.

Within this chapter the algorithm of the model building process will be discussed including the energy data treatment and the selection of suitable forecast methods. The modeling results will be interpreted by statistical tests. The focus of the investigation lies in the application of regression methods and of neural networks for the forecast of the power and heat demand for cogeneration systems. It will be shown that similar methods can be applied to both forecast tasks. The application of the described methods will be demonstrated by the heat and power demand forecast for a real district heating system containing different cogeneration units.

2. Energy data management

2.1 Energy data analysis

Energy management describes the process of managing the generation and the consumption of energy, generally to minimize demand, costs, and pollutant emissions. The energy management has to look for efficient solutions for the challenges of the changing conditions of the international energy economy which are caused by the world wide liberalization of the energy market restricted by limited resources and increasing prices (Doty & Turner, 2009). Computer aided energy management combines applications from mathematics and informatics to optimize the energy generation and consumption process. Information systems represent the basis for controlling and decision activities. Because of the large number of relevant information an efficient data management is to be used. Therefore mathematical analyzing and optimizing methods are to be combined with energy data bases and with the data management of the energy generation process. The detailed analysis of the main input and output data of an energy system is necessary to improve its efficiency. Improving the efficiency of energy systems or developing cleaner and efficient energy systems will slow down the energy demand growth, make deep cut in fossil fuel use and reduce the pollutant emissions.

Much of the energy generated today is produced by large-scale, centralized power plants using fossil fuels (coal, oil, and gas), hydropower or nuclear power, with energy being transmitted and distributed over long distances to the consumers. The efficiency of conventional centralized power systems is generally low in comparison with combined heat and power (CHP) technologies (cogeneration) which produce electricity or mechanical power and recover waste heat for process use. CHP systems can deliver energy with efficiencies exceeding 90%, while significantly reducing the emissions of greenhouse gases and other pollutants (Petchers, 2003). Selecting a CHP technology for a specific application depends on many factors, including the amount of power needed, the duty cycle, space constraints, thermal needs, emission regulations, fuel availability, utility prices and interconnection issues. The tasks and objectives of a local energy provider can be summarized as follows:

- Supply of the power and heat demand of the delivery district (additionally supply of cool and other media as gas and water is possible)
- Logistic management and provision of the primary fuels and of the support materials; dispose of the waste materials
- Portfolio management (i.e. buying and selling power at the power stock exchange)
- Customer relationship management
- Power plant and grid operation

Fig. 1 shows the relationship model of the main input data resources and the data flow of the energy data management. The energy database represents the heart of the energy information system. The energy data management provides information for the energy controlling including all activities of planning, operating, and supervising the generation and distribution process. A detailed knowledge of the energy demand in the delivery district is necessary to improve the efficiency of the power plant and to realize optimization potentials of the energy system.

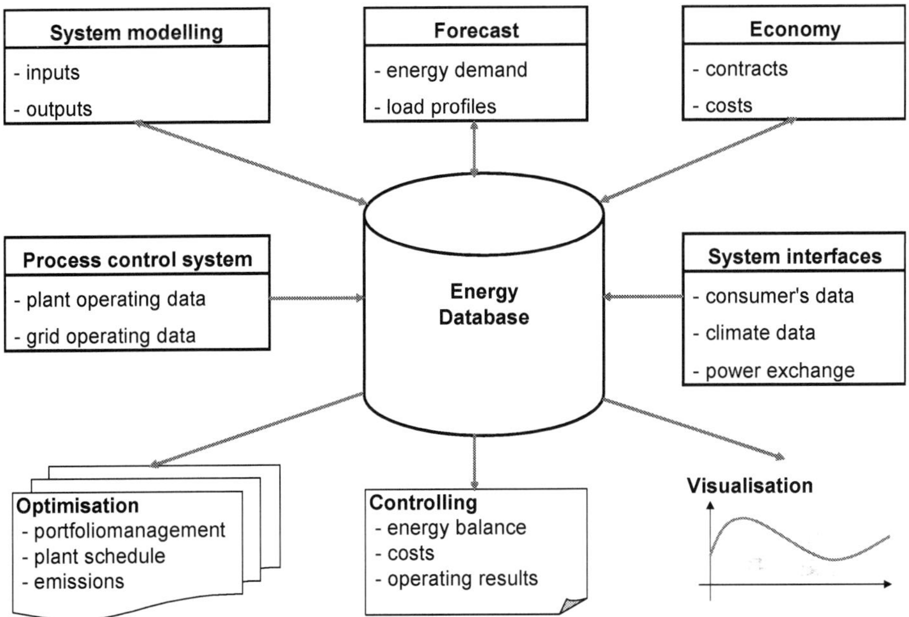

Fig. 1. Energy data management

2.2 Mathematical modeling

With the help of an energy data analysis the relations between the main inputs and outputs of the energy system will be described by mathematical models. The process of the mathematical modeling is characterized by the following properties:

- A mathematical model represents the mapping of a real technical, economical or natural system.
- As in real systems generally many influence parameters are determining, the modeling process must condense and integrate them (section 3.1).
- The mathematical modeling combines abstraction and simplification.
- In the most cases the model is oriented to application, i.e., the model is built up for a special use.

The demands for the modeling process can be summarized to the thesis: The model should be exact as necessary and simple as possible. A wide range of statistical modeling algorithms is used in the energy sector. They can be classified according to these three criteria:

- type of the model function (linear / non-linear)
- number of the influence variables (univariate / multivariate)
- general modeling aspect (parametric / non-parametric)

The separation between linear and non-linear methods depends on the functional relationship. A model is called univariate if only one influence factor will be regarded; otherwise it is of the multivariate type. Parametric models contain parameters besides the input and output variables. The best known linear univariate parametric model is the classical single linear regression model (section 3.4). Non-parametric models as artificial neural networks (section 3.5) don't use an explicit model function.

An explicit algebraic relationship between input and output can be described by the model

$$y = F(x,p) \tag{1}$$

where the function F describes the influence of the input vector x on the output variable y. The function F and the parameter vector p determine the type of the model. Regarding (1) there are two typically used modeling tasks:

Simulation:

Calculate the outputs y for given inputs x and fixed parameters p, and compare the results.

Parameter estimation (inverse problem):

For given measurements of the input x and the output y calculate the parameters p so that the model fits the relation between x and y in a "best" way.

The numerical calculation of the parameters of the regression model described in section 3.4 represents a typical parameter estimation problem.

2.3 Energy demand analysis

The energy consumption of the delivery district of a power plant depends on many different influence factors (fig. 2). Generally the energy demand is influenced by seasonal data, climate parameters, and economical boundary conditions. The heat demand of a district heating system depends strongly on the outside temperature but also on additional climate factors as wind speed, global radiation and humidity. On the other side seasonal factors influence the energy consumption. Usually the power and heat demand is higher on working days than at the weekend. Furthermore vacation and holidays have a significant impact on the energy consumption. Last but not least the heat and power demand in the delivery district is influenced by the operational parameters of enterprises with large energy demand and by the consumer's behavior. Additionally the power and heat demand follow a daily cycle with low periods during the night hours and with peaks at different hours of the day.

The quality of the energy demand forecast depends significantly on the availability of historical consumption data and on the knowledge about the main influence parameters on the energy demand. The functional relationship is non-linear and there are more or less complex interactions between different data types. Because of the large number of influence factors and their uncertainty it is impossible to build up an 'exact' physical model for the energy demand. Therefore the energy demand is calculated on the basis of mathematical models simplifying the real relationships as described in the previous section. Since no simple deterministic laws that relate the predictor variables (seasonal data, meteorological data and economic factors) on one side and energy demand as the target variable on the other side exist, it is necessary to use statistical models. A statistical model learns a quantitative relationship from historical data. During this training process quantitative relationships between the target variables (variables that have to be predicted) and the predictor variables are determined from historical data. Training data sets must be provided for known predictor target variables. From these example data the mathematical model is determined. This model can then be used to compute the values of the target variables as a function of the predictor variables for periods for which only the predictor variables are known. Using meteorological data as predictor variables forecasts for those meteorological variables are needed (Fischer, 2008).

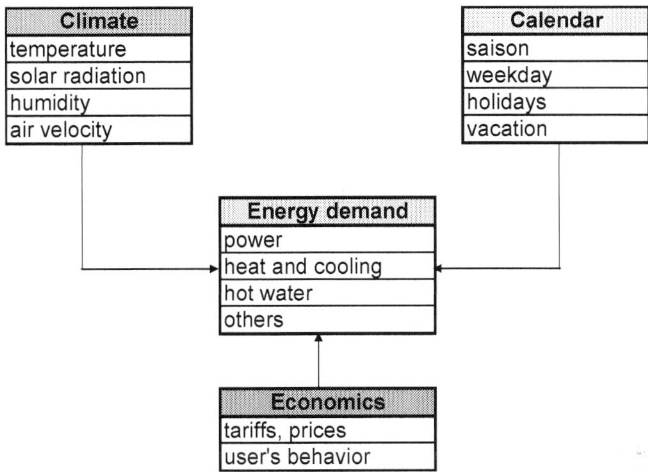

Fig. 2. Relationship model of the energy demand

The analysis of the relationships between energy consumption and climate factors includes the following activities:
- energy balancing (distribution of the demand)
- analysis of the main influence factors (fig. 2)
- design of the mathematical model
- analysis and modeling of typical demand profiles

The daily cycle of the power and heat consumption can be described by time series methods (see 3.3). For non-interval metered customers "Standard load profiles" (SLP) can be used. They describe the time dependent load of special customer groups, e.g. residential buildings, small manufactories, office buildings, etc. (VDEW, 1999).

2.4 Energy controlling and optimization

The power generation system of the provider generally consists of several power plants including distributed units as cogeneration systems, wind turbines, and others (fig. 3). The provider is faced with the task to find the optimal combination (schedule) of the different generation units to satisfy the power and heat demand of the customers. Because of the unbundled structure of the generation, distribution and selling of electricity a lot of technical relations and economical conditions are to be modeled.

As the architecture of the future electricity systems can be characterized by a combination of conventional centralized power plants with an increasing number of distributed energy resources, the generation scheduling optimization becomes more and more important. The schedule selects the operating units and calculates the amount to generate at each online unit in order to achieve the minimum production cost. This generation scheduling problem requires determining the on/off schedules of the plant units over a particular time horizon. Apart from determining the on/off states, this problem also involves deciding the hourly power and heat output of each unit. Thus the scheduling problem contains a large number of discrete (on/off status of plant units) and continuous (hourly power and heat output) variables.

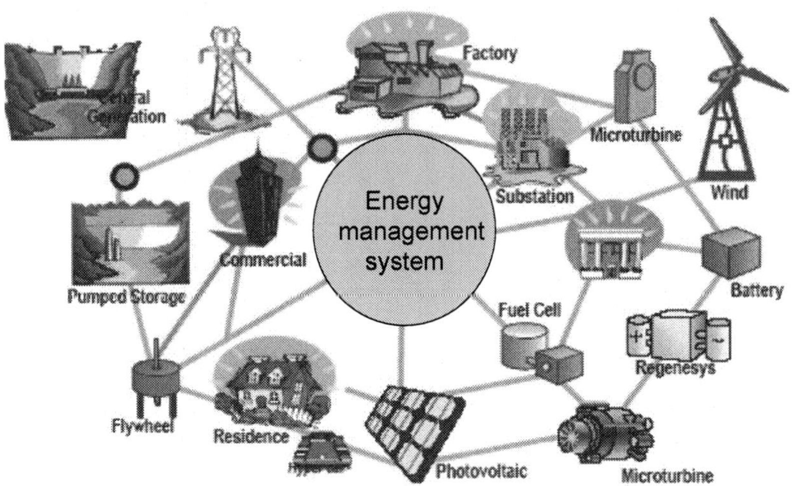

Fig. 3. Distributed energy system (Maegard, 2004)

The objectives of the schedule optimization can be summarized as:
- minimization of the fuel and operating costs
- minimization of the distribution costs
- reduction of CO_2 emissions
- optimization of the power trading

The most important restrictions and boundary conditions of the optimization problem are given by (Schellong, 2006):
- The generation system must satisfy the power and heat demand of the delivery district.
- The power generation in a cogeneration system depends on the heat generation. The mathematical relations can be described in a similar way as described in 2.2.
- There are a lot of boundary restrictions referring the capacity and the operating conditions of the generation units.
- The operating schedule depends on the availability of the single generation units.
- The system is influenced by constraints of the district heating network as well as of the electrical grid.
- The generation system has to fulfill legal constraints referring emissions.
- The optimization system is influenced by the delivery contracts and actual conditions of the energy trading at the energy stock exchange.

Thus the related mathematical optimization model has a very complex structure. Following the ideas described in section 2.2 the generation scheduling problem can be solved as a mixed integer linear optimization problem. The optimization results in an optimal schedule of the generation units using an optimal fuel mix and satisfying all restrictions. To realize this schedule the generation process must be supervised by the energy control system using the data management illustrated in fig. 1.

It is obviously that these processes require the detailed knowledge of the energy demand of the delivery system. Especially for cogeneration systems it is important to know the coincidence of the power and heat demand. CHP units are only able to generate electricity efficiently, when the produced heat is simultaneously used on the demand side.

3. Energy demand forecast methods

3.1 General modeling aspects

As described in the previous section the quality of the forecast methods mainly depends on the available historical data as well as on the knowledge about the factors influencing the energy demand. With the help of the energy data analysis (see 2.1) the necessary data for the training, test, and validation sets are provided to realize the modeling process (see 2.3). The historical energy consumption data are divided into clusters depending on seasonal effects. Thus the modeling process must be specified for each cluster. Furthermore the time horizon of the forecast determines the type of the applied method. Short- term forecasting calculates the power demand for the period of the next view minutes. This task plays an important role for the generation process, but also for the implementation of peak shifting applications at the consumer's side. The forecast of the day-ahead and of the weekly energy demand will be realized by medium-term methods. Based on the day-ahead forecast the operation schedule of the power plant units will be optimized (see 2.4). Finally long-term forecast tools estimate the future demand for periods of several month or years. These methods are necessary for the portfolio management and for the energy logistic (fig. 1).

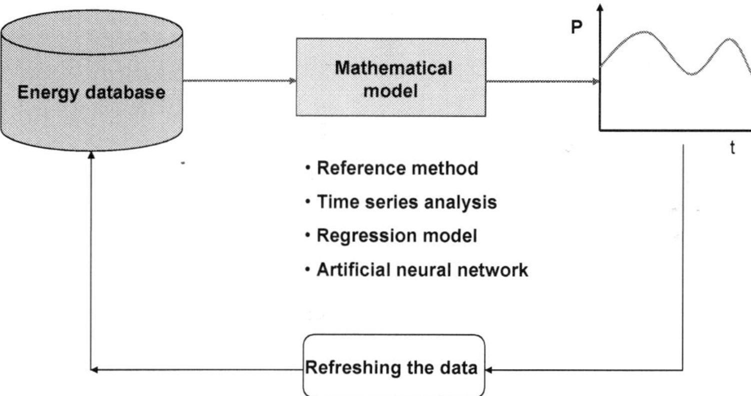

Fig. 4. Forecast methods

Fig. 4 shows an overview of the most common used forecast methods which are described in this section. The methods can be divided into the following three branching pairs: empirical and model-based, extrapolation and causal, and static and dynamic (Fischer, 2008). Empirical methods are useful when only few or no historical data are available, when the past does not significantly affect the future, or when explanation and sensitivity analysis are not required. A popular approach is that of historical analogies implemented in the reference method (see 3.2). Model-based methods use well-specified algorithms to process and analyze data. Extrapolation and causal methods are included in this category. Extrapolation methods are numerical algorithms that help forecasters find patterns in time-series observations of a quantitative variable. These are popular for short-range forecasting. This method is based on the assumption that a stable, systematic structure can describe the future energy demand. These models are characterized by the criteria described in section 2.2. A static forecast is used to predict the energy demand into the near future on the basis of actual data for the variables in the past or the present. On the other hand, a dynamic forecast can be used to make long term projections considering changes of the framework conditions during the forecast period.

3.2 Reference method

The pure reference method works without a mathematical model. The basic idea of this simple method is to find a situation in an energy data base of historical data that is similar to the one that has to be predicted. A set of explanatory variables is defined and similarity between situations is measured by these variables. The method will be described by an example: To calculate the heat or power demand for a Monday, with a mean predicted temperature of +5 deg C the algorithm is simply looking in a data base for another Monday with a mean temperature close to +5 deg C. Thus the historical consumption data for that day are used as the prediction. For a long time this method has been the reference method for energy demand predictions especially for local energy providers, and surprisingly it is still widely used. The advantage of the method is that it is simple to implement. The results are easily to be interpreted. However the disadvantages are numerous. Although the implementation of the method seems to be straightforward, it becomes complicated if the number of criterions increases. If for instance hourly temperatures are used instead of daily mean temperature the measures of similarity are no longer so obvious. With an increasing number of explanatory variables, the probability to find no data set that is similar according to all criteria increases (Fischer, 2008).

In practical applications the reference method is used in combination with some other adaptation criteria depending on the behavior of the energy consumption in the past. Additionally the reference method is supported by a regression model describing the climate influence factors and/or time dependent energy consuming impacts caused by production factors in industrial enterprises. On the other side the knowledge of the energy consumption of selected historical reference days can improve the quality of model based methods as will be described in section 4.

3.3 Time series analysis

This method belongs to the category of the non-causal models of demand forecasting that do not explain how the values of the variable being projected are determined. Here the variable to be predicted is purely expressed as a function of time, neglecting other influence factors. This function of time is obtained as the function that best explains the available data, and is observed to be most suitable for short-term projections. A time series is often the superposition of the following terms describing the energy demand as time dependent output $y(t)$:

- Long-term trend variation (T)
- Cyclical variation (C)
- Seasonal variation (S)

- Irregular variation (R)

The trend variation T describes the gradual shifting of the time series, which is usually due to long term factors such as changes in population, technology, and economy. The cyclical component S represents multiyear cyclical movements in the economy. The periodic or seasonal variation in the time series is, in general, caused by the seasonal weather or by fixed seasonal events. The irregular component contains the residual of the time series if the trend, cyclical and seasonal components are removed from the time series. These terms can be combined to mixed time series model:

Additive model: $\qquad$ $y(t) = T(t) + S(t) + C(t) + R(t)$ $\qquad$ (2)

Hybrid model: $\qquad$ $y(t) = T(t) \times S(t) + R(t)$ $\qquad$ (3)

In addition to the univariate time series analysis, autoregressive methods provide another modeling approach requiring only data on the previous modeled variable. Autoregressive models (AR) describe the actual output y_t by a linear combination of the previous time series $y_{t-1}, y_{t-2}, \ldots, y_{t-p}$ and of an actual impact a_t:

$$y_t = \varphi_1 y_{t-1} + \varphi_2 y_{t-2} + \ldots + \varphi_p y_{t-p} + a_t \qquad (4)$$

The autoregressive coefficients have to be estimated on the basis of measurements. The AR-models can be combined with moving average models (MA) to ARMA models which have been firstly investigated by Box and Jenkins (Box & Jenkins, 1976).

The time series method has the advantage of its simplicity and easy use. It is assumed that the pattern of the variable in the past will continue into the future. The main disadvantage of this approach lies in the fact that it ignores possible interaction of the variables. Furthermore the climate impacts and other influence factors are neglected.

3.4 Regression models

Regression models describe the causal relationship between one or more input variable(s) and the desired output as dependent variable by linear or nonlinear functions. In the simplest case the univariate linear regression model describes the relationship between one input variable x and the output variable y by the following formula:

$$y = f(x,a_0,a_1) = a_0 + a_1 x \qquad (5)$$

Thus geometrically interpreted a straight line describes the relationship between y and x. The shape of the straight line is determined by the so called regression parameters a_0 and a_1. For given measurements $x_1, x_2, \ldots, x_n$ and $y_1, y_2, \ldots, y_n$ of the variables x and y the parameters are calculated such that the mean quadratic distance between the measurements y_i (i=1, . . . ,n) and the model values $\hat{y}_i$ on the straight line is minimized. That means the following optimization problem is to be solved:

$$Q(a_0,a_1) = \sum_{i=1}^{n} (y_i - f(x_i,a_o,a_1))^2 \to \underset{a_0,a_1}{Min} \qquad (6)$$

The calculated regression parameters represent a so called least squares estimation of the fitting problem (Draper & Smith, 1998).

The regression model can be extended to a multivariate linear relationship where the output variable y is influenced by p inputs $x_1, x_2, \ldots, x_p$:

$$y = f(x,a) = a_0 + a_1x_1 + a_2x_2 + \ldots + a_px_p \tag{7}$$

We define the following notations:

$$y = \begin{bmatrix} y_1 \\ y_2 \\ . \\ y_n \end{bmatrix} \quad a = \begin{bmatrix} a_1 \\ a_2 \\ . \\ a_p \end{bmatrix} \quad X = \begin{bmatrix} 1 & x_{11} & . & x_{1p} \\ 1 & x_{21} & . & x_{2p} \\ . & . & . & . \\ 1 & x_{n1} & . & x_{np} \end{bmatrix} \tag{8}$$

where the vector y contains the measurements of the output variable, a represents the vector of the regression parameters, and the matrix X contains the measurements x_{ij} of the ith observation of the input x_j. Thus the least squares estimation of the multivariate linear regression problem will be obtained by solving the minimization task:

$$Q(a_0, a_1, \ldots, a_p) = \sum_{i=1}^{nn} (y_i - a_o - a_1x_{i1} - a_2x_{i2} - \ldots - a_px_{ip})^2 = (y - Xa)^T (y - Xa) \to \underset{a_0, a_1, \ldots, a_p}{Min} \tag{9}$$

The least squares estimation of the regression parameter vector a represents the solution of the normal equation system referring to the minimization problem (9):

$$X^T Xa = X^T y \tag{10}$$

Regarding the special structure of this linear system, adapted methods like Cholesky or Housholder procedures are available to solve (10) using the symmetry of the coefficient matrix (Deuflhard & Hohmann, 2003). The model output can be described as

$$\hat{y} = X\hat{a} \tag{11}$$

where the vector $\hat{y}$ contains the model output values $\hat{y}_i$ (i=1, . . . , n) and $\hat{a}$ represents the vector of the estimated regression coefficients a_j (j=1, . . . , p) as the solution of (10).

The results of the regression analysis must be proofed by a regression diagnostic. That means we have to answer the following questions:

- Does a linear relationship between the input variables $x_1, x_2, \ldots, x_p$ and the output y really exist?
- Which input variables are really relevant?
- Is the basic data set of measurements consistent or are there any "out breakers"?

With the help of the coefficient of determination B we can proof the linearity of the relationship.

$$B = 1 - \frac{\sum_{i=1}^{n} (y_i - \hat{y}_i)^2}{\sum_{i=1}^{n} (y_i - \overline{y})^2} = \frac{SSR}{SSY} \quad , \tag{12}$$

where $\hat{y}_i$ represent the calculated model values given by (11) and $\overline{y}$ is the arithmetic mean value of the measured outputs y_i. B ranges from 0 to 1. Values of B in the near of 1 indicate, that there exists a linear relationship between the regarded input and output. To identify the most significant input variables the modeling procedure must be repeated by leaving one of the variables from the model function within an iteration process. The coefficient of determination and the expression $s^2 = SSR/(n-p-1)$ indicate the significance of the left variable. s^2 represents the estimated variance of the error distribution of the measured values of y. Finally the analysis of the individual residuals $r_i = y_i - \hat{y}_i$ gives some hints for the existence of "out breakers" in the basic data set.

Multivariate linear regressions are widely used in the field of energy demand forecast. They are simple to implement, fast, reliable and they provide information about the importance of each predictor variable and the uncertainty of the regression coefficients. Furthermore the results are relatively robust. Nonlinear regression models are also available for the forecast. But in this case the parameter estimation becomes more difficult. Furthermore the nonlinear character of the influence variable must be guaranteed. Regression based algorithms typically work in two steps: first the data are separated according to seasonal variables (e.g. calendar data) and then a regression on the continuous variables (meteorological data) is done. That means a regression analysis must be done for each seasonal cluster following the algorithm:

Step 1. Analysis of the available energy data
Step 2. Splitting the historical energy consumption data into seasonal clusters
Step 3. Identifying the main meteorological factors on the energy demand as described in section 2.3
Step 4. Regression analysis as described above
Step 5. Validation of the model (regression diagnostic)
Step 6. Integration of the sub models

The application of regression methods to the heat demand forecast for a cogeneration system will be described in section 4.

3.5 Neural networks

Neural networks (NN) represent adaptive systems describing the relationship between input and output variables without explicit model functions. NN are widely used in the field of energy demand forecast (Schellong & Hentges, 2007). The basic elements of neural networks (NN) are the neurons, which are simple processing units linked to each other with directed and weighted connections. Depending on their algebraic sign and value the connections weights are inhibiting or enhancing the signal that is to be transferred. Depending on their function in the net, three types of neurons can be distinguished: The units which receive information from outside the net are called input neurons. The units which communicate information to the outside of the net are called output neurons. The remaining units are called hidden neurons because they only send and receive information from other neurons and thus are not visible from the outside. Accordingly the neurons are grouped in layers. Generally a neural net consists of one input and one output layer, but it can have several hidden layers (fig. 5).

The pattern of the connection between the neurons is called the network topology. In the most common topology each neuron of a hidden layer is connected to all neurons of the preceding and the following layer. Additionally in so-called feedforward networks the signal is allowed to travel only in one direction from input to output (Fine, 1999).

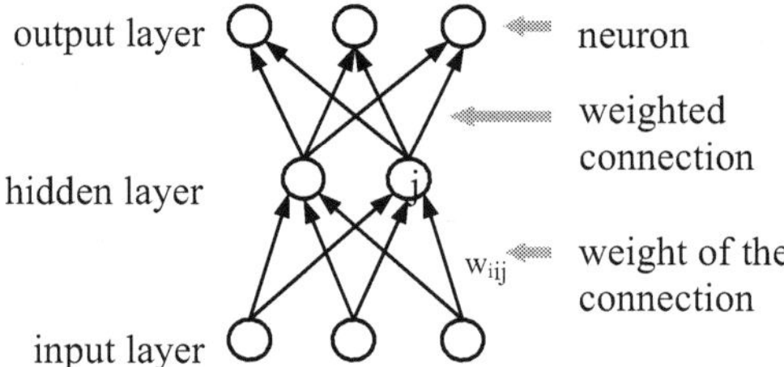

Fig. 5. Structure of a neural network

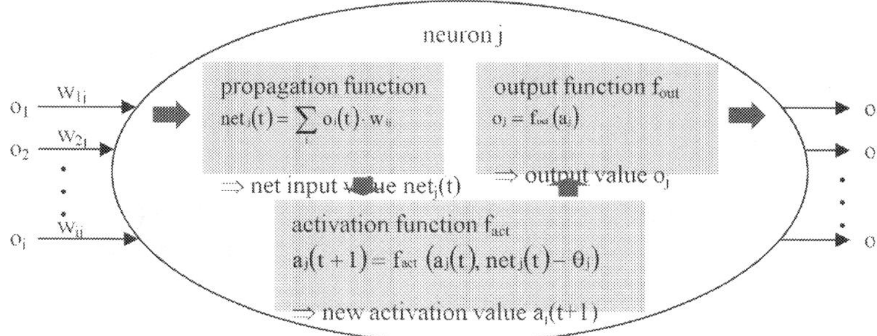

Fig. 6. Structure of a neuron

To calculate its new output depending on the input coming from the preceding units (or from outside) a neuron uses three functions (Galushkin, 2007): First the inputs to the neuron j from the preceding units combined with the connection weights are accumulated to yield the net input. This value is subsequently transformed by the activation function f_{act}, which also takes into account the previous activation value and the threshold θ_j (bias) of the neuron to yield the new activation value of the neuron. The final output o_j can be expressed as a function of the new activation value of the neuron. In most of the cases this function f_{out} is not used so that the output of the neurons is identical to their activation values (fig. 6).

Three sigmoid (S-shaped) activation functions are usually applied: the logistic, hyperbolic tangent and limited sine function. The formulas of the functions are given by:

$$f\log(x)=\frac{1}{1+e^{-x}} \quad f\tanh(x)=\tanh(x)=\frac{e^x-e^{-x}}{e^x+e^{-x}} \quad f\sin(x)=\begin{cases} 1 & \text{for } x>\pi/2 \\ \sin(x) & \text{for } -\pi/2\leq x\leq\pi/2 \\ -1 & \text{for } x<\pi/2 \end{cases} \quad (13)$$

A neural network has to be configured such that the application of a set of inputs produces the desired set of outputs. This is obtained by training, which involves modifying the connection weights. In supervised learning methods, after initializing the weights to random values, the error between the desired output and the actual output to a given input vector is used to determine the weight changes in the net. During training, input pattern after input pattern is presented to the network and weights are continually adapted until for any input the error drops to an acceptable low value and the network is not overfitted. In the case that a network has been adjusted too many times to the patterns of the training set, it may in consequence be unable to accurately calculate samples outside of the training set. Thus by overlearning the neural network loses its capability of generalization. One way to avoid overtraining is by using cross-validation. The sample set is split into a training set, a validation set and a test set. The connection weights are adjusted on the training set, and the generalization quality of the model is tested, every few iterations, on the validation set. When this performance starts to deteriorate, overlearning begins and the iterations are stopped. The test set is used to check the performance of the trained neural network (Caruana et al., 2001). The most widely used algorithm for supervised learning is the backpropagation rule. Backpropagation trains the weights and the thresholds of feedforward networks with monotonic and everywhere differentiable activation functions.

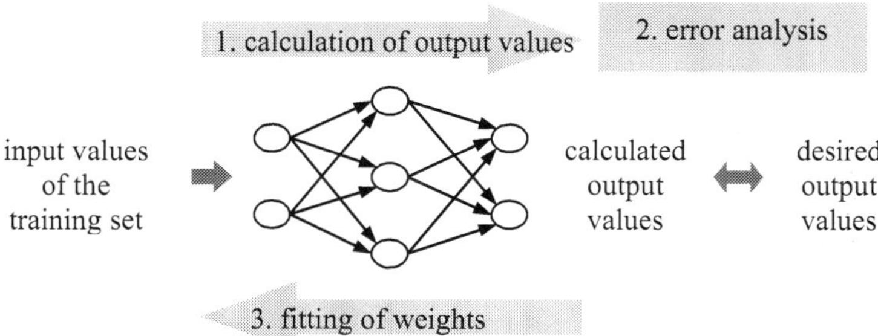

Fig. 7. Backpropagation learning rule

Mathematically, the backpropagation rule (fig. 7) is a gradient descent method, applied on the error surface in a space defined by the weight matrix. The algorithm involves changing each weight by the partial derivative of the error surface with respect to the weight (Rumelhart et al., 1995). Typically, the error E of the network that is to be reduced is calculated by the sum of the squared individual errors for each pattern of the training set. This error depends on the connection weights:

$$E(W) = E(w_{11}, w_{12}, ..., w_{nn}) = \sum_p E_p \quad \text{with} \quad E_p = \frac{1}{2}\sum_j (t_{pj} - o_{pj})^2 \tag{14}$$

where E_p is the error for one pattern p, t_{pj} is the desired output from the output neuron j and o_{pj} is the real output from this neuron.

The gradient descent method has different drawbacks, which result from the fact that the method aims to find a global minimum with only information about a very limited part of the error surface. To allow a faster and more effective learning the so-called momentum term and the flat spot elimination are common extensions to the backpropagation method. These prevent, for example, the learning process from sticking on plateaus where the slope is extremely slight, or being stuck in deep gaps by oscillation from one side to the other (Reed et al., 1998).

Although the algorithm of NN is very flexible and can be used in a wide range of applications, there are also some disadvantages. Generally the design and learning process of neural networks takes a large amount of computing time. Due to the capacity of computational time it is in most cases not possible to re-train a model in operational mode every day. Furthermore it is difficult to interpret the modeling results.

In order to use neural networks for the energy demand forecast the following algorithm must be realized:

Step 1. Preliminary analysis of the main influence factors on the energy demand as described in section 2.3

Step 2. Design of the topology of the NN

Step 3. Splitting the basic data into a training set, a validation set and a test set

Step 4. Test and selection of the best suitable activation function

Step 5. Application of the backpropagation learning rule with momentum term and flat spot elimination

Step 6. Validation and comparison of the modeling results

Step 7. Selection of the best suitable network

The application of neural networks to the heat and power demand forecast for a cogeneration system will be described in section 4.

4. Heat and power demand forecast for a cogeneration system

4.1 The cogeneration system

The cogeneration system consists of two cogeneration units and two additional heating plants (fig. 8). The first cogeneration unit represents a multi-fuel system with hard coal as primary input. Additionally gas and oil are used. The second unit works as incineration plant with waste as primary fuel. The heating plants use mainly gas as fuel. The cogeneration system provides power and heat for a district heating system. The heating system consists of 3 sub networks connected by transport lines. About 3.000 customers from

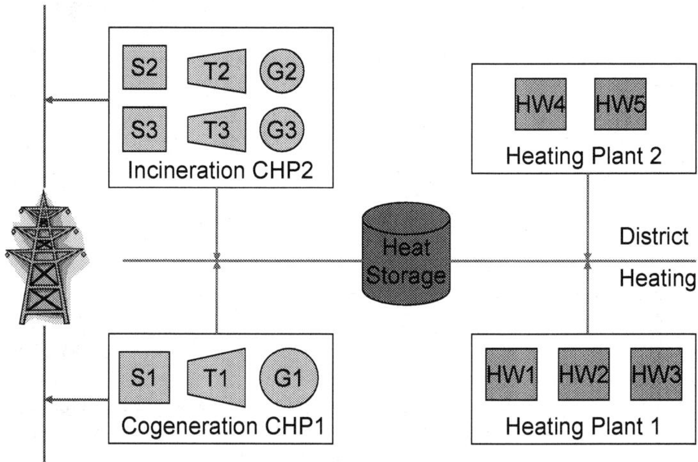

Fig. 8. Cogeneration system

industry, office buildings, and residential areas are delivered by the system. Thus the consumption behavior is characterized by a mixed structure. But the main part of the heat consumption is used for room heating purposes. The annual heat consumption amounts to about 460 GWh, and the power consumption to 6.700 GWh (Schellong & Hentges, 2007). Thus the power demand can not be completely supplied by the cogeneration plant. The larger part of the demand must be bought from other providers and at the European energy exchange (EEX). Therefore the forecast tool for the power demand is not only necessary for the operating of the cogeneration plant but also for the portfolio management.

Generally the power plant of a district heating system is heat controlled, because the heat demand of the area must completely be supplied. Although in the system a heat accumulator is integrated, the heat demand must be fulfilled more or less 'just in time'. But as in the cogeneration plant 3 extraction condensing turbines are involved (fig. 8), the system is also able to follow the power demand.

4.2 Data analysis

As described in section 2.3 the energy consumption of the district delivery system depends on many different influence factors (fig. 3). Generally the energy demand is influenced by seasonal data, climate parameters, and economical boundary conditions. The heat demand of the district heating system depends strongly on the outside temperature but also on additional climate factors as wind speed, global radiation and humidity. On the other side seasonal factors influence the energy consumption. As a result of a preliminary analysis, the strongest impact among the climate factors on the heat demand has the outdoor temperature. Additionally the temperature difference of two sequential days represents a significant influence factor, describing the heat storage effects of buildings and heating systems. Concerning the power forecast, the influence of the power consumption measured in the previous week proved to be an interesting factor. These influence factors represent the basis of the model building process. For the forecast calculations, the power and the heat consumption data are divided into three groups depending on the season:

- winter
- summer
- transitional period containing spring and autumn

In each cluster the consumption data of a whole year are separately modeled for working days, weekend and holidays.

4.3 Heat demand forecast by regression models

Following the modeling strategy of section 2.3 the heat demand Q_{th} of a district heating system can be simply described by a linear multiple regression model (RM):

$$Q_{th} = a_0 + a_1 t_{out} + a_2 \Delta t_{out} \tag{15}$$

where t_{out} represents the daily average outside temperature and Δt_{out} describes the temperature difference of two sequential consumption days.

The model (15) can be extended by additional climate factors as wind, solar radiation and others. But in order to get a model based on a simple mathematical structure and because of the dominating impact of the outdoor temperature among the climate factors only the two regression variables are used in (15). The results of the regression analysis for each cluster depending on the season and on the type of the day are checked by the correlation coefficients and by a residual analysis. Corresponding to the modeling aspects described in chapter 2.2 for each season and each weekday a regression model (see equation 1) is calculated. The models describe the dependence of the daily heat demand on the outdoor temperature and the temperature difference of two sequential days. In order to estimate the regression parameters of the model (15) the database of the reference year is split up into the training set and the test set. The regression parameters are calculated by solving the corresponding least squares optimization (see section 3.4) on the basis of the training set. The quality of the model is checked by the comparison between the forecasted and the real heat consumption for the test dataset.

The correlation coefficients and the mean prediction errors (see table 1) are used as quality parameter. The mean error is calculated for each model by:

$$\varepsilon = \frac{1}{n}\sum_{i=1}^{n}\frac{|Q_{th}-Q_{real}|}{Q_{real}} \cdot 100\% \text{ , where n represents the number of test data} \qquad (16)$$

For the reference year the correlation coefficients range from 0.81 for the summer time to 0.93 for the winter season. The quality of the regression models of the heat consumption strongly depends on seasonal effects. The modeling results show that the quality of the models for the summer and transitional seasons is worse in comparison with the winter time (Schellong & Hentges, 2007). The large errors in the summer and transitional periods are caused by the fact that during the 'warmer' season the heat demand does not really depend on the outside temperature. In this case the heat is only needed for the hot water supply in the residential areas.

season	summer		transitional period		winter	
day type	workdays	weekend	workdays	weekend	workdays	weekend
ε	16.0	12.0	12.9	19.8	5.5	5.6

Table 1. Mean errors for the daily heat demand forecast calculated by RM

4.4 Heat and power demand forecast by neural networks
4.4.1 Methodology
In order to calculate the forecast of the heat and power demand, feedforward networks are used with one layer of hidden neurons connected to all neurons of the input and output layer. The applied learning rule is the backpropagation method with momentum term and flat spot elimination (see section 3.5). The optimal learning parameters are defined by testing different values and retaining the values which require the lowest number of training cycles.

In order to find the most accurate model, several types of neural networks are trained and their prediction error for the test set is compared corresponding to formula (16). Networks with different numbers of hidden neurons are used with three sigmoid (S-shaped) activation functions: the logistic, hyperbolic tangent and limited sine function. Each neural net is trained three times up to the beginning overlearning phase and then the net with the best forecast is retained (Schellong & Hentges, 2011).

Corresponding to the preliminary data analysis described in section 4.1 the power and the heat consumption data are divided into three groups depending on the season: winter, summer, and the transitional period. In each cluster the consumption data are separately modeled for working and for holidays. Thus overall 18 networks have to be tested for the heat and power demand models. For each network the topology varies from 3 to 8 neurons in the hidden layer.

Following the mathematical modeling strategies of section 2.2 such models are preferred which have a simple structure. Thus overlearning effects can be avoided, and the adaptation properties of the model will be better than for more complex structures. Furthermore computing time can be reduced.

4.4.2 Heat demand model

As analyzed in section 4.2 the heat demand depends strongly on the outside temperature. Additionally the temperature difference of two sequential days has an effect on the heat consumption. Thus the daily heat demand can be described by the network shown in fig. 9.

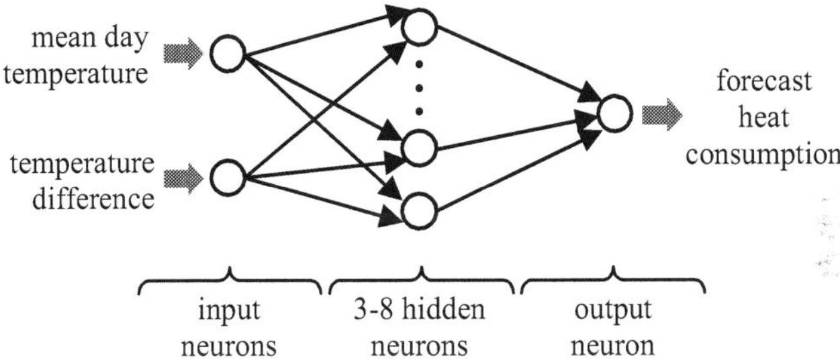

Fig. 9. Network for the daily heat demand

For the daily heat forecast the comparison of the mean prediction error for the 6 categories in which the days are divided (workdays and weekend in winter, summer or in the transitional period) shows that neural nets with a logistic activation function and 6 neurons in the hidden layer deliver the best forecast results (Schellong & Hentges, 2007). As an example fig. 10 demonstrates the network for the heat demand of workdays in the winter period with calculated weights:

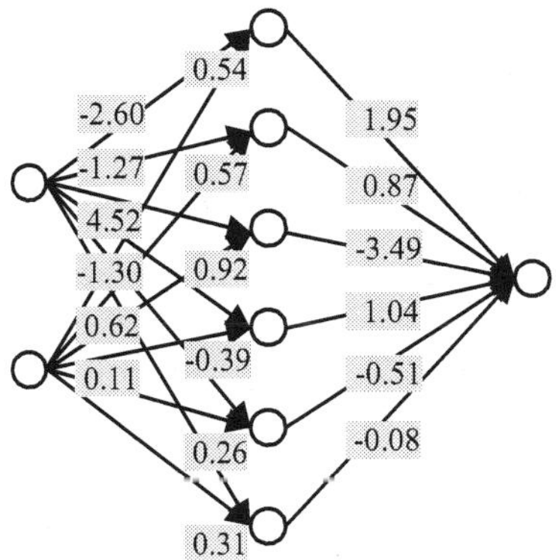

Fig. 10. Network for the daily heat forecast of workdays in winter

Table 2 contains the mean prediction errors corresponding to formula (16). For the winter period we achieve the same quality of modeling results in comparison with RM (table 1).

season	summer		transitional period		winter	
day type	workdays	weekend	workdays	weekend	workdays	weekend
ε	16.1	12.0	15.0	15.8	5.6	5.7

Table 2. Mean errors for the daily heat demand forecast calculated by NN

4.4.3 Power demand model

For the power forecast two different neural networks were used (Schellong & Hentges, 2011). The first considered network receives as only information on one input neuron the coded time (quarter of an hour). The subsequently calculated forecasted power consumption is presented on one output neuron. The second considered network has two input neurons. Additionally to the coded time this network calculates the forecasted power consumption using the consumption measured in the previous week. If the considered day was a holiday the respective previous Sunday is used as comparative day. On the other hand if for a given working day the comparative day of the previous week was a holiday then the according day from the preceding week is used. The prediction accuracies of very small networks with 1 neuron in the hidden layer up to bigger nets with 8 hidden neurons are compared. Fig. 11 shows the structure of the second type of networks.

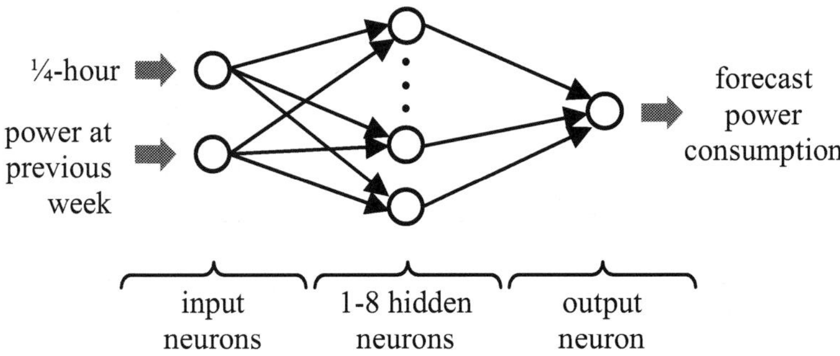

Fig. 11. Network for the power demand

The optimal parameter values identified for the backpropagation learning rule with momentum term α and flat spot elimination term c are similar for both networks. For the power forecast without using a comparative day the analysis of the above defined 24 networks (nets with 1-8 hidden neurons and 3 different activation functions) shows that nets with a logistic activation function and 4 hidden neurons yields the best forecast results. The corresponding comparison of the forecast results, using the power at previous week as additional input, demonstrates that networks with a logistic activation function and 5 neurons in the hidden layer calculate the most accurate forecasts (see fig. 12).

Fig. 13 shows the mean prediction error for the power demand forecast without (blue) and with (orange) comparative day corresponding to formula (16).

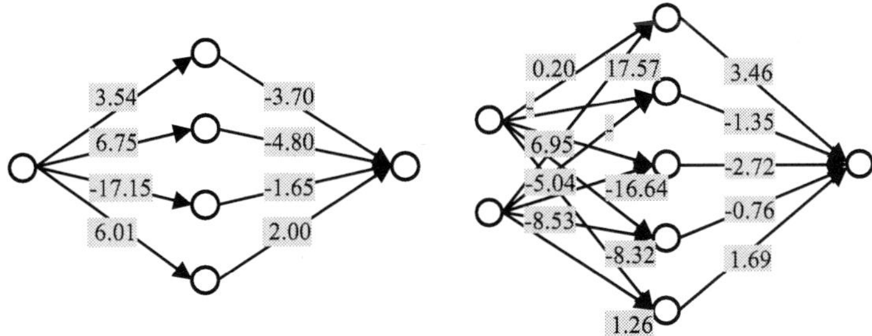

Fig. 12. Networks for the power demand of workdays in winter

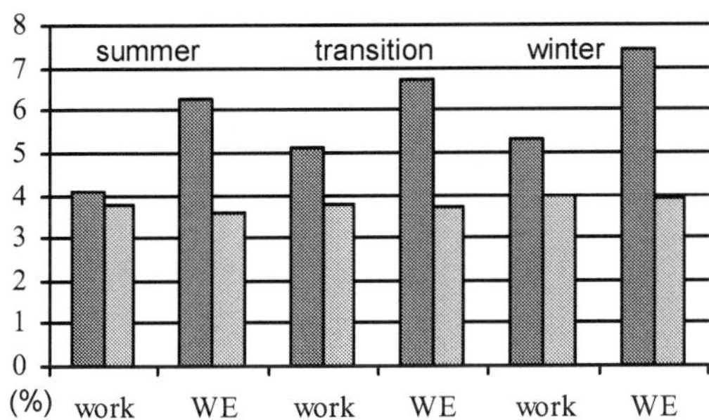

Fig. 13. Mean prediction errors for the power demand

5. Conclusion

The analysis and the forecast of the energy demand represent an essential part of the energy management for sustainable systems. The energy consumption of the delivery district of a power plant is influenced by seasonal data, climate parameters, and economical boundary conditions. Within this chapter the algorithm of the model building process was discussed including the energy data analysis and the selection of suitable forecast methods. It was shown that the quality of the demand forecast tools depends significantly on the availability of historical consumption data as well as on the knowledge about the main influence parameters on the energy consumption. The energy data management must provide information for the energy controlling including all activities of planning, operating, and supervising the generation and distribution process. A detailed knowledge of the energy demand in the delivery district is necessary to improve the efficiency of the power plant and to realize optimization potentials of the energy system.

In this chapter the application of regression methods and of neural networks for the forecast of the power and heat demand for a cogeneration system was investigated. It was shown that similar methods can be applied to both forecast tasks. Generally the energy consumption data must be divided into seasonal clusters. For each of them the forecast models were developed. The heat demand could be calculated by relatively simple regression models based on the outside temperature as the main impact. Involving the temperature difference between two sequential days into the model improved the quality of the forecast.

Additionally feedforward networks were used with one layer of hidden neurons connected to all neurons of the input and output layer in order to calculate the forecast of the heat and power demand. The backpropagation method with momentum term and flat spot elimination was applied as learning rule. Neural networks using the coded time and the consumption measured in the previous week as inputs produced good forecast results for the power demand. Thus the quality of the power and heat forecast could be improved by using information of the 'near' past.

6. References

Box, G. & Jenkins, G. (1976). *Time series analysis, forecasting and control*. Prentice Hall, NY, USA, ISBN 0-130-60774-6

Caruana, R.; Lawrence, S. & Giles, C. (2001). Overfitting in Neural Nets: Backpropagation, Conjugate Gradient, and Early Stopping. *Advances in Neural Information Processing Systems*, Vol 13, MIT Press, Cambridge MA ,pp. 402-408, ISBN 100-262-12241-3

Deuflhard, P. & Hohmann, A. (2003). *Numerical Analysis in Modern Scientific Computing*. Springer Verlag, New York, ISBN 0-387-95410-4

Doty, S. & Turner, W, (2009). *Energy management handbook*. The Fairmont press, Inc., Lilburn, USA, ISBN 0-88173-609-0

Draper, N. & Smith, H. (1998). *Applied Regression Analysis*. Wiley Series in Probability and Statistics, New York, ISBN 0-471-17082-8

Fine, T. L (1999). *Feedforward Neural Network Methodology*. Springer Verlag, New York, ISBN 978-0-387-98745-3

Fischer, M. (2008). Modeling and Forecasting energy demand: Principles and difficulties, In *Management of Weather and Climate Risk in the Energy Industry*, Troccoli, A. (Ed.), pp. 207-226, Springer Verlag, ISBN 978-90-481-3691-9, Dordrecht, The Netherlands.

Galushkin, A. (2007). *Neural Networks Theory*. Springer Verlag, New York, ISBN 978-3-540-48124-9

Hahn, H.; Meyer-Nieberg, S. & Pickl, S. (2009). Electric load forecasting methods: tools for decision making. *European Journal of Operational Research*, Vol.199, No.3, pp. 902-907, ISSN 0377-2217

Maegaard, P. & Bassam, N. (2004). *Integrated Renewable Energy for Rural Communities, Planning Guidelines, Technologies and Applications*. Elsevier, ISBN 0-444-51014-1

Petchers, N. (2003). Combined heating, cooling and power handbook. The Fairmont press, Inc., Lilburn, USA, ISBN 0-88173-4624

Reed, R. & Marks, R. (1998). Neural Smithing: Supervised Learning in Feedforward Artificial Neural Networks. MIT Press, Cambridge MA, ISBN-10:0-262-18190-8

Rumelhart, D.; Durbin, R.; Golden, R. & Chauvin, Y. (1995). Backpropagation: The basic theory. In *Backpropagation: Theory, architectures, and applications*, Chauvin, Y. & Rumelhart, D. (Ed.), pp. 1-34., Lawrence Erlbaum, Hillsdale New Jersey, ISBN-10: 0805812598

Schellong, W. (2006). Integrated energy management in distributed systems. *Proc. Conf. Power Electronics Electrical Drives, Automation and Motion, SPEEDAM 2006*, pp. 492-496, ISBN 1-4244-0193-3 , Taormina, Italy, 2006

Schellong, W. & Hentges, F. (2007). Forecast of the heat demand of a district heating system. *Proc. 7th Conf. on Power and Energy Systems*, pp. 383-388, ISBN 978-0-88986-689-8, Palma de Mallorca, Spain, 2007

Schellong, W. & Hentges, F. (2011). Energy Demand Forecast for a Cogeneration System. *Proc. 3rd Conf. on Clean Electrical Power*, Ischia, Italy, 2011

VDEW (1999). *Standard load profiles*. VDEW Frankfurt (Main), Germany

Part 2

Energy Systems: Applications, Smart Grid Management

6

Energy Management for Intelligent Buildings

Abiodun Iwayemi[1], Wanggen Wan[2] and Chi Zhou[1]
[1]Illinois Institute of Technology
[2]Shanghai University
[1]USA
[2]PRC

1. Introduction

The increasing availability and affordability of wireless building and home automation networks has increased interest in residential and commercial building energy management. This interest has been coupled with an increased awareness of the environmental impact of energy generation and usage. Residential appliances and equipment account for 30% of all energy consumption in OECD countries and indirectly contribute to 12% of energy generation related carbon dioxide (CO_2) emissions (International Energy Agency, 2003). The International Energy Association also predicts that electricity usage for residential appliances would grow by 12% between 2000 and 2010, eventually reaching 25% by 2020. These figures highlight the importance of managing energy use in order to improve stewardship of the environment. They also hint at the potential gains that are available through smart consumption strategies targeted at residential and commercial buildings. The challenge is how to achieve this objective without negatively impacting people's standard of living or their productivity.

The three primary purposes of building energy management are the reduction/management of building energy use; the reduction of electricity bills while increasing occupant comfort and productivity; and the improvement of environmental stewardship without adversely affecting standards of living.

Building energy management systems provide a centralized platform for managing building energy usage. They detect and eliminate waste, and enable the efficient use electricity resources. The use of widely dispersed sensors enables the monitoring of ambient temperature, lighting, room occupancy and other inputs required for efficient management of climate control (heating, ventilation and air conditioning), security and lighting systems.

Lighting and HVAC account for 50% of commercial and 40% of residential building electricity expenditure respectively, indicating that efficiency improvements in these two areas can significantly reduce energy expenditure. These savings can be made through two avenues: the first is through the use of energy-efficient lighting and HVAC systems; and the second is through the deployment of energy management systems which utilize real time price information to schedule loads to minimize energy bills. The latter scheme requires an intelligent power grid or smart grid which can provide bidirectional data flows between customers and utility companies.

The smart grid is characterized by the incorporation of intelligenceand bidirectional flows of information and electricity throughout the power grid. These enhancements promise to revolutionize the grid by enabling customers to not only consume but also supply power.

Utilities will be able to provide customers with real time pricing (RTP) information and enable their active participation in demand response (DR) programs to reduce peak electricity demand. The smart grid will also facilitate greater incorporation of renewable energy sources such as wind and solar energy, resulting in a cleaner power grid.

The smart grid must however, be allied with smart consumption in order to realize its full potential. The extension of the smart grid into the home via smart meters, home automation networks (HAN's) and advanced metering infrastructure (AMI) enables the provision of real-time pricing information and other services to consumers. This facilitates services such as residential DR. DR is the modification of user electricity consumption patterns due to price variations or incentives from the utility, and its objective is to reward behaviour which reduces energy utilization during peak pricing periods. Smart grid DR provides a means of stretching current power infrastructure and delaying the need to build new power plants. It also reduces the rate of greenhouse gas emission by limiting the need for costly and dirty coal-fired peaker plants.

In this work, we focus on two of the largest electricity consumers in buildings – appliances and lighting. Efficient management of these two load categories will result in substantial savings in electricity expenditure and energy use. In order to achieve the three energy management goals discussed above, we require insight into appliance usage patterns and individual appliance energy use. This is achieved by means of distributed and single-point sensing schemes. We therefore survey the various approaches and detail their advantages and disadvantages. We also survey intelligent lighting schemes which utilize networked ambient intelligence to balance energy conservation with occupant comfort. The combination of appliance energy monitoring and control, with intelligent lighting can result in energy savings greater than 15% in residences alone.

We begin by defining intelligent buildings and discuss building and home automation networks, as they provide the framework for intelligent environments. We then discuss appliance energy management and follow this with intelligent lighting control. We conclude with a discussion of the privacy and security threats that must be addressed in smart environments in order to guarantee widespread adoption of these technologies.

2. Intelligent buildings

The Intelligent Building Institute defines an intelligent building as: """.... one that provides a productive and cost-effective environment through optimization of its four basic elements – structure, systems, services and management – and the interrelationships between them. Intelligent buildings help building owners, property managers and occupants realise their goals in the area of cost, energy management, comfort, convenience, safety, long term flexibility and marketability." (Caffrey 1985). These buildings are characterized by three features (Wong et al.,2005):

- Automated control
- The incorporation of occupant preferences and feedback
- Learning ability (performance adjustment based on environmental and occupant changes)

Such environments are distinguished by a tight coupling of HVAC, security, lighting, and fire protection systems. They are sensor rich and produce large amounts of data which can be analysed to predict occupant behaviour and detect equipment faults. They can automatically sense, infer and act in order to balance user comfort and energy efficiency

(Paradiso et al., 2011), a concept also known as ambient intelligence or pervasive computing.

Pervasive computing is the networking of everyday devices, objects and materials using embedded computers equipped with networking, sensing and actuation capabilities. As networked embedded computers reduce in size and cost, they will proliferate at even greater rates. This development, combined with smaller and cheaper sensors and actuators, will result in the availability of networked processing power in smaller and smaller packages. The result is the permeation of pervasive computing into homes, offices, factories, automobiles, airplanes and every area that humans occupy.

Intelligent embedded agents are software programs which run on embedded computers and mimic some of the attributes associated with intelligence – they reason, plan and learn from occupant behaviour. This intelligence is useful as it reduces the burden and complexity of managing and programming large numbers of agents; it enables the agents to adapt to changing occupant needs or environments; and it frees occupants from requiring in-depth understanding of the system or having to make complex decisions (Callaghan et al., 2009). This is because the agents filter the information received by sensing and observation of building occupants, and make decisions or inferences about what the occupants are trying to achieve. This frees building occupants to concentrate on more productive or important tasks. The benefits of these systems include environmentally friendly buildings; increased occupant comfort, health ,security and quality of life; and significant increases in energy efficiency.

The intelligence and sensing capabilities required to support such environments are provided by wireless sensor and actuator networks (WSAN's). WSANs consist of large numbers of tiny, networked sensor or actuator-equipped, power-constrained wireless devices with limited amount of memory and processing power. These devices are the building blocks for the modern day building and home automation networks which we discuss below.

2.1 Building automation and home automation networks

Building automation systems provide centralized management of climate control, lighting, and security systems in order to improve energy efficiency and occupant comfort. These systems reduce energy waste and costs, while boosting occupant productivity (A. C. W. Wong & So, 1997; Kastner et al., 2005). They also facilitate or remote building management as well as improved occupant safety and security (Gill et al., 2009; Newman & Morris, 1994).

Building automation systems have a hierarchical structure consisting of field, automation and management layers (Kastner et al., 2005b) as shown in figure 1. The field layer comprises of temperature, humidity, light level, and room occupancy sensors. The actuators are made up of automated blinds, light switches, flow valves etc. The automation layer consists of direct digital controllers (DDC's) which provide precise automated control of building processes using digital devices (Newman & Morris, 1994), while the management layer provides centralized management of the entire system. It provides a view of the whole building, facilitating centralized control, data collection and analysis.

A primary function of building automation systems is energy management. This goal is achieved by means of schemes such as the duty-cycling of loads to conserve energy; peak load management to regulate total power consumption during peak hours; scheduled start/stop of building HVAC systems at the beginning and end of each day; and real time control of building systems in response to occupancy detection (Merz et al., 2009). The use of BAS' has enabled buildings to dynamically respond to current weather conditions, room occupancy, time of day and various other inputs, resulting in significant reductions in building energy usage.

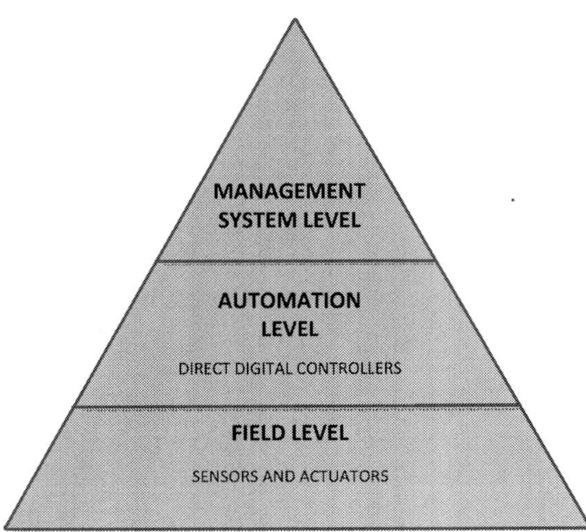

Fig. 1. Building automation system hierarchy (Kastner et al., 2005a)

Sensors and actuators are an integral part of home and building automation networks. These devices serve as the eyes, ears, hands and feet of the system. Unfortunately, wiring costs frequently exceed the cost of sensors (Gutierrez, 2004), so the availability of low-cost wireless communication schemes such as Zigbee (Zigbee Alliance, 2008) enables cost effective and rapid deployment of wireless sensors and actuators throughout a building. Wireless nodes also provide flexibility, easy re-deployment and reconfiguration, all of which are very important features for commercial buildings as they are often re-partitioned and modified to meet differing occupant requirements.

Wireless sensor and actuator networks (WSAN) are defined as a group of sensors and actuators connected by wireless medium to perform distributed sensing and actuation tasks (Dengler et al., 2007). These sensors tend to have the following features: battery powered; low-cost; low-energy consumption; short range communication facilities; limited sensing and computation capabilities. Actuators tend to have greater capabilities than sensing nodes and are often mains-powered, thereby reducing their power and processing constraints. WSANs can observe their physical environment, process sensed data, make decisions based on observations, and utilize their actuators to take appropriate action.

HAN's comprise of smart appliances which can communicate with one another or a Home Energy Controller (HEC) to enable residents to automatically monitor and control home lighting, safety and security systems, and manage home energy usage. The widespread availability of low-cost wireless technologies such as Zigbee has accelerated the deployment of HAN's by facilitating the addition of communication capabilities to household appliances. Figure 2 shows a typical HAN architecture.

Smart appliances are home appliances which combine embedded computing, sensing and communication capabilities to enable intelligent decision-making. Sensing capabilities enable these devices to measure and report their energy consumption to the HEC, while their actuation abilities enable them to respond to commands from the HEC. These commands can be simple on/off signals, or a DR command to operate in energy saving mode. Their communication capabilities also enable them to report their current operating state to the HEC, which then determines their level of participation in any DR activities. For example delaying the current operation of a washer in the middle of a wash cycle may result in the use of more energy than if it is allowed to finish its operation.

Smart appliances support DR signals in one of two ways. They can operate in energy saving modes when electricity prices are high, or they can delay their operation till prices drop below a specified threshold. Examples include smart dishwashers which can receive DR signals and delay wash cycles till off-peak periods; Microwave ovens which automatically reduce their power levels during peak periods or refrigerators which can delay their defrost cycle till off-peak periods ("GE 'Smart' Appliances). Legacy devices such as water heaters, pool pumps or lighting fixtures which do not contain embedded controllers or communication abilities of their own can be controlled via smart plugs. Smart plugs are intelligent power outlets with measurement and communication capabilities which enable device energy monitoring.and remote device shut off. We have discussed the architecture required for appliance management and now proceed to show how this architecture can be leveraged to manage building energy use.

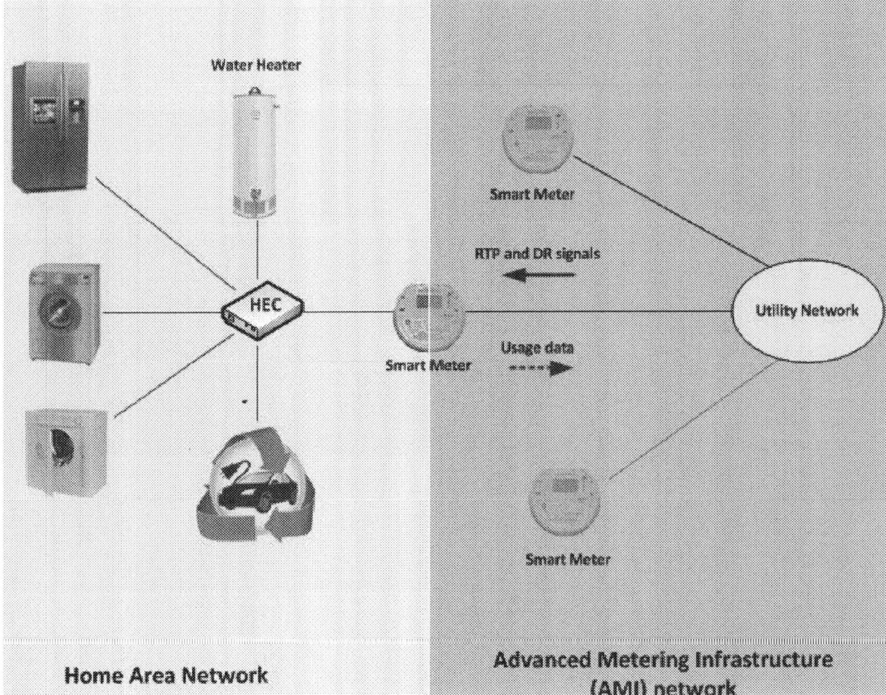

Fig. 2. Home automation network

3. Appliance management

Visibility into load or appliance energy usage is essential for energy-efficient management of building loads. Froehlich et al (Froehlich et al., 2011) note that the greatest reductions in energy usage are made when users are provided with disaggregated energy use data for each appliance, rather than just aggregate energy use data. Therefore in order to determine appropriate energy management strategies, building managers and residents require knowledge of their largest loads, peak usage times and their usage patterns.

Energy-efficient appliance management requires energy sensing/measurement, appliance control, and data analysis (recommendations and predictions based on energy usage patterns). In this section we discuss appliance energy consumption, the various energy sensing schemes and conclude with a discussion of how these schemes can be incorporated into the next generation of smart meters.

3.1 Appliance energy consumption

Residential and commercial electricity usage accounts for 75% of US electricity consumption (US Department of Energy, 2009). As can be seen in figure 3a, all appliances (excluding refrigerators) and lighting account for 60% of residential energy usage. The primary electrical loads in commercial buildings are lighting and cooling, which comprise almost 50% of all electricity usage (figure 3b and 3c) and the bulk of commercial electricity bills. It is estimated that a 10-15% reduction in residential electricity use will result in energy savings of 200 billion kWh, equivalent to the output of 16 nuclear power plants (Froehlich et al., 2011). These statistics demonstrate the importance of appliance energy management, along with the potential savings that can be achieved by means of energy efficiency schemes.

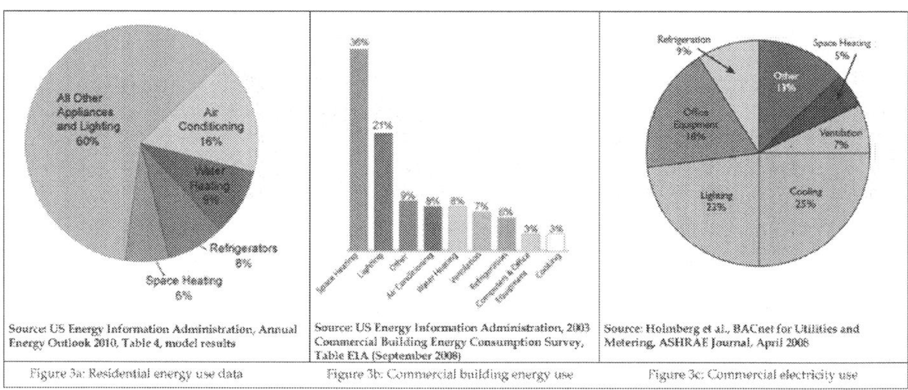

Fig. 3. Residential and Commercial energy usage data

3.2 Energy sensing, measurement and control

As earlier discussed, visibility or feedback into energy use is the first step for energy management. Energy usage measurement schemes fall into two classes – direct or distributed sensing and single point sensing schemes. The choice of schemes used is a function of system cost, the size of the system to be measured and ease of installation.

3.2.1 Distributed or direct sensing

This is the most accurate scheme for obtaining disaggregated appliance energy use data. Each of the sensed devices is connected to the mains through a smart plug or sensor which measures appliance energy usage. The smart plug either displays device energy usage directly or it transmits readings to a central controller. An important feature of these devices is the ability to control attached appliances and switch them on or off. Examples include the University of California, Berkeley's Acme (Jiang et al., 2009) and the Plogg ("Plogg Smart Meter Plug,). The features of these devices are:

- Highly accurate measurements
- Simple device tagging/identification
- The ability to control the sensed device
- Requires the deployment of a large number of nodes
- High system and installation cost

3.2.2 Single point sensing

Single point sensing addresses the cost and convenience issues associated with distributed sensing schemes. In this method, disaggregated energy use data is obtained from a single point in the household or building. This provides a cost-effective and easily deployed solution with fewer points of failure than a distributed solution. It is especially attractive in large building and commercial environments where a large number of devices are to be sensed. This scheme is known as non-intrusive load monitoring (NILM) or non-intrusive appliance load monitoring (NALM). Aggregate power measurements are monitored and are converted into feature vectors that can be used to disaggregate individual devices by identifying signatures unique to each monitored device.

Single-point sensing involves feature extraction, event detection (e.g. device turn on/off) and event classification. The features of this scheme are:

- Lower cost and easier installation
- No device control
- Training required to identify/tag
- Some schemes can only sense appliance activity but not measure energy use

This sensing scheme can be divided into two classes – low and high sampling frequency methods, with the sampling frequency requirements being a function of the selected feature vector components. The shorter the duration of the events we are trying to detect, the higher the sampling frequency requirements. Low-sampling frequency schemes are cheap and simple, making them ideal for residential environments with a small number of high-power loads. On the other hand, high-sampling frequency schemes provide greater versatility in detecting and disaggregating loads, but this comes at the price of higher system cost and computational complexity.

3.2.2.1 Low-sampling frequency schemes

The first NALM scheme was developed by Hart et al in the late 1980's (Hart, 1992). It utilizes aggregate complex power (i.e. real and reactive power) to identify step changes in a real vs. reactive power (P-Q) space.

Hart classified loads into 3 groups in order of complexity – on/off, finite state machine and continuously variable loads. Examples of on/off loads are light switches and other devices with only two operating states. Finite state machines are appliances with different operating modes e.g. a washing machine with wash, rinse and spin cycles; while continuously variable loads include power tools and motor loads whose electricity draw varies continuously. Hart's scheme worked quite well for the first two categories but was unable to disaggregate the last group. His ingenious scheme involved noting step changes in energy use in the P-Q plane, and mapping these step changes to appliance state changes. This enabled the identification of loads along with their energy usage. It however only functioned well in home environments, as it was unable to detect and classify loads smaller than 100W, or continuously varying loads. It was also unable to distinguish between loads of the same type – e.g. two identical light bulbs.

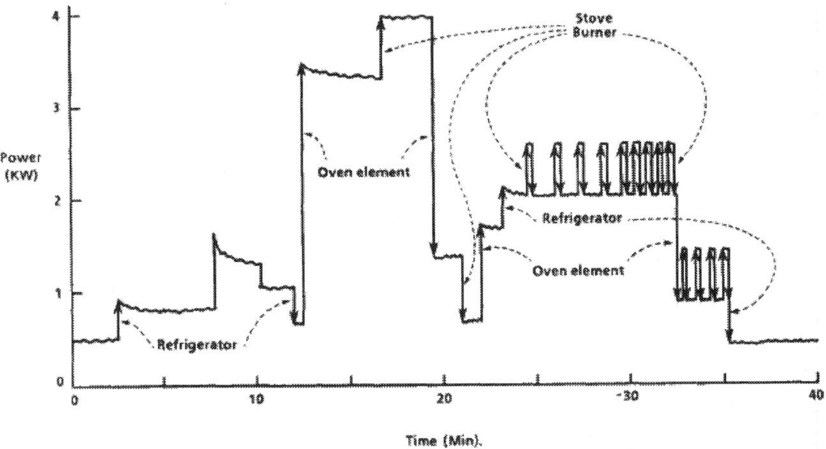

Fig. 4. Energy disaggregation (Hart, 1992)

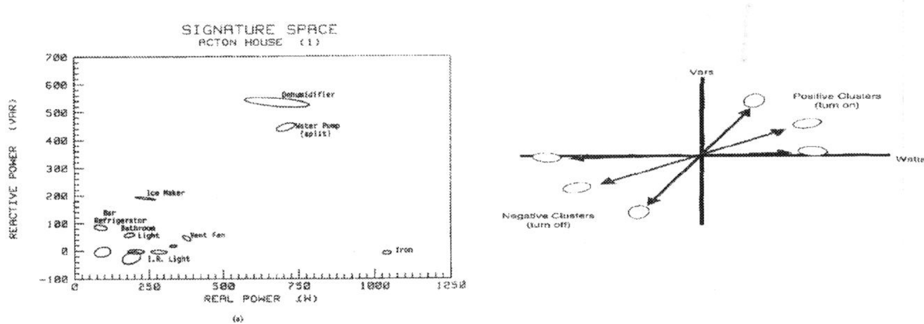

Fig. 5. NALM (Drenker & Kader, 1999)

3.2.2.2 High-sampling frequency schemes

3.2.2.2.1 NALM combined with harmonics and transients

Hart's NALM work was extended by his colleagues to utilize a feature vector consisting of harmonics and transients, in addition to complex power (Laughman et al., 2003). This extension enabled the detection of continuously varying loads as well as the resolution of low

power devices, thereby overcoming the primary deficiencies of basic NALM. These deficiencies were present because the original NALM scheme was based on three assumptions which did not always hold, especially in commercial buildings. The assumptions were:
1. Each load can be uniquely identified in the P-Q space
2. After a brief transient period, load power consumptions settle to a steady state value
3. Energy data would be batch processed at the end of the day
It was found that different loads could have almost identical loci on the P-Q plane, leading to inaccurate load classification. Analysis of aggregate power in commercial buildings also showed that in buildings with large numbers of variable speed loads, steady state power draws were never achieved. Finally, the original NALM scheme was designed with batch processing in mind, limiting its utility for real or near real-time energy data analysis.

In this scheme, the aggregate current waveform is sampled at 8 kHz or greater, and the Fourier transform of the sampled waveform is used to obtain spectral envelope of the signal. The spectral envelope is the summary of the harmonic content of the line current and is used to obtain estimates of the real, reactive and higher frequency content of the current. The combination of real and apparent power with harmonic content enables disaggregation of loads which would be indistinguishable using only P-Q information. The spectral envelope is given by equation 1 where $a_m(t)$ is proportional to real power, and $b_m(t)$ is proportional to reactive power.

$$a_m(t) = \frac{2}{T} \int_{t-T}^{t} i(\tau) \sin(m\omega\tau)\, d\tau$$

$$(1)$$

$$b_m(t) = \frac{2}{T} \int_{t-T}^{t} i(\tau) \cos(m\omega\tau)\, d\tau.$$

Transient events are learned and used to create signatures which detect appliance events, hence loads are detected via their unique transient profiles, and these profiles can also be used for device diagnosis. They can also detect continuously varying loads such as variable speed drives, and the use of transients to detect device start-up/shutdown is shown in figure 6 below:

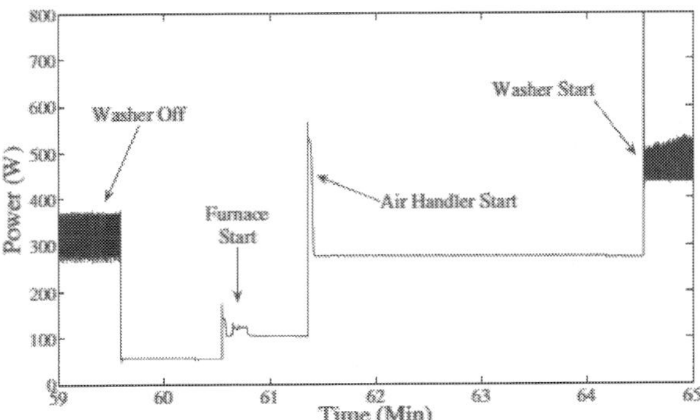

Fig. 6. Transient event detection (Sawyer et al. , 2009)

The use of harmonics enables the disaggregation of loads that appear almost identical in the P-Q plane. This is apparent in figure 7, where the 3rd harmonic is used in conjunction with the real and reactive power respectively to disaggregate an incandescent bulb and a computer.

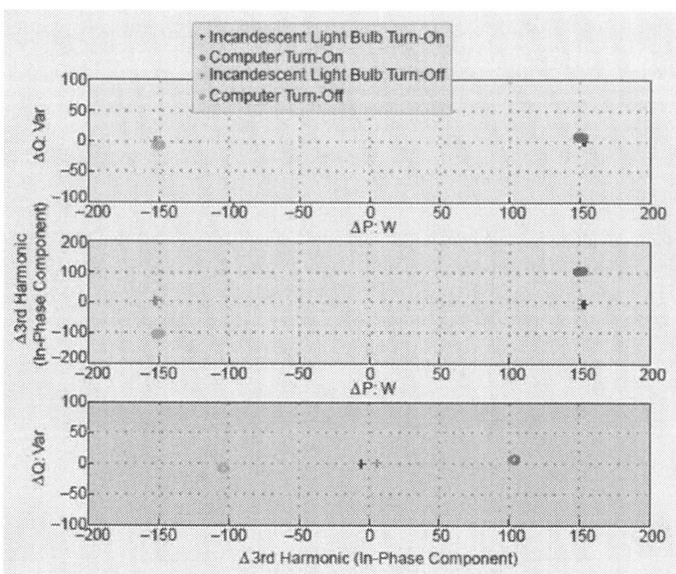

Fig. 7. Complex power and harmonic device signatures (Laughman et al., 2003)

3.2.2.2.2 Noise as an appliance feature

An innovative approach to appliance disaggregation was developed by Patel (Patel et al., 2007). Their scheme utilized transient noise as the feature vector for appliance detection. Real-time event detection and classification were performed via transient noise analysis of device turn on or off events. The novelty of their scheme was the ability to perform single point sensing from any power outlet in the home, obviating the need for professional installation or any work at the meter or junction box. Another advantage is the fact that appliances of the same type have unique features due to their mechanical characteristics and the length of their attached power line. This enables their scheme to not only detect that a light has been switched on, but also which light. Transient noise only lasts for a few milliseconds but is rich in harmonics in the range of 10Hz-100 kHz depending on the device, therefore this scheme requires high sampling rates (1-100MHz).

3.2.2.2.3 Continuous voltage noise signature

Rather than looking at transient noise, Froehlich et al (Froehlich et al., 2011) utilize the steady state noise generated by all electrical appliances as a feature vector. Appliances produce steady state noise during operation, and introduce this noise into the home power wiring. Most appliances (laptop chargers, CFL bulbs, TV's etc.) use switched mode power supplies (SMPS') and it has been found that these units emit unique continuous noise signatures which vary between device types. As with their earlier work, this scheme permits single point sensing from any point in the home. Steady state noise events have longer periods than transient noise, so the sampling rate required is significantly lower than that for transient noise detection. As a result, sampling rates of only 50-500 kHz are required. A spectrogram of steady state noise signatures is shown in fig 8.

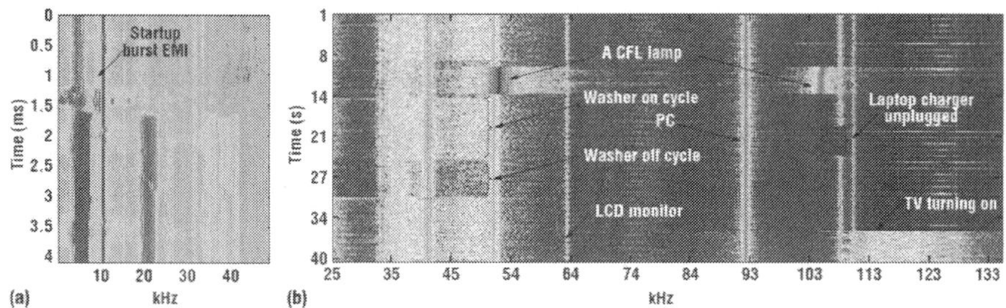

(a) Transient voltage noise of a light switch being turned on. Colors indicate amplitude at each frequency. (b) Steady-state continuous voltage noise signatures of devices during various periods of operation

Fig. 8. Spectrogram of steady state noise signatures (Froehlich et al., 2011)

3.3 Open issues

The primary question is how can we use existing HAN infrastructure to perform NALM? The high sampling rates required for noise and harmonic signature-based schemes preclude their use in Smart meters which only sample electricity at 1Hz, while conventional NALM schemes have lower sampling rates but are too processor intensive to be incorporated into smart meters. The constraints to widespread NALM adoption are:

* Meter sampling rate
* Meter processing power
* Installation cost
* Consumer privacy concerns

Open issues include finding feature vectors which can be obtained with a sampling rate of 1Hz or less, while providing accurate disaggregation, as this will allow us to harness the smart meter for energy usage measurement without installing additional measurement hardware. The Home Energy Controller can then be leveraged to collect raw power data from the smart meter via wireless links. It can then perform signal processing on the aggregate data and disaggregate energy usage. The HEC can also be used to schedule home appliances in order to reduce residential energy cost.

The greatest cost savings are achieved when energy usage is correlated with real-time pricing, hence the synergy between appliance energy management and the smart grid. Unfortunately, the usage of the smart grid introduces security and privacy concerns which need to be addressed. These issues are related to the visibility into appliance energy usage and the availability of information which enables the profiling of occupant habits and behaviour, we therefore address this issue in detail in section 5 of this paper.

4. Intelligent lighting

Lighting accounts for 28% of all commercial building electricity expenditure (US Department of Energy,) and represents a potential source of energy savings. These systems also directly influence workplace comfort and occupant productivity. Improvements to lighting systems promise significant energy and cost savings (Rubinstein et al., 1993), as well as improved occupant comfort (Fisk, 2000).

A substantial amount of research has been done on energy efficient lighting e.g. CFL's etc., now the next challenge is the addition of intelligence and communication capabilities. The objective is to drive down energy usage even further, while enhancing occupant comfort and productivity. The integration of WSAN's into lighting systems permits granular control of lighting systems, permitting personalized control of workspace lighting.

The functions of a lighting control system are workspace illumination, ambience and security. They directly affect workspace safety and occupant productivity, but are also one of the largest consumers of electricity. A system diagram of an intelligent lighting control system is provided in Figure 9.

Lighting systems consist of ballasts and luminaires or lighting fixtures. Ballasts provide the start-up voltages required for lamp ignition, and regulate current flow through the bulb. Newer ballasts enable fluorescent dimming using analogue or digital methods, enabling granular control of lighting output. It has been discovered that the human eye is insensitive to dimming of lights by as much as 20%, as long as the dimming is performed at a slow enough rate (Akashi & Neches, 2004), thereby permitting significant savings in energy use.

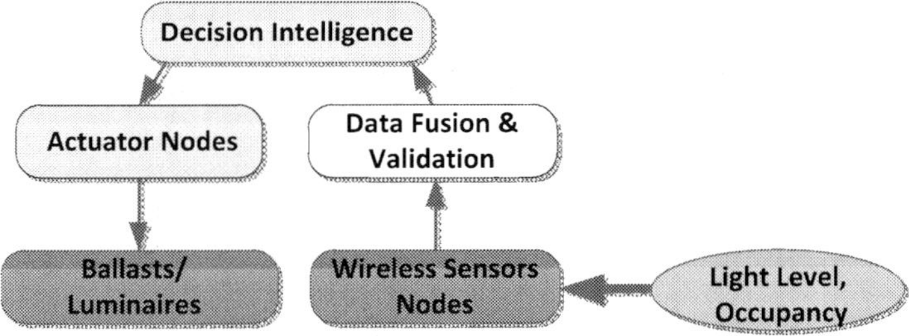

Fig. 9. Intelligent lighting system

4.1 Sensors

Sensors serving as the eyes and ears of the intelligent environmental control system allow the system to detect and respond to events in its environment. The most commonly utilized sensors are occupancy and photo sensors, although some systems incorporate the use of smart tags to detect and track occupants. However, these smart tag based schemes are yet to gain widespread acceptance due to privacy concerns.

Occupancy sensors are used in detecting room occupancy and are utilized in locations with irregular or unpredictable usage patterns such as conference rooms, toilets, hallways or storage areas (DiLouie, 2005). The primary technologies used in occupancy sensors are ultrasonic and Passive Infra-red (PIR) sensors.

Photo sensors detect the amount of ambient light, and use this information to determine the amount of artificial lighting required to maintain total ambient lighting at a defined value. Therefore, photo sensors are an integral component of daylight harvesting systems.

4.2 Lightning control modes

Basic lighting control modes include on/off control, scheduling, occupancy detection, and dimming. More advanced schemes include daylight harvesting, task tuning and demand response. Daylight harvesting involves measurement of how much ambient light is present, and harnessing ambient light to reduce the amount of artificial lighting required to keep the amount of light at a pre-set level. Task tuning involves adjusting the light output in accordance with the function or tasks which will be performed in the lighted area. Demand response is the dimming of lighting output in response to signals from the utility. As discussed earlier this dimming is often un-noticeable to building occupants.

Intelligent lighting control systems combine digital control with computation and communications capabilities. The result is a low cost, yet highly flexible lighting system. These systems were surveyed in (Iwayemi et al., 2010) and a taxonomy of the schemes is provided in figure 10.

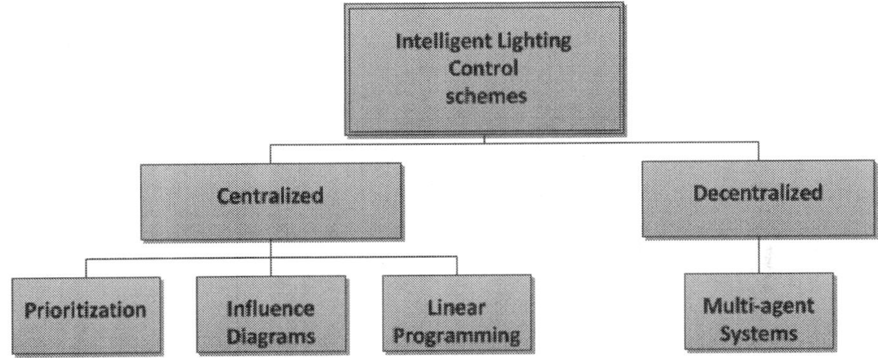

Fig. 10. Taxonomy of intelligent wireless lighting control (Iwayemi et al., 2010)

Centralized intelligent lighting schemes deliver faster performance and lower convergence times than de-centralized schemes, but this comes at the cost of scalability and single-point of failure issues. An overview of the various schemes is provided in Table 1.

4.3.1 Prioritization

This is the most basic intelligent lighting scheme, where conflicting occupant lighting requirements are resolved by the assignment of user rankings or priorities. In this system, area lighting settings are determined by the occupant with the highest ranking. Such a scheme was deployed by Li (S.-F. Li, 2006) and used a WSAN-based lighting monitoring and control test bed with pre-assigned user priorities.

4.3.2 Influence diagrams

An influence diagram is a graphical representation of a decision problem and the relationship between decision variables. The relationship between decision variables is determined by means of marginal and conditional probabilities, enabling the use of Bayes rule for non-deterministic decision-making and inference (Granderson et al., 2004).

Influence diagrams are directed acyclic graphs made up of three node types, namely state, decision and value nodes. Decision nodes are denoted by rectangles and represent the control actions available to controllers/actuators within the system. State nodes are denoted by ellipses, and represent uncertain events over which we have no control, while value nodes represent the cost functions we seek to minimize or maximize. They are represented

	Prioritization	Influence Diagrams	Linear Programming	Multi-agent Systems
Overview	Conflicts resolved by deferring to the highest priority user present	Complex interrelationships formulated using simple graphs. Non-deterministic decision-making	Effective optimization scheme for modeling and satisfying competing objectives	Ideal for environments where learning and prediction are essential while interrelationships between system parameters are either unknown or not well-defined
Approach	Node prioritization	Bayesian probabilities	Linear optimization, scalarization,	Artificial Intelligence - Neural networks, expert systems
Response time	Fastest	Rapid response	Rapid response	Medium
Scalability	Centralized architecture which limits scalability and produces single-point failures			Highly scalable due to distributed architecture
Weaknesses	Can only guarantee comfort for a single occupant	Probabilities must be determined via experimentation	Optimization problem formulation is a non-trivial task	No wireless scheme currently deployed due to complexity of the problem

Table 1. Comparison of intelligent lighting control schemes

by hexagons. These nodes rank the different options available to the system controller based on the current system state, with the optimal decision being the choice that maximizes (or minimizes) the selected cost function. Arcs represent the interrelationships between system nodes. Input arcs (arcs from state nodes to decision nodes) represent the information available to decision nodes or controllers at decision time, while arcs from decision nodes to state nodes indicate causal relationships. An influence diagram for intelligent lighting control is shown in fig 11 and displays the various states, decision nodes and inputs.

Granderson (Jessica Granderson, 2007; Jessica Granderson et al., 2004), and Wen (Wen, J. Granderson, & A.M. Agogino, 2006) utilize influence diagrams to provide intelligent decision-making capabilities for WSAN-based lighting schemes. Their systems utilized dimmable ballasts and were able to satisfy conflicting occupant preferences in shared workspaces.

4.3.3 Linear optimization

This is the most common scheme for minimizing lighting energy consumption subject to the constraint of satisfying conflicting user requirements. It seeks to maximize or minimize an objective function subject to constraints, and there is a rich collection of work in this area (Akita et al.,2010; Kaku et al., 2010; M. Miki et al., 2004; Pan, et al., 2008; Park et al., 2007; Singhvi et al.,2005; S. Tanaka, M. Miki et al., 2009; Tomishima et al., 2010; Yeh et al., 2010). For example, Wen (Wen & Alice M. Agogino, 2008) created an illuminance model of the room to be lighted, and this model captured the effect of each individual luminaire on work surface lighting. Their objective was the minimization of work surface illuminance levels subject to the satisfaction of lighting preferences of current room occupants. Their system calculates the optimal linear combination of individual illuminance models and lighting levels which minimizes energy usage.

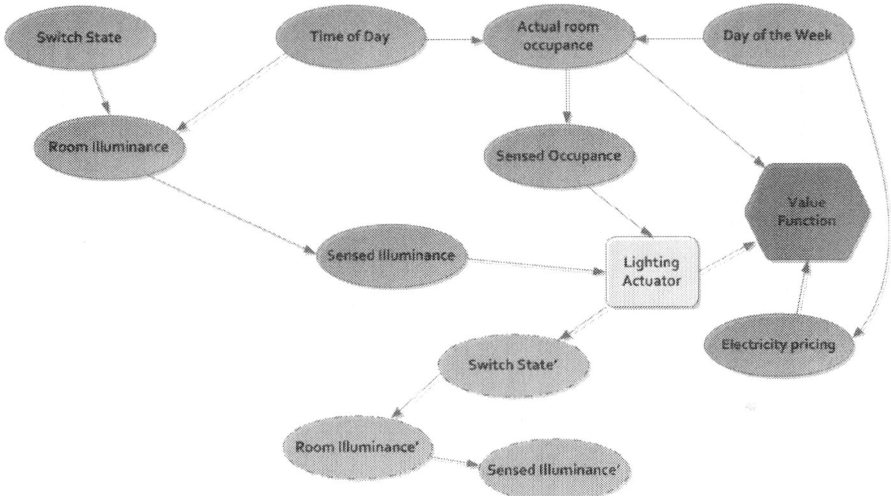

Fig. 11. Inference Diagram for Intelligent lighting control

4.3.4 Multi-agent systems

Multi-agent systems utilize large numbers of autonomous intelligent agents which cooperate to provide decentralized control of complex tasks. These schemes incorporate the advantages of inference diagram based lighting control schemes, without requiring centralized control. Their advantages include scalability, self-configuration and adaptation by means of machine learning techniques. A theoretical framework for such as system was proposed by Sandhu (Sandhu, 2004).

5. Smart grid security

Smart environments promise great convenience through the use of autonomous intelligent agents which learn and predict occupant desires, and smart appliances which monitor and automatically regulate energy use. As noted by Cavoukian et al (Cavoukian et al., 2010), these environments generate and observe tremendous amounts of detailed data about their occupants, providing them with information to control their energy consumption and electricity bills; reduce greenhouse gas emissions; and improve occupant comfort and quality of life. The benefits to the utility (via smart metering and other smart grid technologies) are the provision of real-time billing, customer energy management, and highly accurate system load prediction data.

Unfortunately the use of these technologies poses tremendous security and privacy risks due to the type and quality of the data they capture (Callaghan et al., 2009). The home is man's most private place, and analysis of fine-grained smart metering data by NALM enables the utility to learn occupant habits, behaviours and lifestyles (Bleicher, 2010; Quinn, 2009) . The smart grid increases the amount and quality of personally identifiable information available, and there is significant concern that this information will be used for applications beyond the purposes for which it was originally collected. This information is extremely valuable to third parties such as advertisers, government agencies or criminal elements, and has led to the fear that users can be spied upon by their meters, negatively impacting smart meter deployment (Bleicher, 2010). In addition, the networking of smart meters with the electricity grid also raises the spectre of smart meter fraud, and increases the vulnerability of these devices to malicious attacks such as Denial of service (DoS) attacks. We discuss these issues in more detail below, along with some proposed solutions.

5.1 Privacy issues

The use of earlier discussed non-intrusive appliance load monitoring technology (NALM) has enabled the identification of appliances by means of their unique fingerprint or "appliance load profiles." Data mining and machine learning tools enable utilities to determine which appliances are in use and at what frequency. This provides access to information including the types of appliances a resident possesses, when he/she has their shower each day (by monitoring extended usage of the heater), how many hours they spend using their PC, or whether they cook often or eat microwave meals. This has led to the very valid fear that customers can be profiled, and monitored by means of their smart meter (Hansen, 2011). In addition, improper access to such data can lead to violations of privacy or even make one open to burglary by determining the times the house is empty. As with internet advertising companies that track users and build profiles based on browsing histories, utilities will be in a position to create detailed profiles of their customers which they can mine or sell to third parties.

In order to address this privacy concerns, we need to determine the type, amount and quality of information required by utilities to provide real-time billing and other smart grid services. Utilities require insight into electricity usage patterns in order to optimize their operations and scale them appropriately, while residents desire the benefits of the smart grid but do not want to exchange them for their privacy. Therefore a balance between these two extremes is required.

Smart grid security issues can only be solved by a combination of regulatory and technological solutions. A regulatory framework is required to specify who has access to smart meter data and under which conditions, as well as enforcement of penalties for data misuse (McDaniel & McLaughlin, 2009). Technological solutions focus on anonymization or privacy-preserving methods.

Quinn (Quinn, 2009) suggests aggregating residential data at the neighbourhood transformer and then anonymizing it by stripping it of its source address before transmitting it to the utility. Kalogridis et al (Kalogridis et al., 2010) provide privacy by obscuring load signatures by means of a rechargeable battery as an alternate power source, a process they term "load signature moderation". In this scheme, a power router and a rechargeable battery are added to the HAN network. The power router determines the amount of electricity required by an appliance and 'routes' the power to the appliance via various sources. For example, a fridge could be supplied by a combination of utility power, a solar cell and rechargeable battery. This power mixing is performed in conjunction with battery recharge events which obscure load signatures and prevent their disaggregation by means of NALM (figure 12).

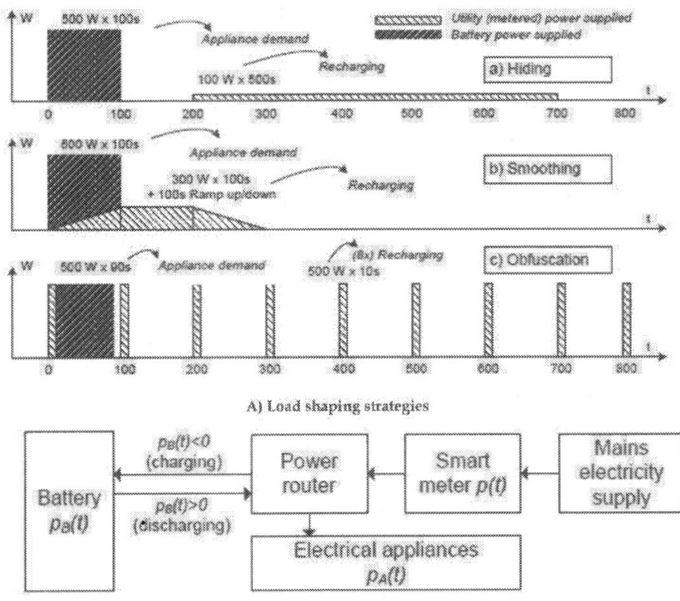

Fig. 12. a) Load shaping and b) battery power mixing (Kalogridis et al., 2010)

Other schemes include Rial and Danezis' (Rial & Danezis, 2010) privacy preserving smart metering scheme. In this scheme, the smart meter provides certified readings to the user who then combines with a certified tariff policy to generate a final bill. The bill is then transmitted to the utility along with a zero-knowledge proof which confirms that the billing calculation is correct. The advantage of this scheme is that no additional information is sent to the utility apart from what is required for billing purposes. However, their scheme also permits the disclosure of individual or aggregate readings to facilitate load prediction.

We propose a digital rights management system (DRMS) based scheme which extends that proposed in (Fan et al., 2010). Users license permission to the utility to access their data at varying levels of granularity. By default the utility would have access to monthly usage and billing data, but customers have to grant the utility permission to access their data at higher levels of granularity in exchange for rebates or other incentives. Such a system eliminates the need for an intermediary between the utility and the consumer, but requires a means of guaranteeing that the utility cannot access restricted customer data.

5.2 Smart meter fraud

Users can manipulate their smart meter readings in order to reduce their electricity bills, as the desire for lower electricity bills provides a compelling incentive for smart meter fraud. The ability to report inaccurate data to the utility means that customers can reduce their bills by falsely claiming to supply power the grid, or consume less power than their actually do. The commercial availability of smart meter hacking kits means that those with sufficient skill and interest can engage in meter fraud(McDaniel & McLaughlin, 2009).

5.3 Malicious attacks

The internetworking of smart meters makes them especially vulnerable to denial of service attacks in which several meters are hijacked in order to flood the network with data in order to shut down portions of the power grid, or report false information which can result in grid failures.

6. Conclusion

Commercial and residential buildings are the largest consumers of electricity in the United States and contribute significantly to greenhouse gas emissions. As a result, building energy management schemes are being deployed to reduce/manage building energy use; reduce electricity bills while increasing occupant comfort and productivity; and improve environmental stewardship without adversely affecting standards of living. The attainment of these energy management goals requires insight into appliance usage patterns and individual appliance energy use, combined with intelligent appliance operation and control. This is achieved by application of distributed and single-point sensing schemes to provide appliance energy sensing and measurement, and the use of intelligent WSAN-based lighting control schemes. We have therefore surveyed the two schemes which promises the greatest reductions in residential and commercial building energy use – non-intrusive appliance load monitoring, and intelligent lighting.
Our survey of NALM techniques indicates that there is currently no one size fits all solution, and that the schemes with the highest resolution also tend to have the highest processor and sampling rate requirements. An open issue is how to leverage smart meter and HAN infrastructure for NALM as this will provide the cheapest and most convenient approach to widespread NALM deployment.
We have also demonstrated the utility of WSAN-based intelligent lighting for providing substantial energy savings, especially in commercial buildings, and provided a taxonomy of intelligent lighting schemes. In addition, the security and privacy problems inherent to smart grids and pervasive computing environments were discussed and solutions proffered. Building energy management is poised to experience tremendous growth over the next decade as the issues outlined in this work are addressed and resolved, leading to cleaner, more efficient buildings.

7. Acknowledgment

This work was funded by the United States Department of Energy under grant DE-FC26-08NT02875.

8. References

Akashi, Y., & Neches, J. (2004). Detectability and Acceptability of Illuminance Reduction for Load Shedding. *Journal of the Illuminating Engineering Society*, 33, 3-13.

Akita, M., Miki, Mitsunori, Hiroyasu, Tomoyuki, & Yoshimi, M. (2010). Optimization of the Height of Height-Adjustable Luminaire for Intelligent Lighting System. In L. Rutkowski, R. Scherer, R. Tadeusiewicz, L. Zadeh, & J. Zurada (Eds.), *Artifical Intelligence and Soft Computing*, Lecture Notes in Computer Science (Vol. 6114, pp. 355-362). Springer Berlin / Heidelberg. Retrieved from http://dx.doi.org/10.1007/978-3-642-13232-2_43

Bleicher, A. (2010, October). Privacy on the Smart Grid. *IEEE Spectrum*. Retrieved November 1, 2010, from http://spectrum.ieee.org/energy/the-smarter-grid/privacy-on-the-smart-grid

Callaghan, V., Clarke, G., & Chin, J. (2009). Some socio-technical aspects of intelligent buildings and pervasive computing research. *Intelligent Buildings International, 1*(1), 56-74. doi:10.3763/inbi.2009.0006

Cavoukian, A., Polonetsky, J., & Wolf, C. (2010). SmartPrivacy for the Smart Grid: embedding privacy into the design of electricity conservation. *Identity in the Information Society, 3*(2), 275-294. doi:10.1007/s12394-010-0046-y

Dengler, S., Awad, A., & Dressler, F. (2007). Sensor/Actuator Networks in Smart Homes for Supporting Elderly and Handicapped People. *Advanced Information Networking and Applications Workshops, 2007, AINAW '07. 21st International Conference on* (Vol. 2, pp. 863-868). Presented at the Advanced Information Networking and Applications Workshops, 2007, AINAW '07. 21st International Conference on. doi:10.1109/AINAW.2007.325

DiLouie, C. (2005). *Advanced Lighting Controls: Energy Savings, Productivity, Technology and Applications* (1st ed.). Fairmont Press.

Drenker, S., & Kader, A. (1999). Nonintrusive monitoring of electric loads. *Computer Applications in Power, IEEE, 12*(4), 47-51. doi:10.1109/67.795138

Fan, Z., Kalogridis, G., Efthymiou, C., Sooriyabandara, M., Serizawa, M., & McGeehan, J. (2010). The new frontier of communications research: smart grid and smart metering. *e-Energy '10: Proceedings of the 1st International Conference on Energy-Efficient Computing and Networking* (p. 115–118). New York, NY, USA: ACM. doi:http://doi.acm.org/10.1145/1791314.1791331

Fisk, W. J. (2000). HEALTH AND PRODUCTIVITY GAINS FROM BETTER INDOOR ENVIRONMENTS AND THEIR RELATIONSHIP WITH BUILDING ENERGY EFFICIENCY. *Annual Review of Energy and the Environment, 25*(1), 537-566. doi:10.1146/annurev.energy.25.1.537

Froehlich, J., Larson, E., Gupta, S., Cohn, G., Reynolds, M., & Patel, S. (2011). Disaggregated End-Use Energy Sensing for the Smart Grid. *Pervasive Computing, IEEE, 10*(1), 28-39. doi:10.1109/MPRV.2010.74

GE "Smart" Appliances Empower Users to Save Money, Reduce Need for Additional Energy Generation / GE Appliances & Lighting. (n.d.). . Retrieved January 20, 2011, from http://pressroom.geconsumerproducts.com/pr/ge/smart_meter_pilot09.aspx

Gill, K., Shuang-Hua Yang & Fang Yao & , and & Xin Lu (2009). A zigbee-based home automation system. *Consumer Electronics, IEEE Transactions on, 55*(2), 422-430. doi:10.1109/TCE.2009.5174403.

Granderson, Jessica. (2007). *Human-centered sensor-based Bayesian control: Increased energy efficiency and user satisfaction in commercial lighting* (Ph.D. Dissertation). Department of Mechancial Engineering, UC Berkeley.

Granderson, Jessica, Agogino, Alice M, Yao-Jung, W., & Goebel, K. (2004). Towards Demand-Responsive Intelligent Lighting with Wireless Sensing and Actuation. *PROC. IESNA (ILLUMINATING ENGINEERING SOCIETY OF NORTH AMERICA) ANNUAL CONFERENCE, 265--274.*

Gutierrez, J. A. (2004). On the use of IEEE 802.15.4 to enable wireless sensor networks in building automation. *Personal, Indoor and Mobile Radio Communications, 2004. PIMRC 2004. 15th IEEE International Symposium on* (Vol. 3, pp. 1865 - 1869 Vol.3).

Hansen, T. (2011). Personal Information Concerns Smart Grid Developers. *Electric Light & Power*, (2). Retrieved from http://www.elp.com/index/display/article-display/0996375781/articles/electric-light-power/volume-89/issue-2/sections/personal-information-concerns-smart-grid-developers.html

Hart, G. W. (1992). Nonintrusive appliance load monitoring. *Proceedings of the IEEE, 80*(12), 1870-1891. doi:10.1109/5.192069

Iwayemi, A., Peizhong Yi & and & Chi Zhou (2010). Intelligent Wireless Lighting Control using Wireless Sensor and Actuator Networks: A Survey. *EJSE Special Issue: Wireless Sensor Networks and Practical Applications (2010)*, 67-77.

Jiang, X., Dawson-Haggerty, S., Dutta, P., & Culler, D. (2009). Design and implementation of a high-fidelity AC metering network. *Proceedings of the 2009 International Conference on Information Processing in Sensor Networks*, IPSN '09 (p. 253–264). Washington, DC, USA: IEEE Computer Society. Retrieved from http://portal.acm.org/citation.cfm?id=1602165.1602189

Kaku, F., Miki, Mitsunori, Hiroyasu, Tomoyuki, Yoshimi, M., Tanaka, Shingo, Nishida, T., Kida, N., et al. (2010). Construction of Intelligent Lighting System Providing Desired Illuminance Distributions in Actual Office Environment. In L. Rutkowski, R. Scherer, R. Tadeusiewicz, L. Zadeh, & J. Zurada (Eds.), *Artifical Intelligence and Soft Computing*, Lecture Notes in Computer Science (Vol. 6114, pp. 451-460). Springer Berlin / Heidelberg. Retrieved from http://dx.doi.org/10.1007/978-3-642-13232-2_55

Kalogridis, G., Efthymiou, C., Denic, S., Lewis, T., & Cepeda, R. (2010). Privacy for Smart Meters: Towards Undetectable Appliance Load Signatures. *IEEE SmartGridComm.* Presented at the IEEE SmartGridComm, Gaithersburg, Maryland, USA.

Kastner, W., Neugschwandtner, G., Soucek, S., & Newmann, H. M. (2005a). Communication Systems for Building Automation and Control. *Proceedings of the IEEE, 93*(6), 1178-1203. doi:10.1109/JPROC.2005.849726

Kastner, W., Neugschwandtner, G., Soucek, S., & Newmann, H. M. (2005b). Communication Systems for Building Automation and Control. *Proceedings of the IEEE, 93*(6), 1178-1203. doi:10.1109/JPROC.2005.849726

Laughman, C., Kwangduk, L., Cox, R., Shaw, S., Leeb, S., Norford, L., & Armstrong, P. (2003). Power signature analysis. *Power and Energy Magazine, IEEE, 1*(2), 56-63. doi:10.1109/MPAE.2003.1192027

Li, S.-F. (2006). Wireless sensor actuator network for light monitoring and control application. *Consumer Communications and Networking Conference, 2006. CCNC 2006.*

3rd IEEE (Vol. 2, pp. 974-978). Presented at the Consumer Communications and Networking Conference, 2006. CCNC 2006. 3rd IEEE.

McDaniel, P., & McLaughlin, S. (2009). Security and Privacy Challenges in the Smart Grid. *IEEE Security and Privacy, 7*(3), 75-77.

Merz, H., Hansemann, T., & Hübner, C. (2009). *Building Automation: Communication systems with EIB/KNX, LON and BACnet (1st ed.) Springer.*

Miki, M., Hiroyasu, T., & Imazato, K. (2004). Proposal for an intelligent lighting system, and verification of control method effectiveness. *Cybernetics and Intelligent Systems, 2004 IEEE Conference on* (Vol. 1, pp. 520-525 vol.1). Presented at the Cybernetics and Intelligent Systems, 2004 IEEE Conference on. doi:10.1109/ICCIS.2004.1460469

Newman, H. M., & Morris, M. D. (1994). *Direct digital control of building systems: theory and practice.* Wiley-Interscience.

Pan, M.-S., Yeh, L.-W., Chen, Y.-A., Lin, Y.-H., & Tseng, Y.-C. (2008). A WSN-Based Intelligent Light Control System Considering User Activities and Profiles. *Sensors Journal, IEEE, 8*(10), 1710-1721. doi:10.1109/JSEN.2008.2004294

Paradiso, J., Dutta, P., Gellersen, H., & Schooler, E. (2011). Guest Editors' Introduction: Smart Energy Systems. *Pervasive Computing, IEEE, 10*(1), 11-12. doi:10.1109/MPRV.2011.4

Park, H., Burke, J., & Srivastava, M. B. (2007). Design and Implementation of a Wireless Sensor Network for Intelligent Light Control. *Information Processing in Sensor Networks, 2007. IPSN 2007. 6th International Symposium on* (pp. 370-379). Presented at the Information Processing in Sensor Networks, 2007. IPSN 2007. 6th International Symposium on. doi:10.1109/IPSN.2007.4379697

Patel, S. N., Robertson, T., Kientz, J. A., Reynolds, M. S., & Abowd, G. D. (2007). At the flick of a switch: detecting and classifying unique electrical events on the residential power line. *Proceedings of the 9th international conference on Ubiquitous computing,* UbiComp '07 (p. 271–288). Berlin, Heidelberg: Springer-Verlag. Retrieved from http://portal.acm.org/citation.cfm?id=1771592.1771608

Plogg Smart Meter Plug. (n.d.). . Retrieved May 5, 2011, from http://www.plogginternational.com/products.shtml

Quinn, E. L. (2009, February 15). Privacy and the New Energy Infrastructure. *SSRN eLibrary.* Retrieved November 1, 2010, from http://papers.ssrn.com/sol3/papers.cfm?abstract_id=1370731

Rial, A., & Danezis, G. (2010). Privacy-Preserving Smart Metering. *Microsoft Research Technical Report, MSR-TR-2010-150.* Retrieved from http://research.microsoft.com/apps/pubs/?id=141726

Rubinstein, F., Siminovitch, M., & Verderber, R. (1993). Fifty percent energy savings with automatic lighting controls. *Industry Applications, IEEE Transactions on, 29*(4), 768-773. doi:10.1109/28.231992

Sandhu, J. S. (2004). Wireless Sensor Networks for Commercial Lighting Control: Decision Making with Multi-agent Systems. *AAAI Workshop on Sensor Networks, 10,* 131-140.

Sawyer, R. L., Anderson, J. M., Foulks, E. L., Troxler, J. O., & Cox, R. W. (2009). Creating low-cost energy-management systems for homes using non-intrusive energy monitoring devices. *Energy Conversion Congress and Exposition, 2009. ECCE 2009. IEEE* (pp. 3239-3246). Presented at the Energy Conversion Congress and Exposition, 2009. ECCE 2009. IEEE. doi:10.1109/ECCE.2009.5316132

Singhvi, V., Krause, A., Guestrin, C., James H. Garrett, J., & Matthews, H. S. (2005). Intelligent light control using sensor networks. *Proceedings of the 3rd international conference on Embedded networked sensor systems* (pp. 218-229). San Diego, California, USA: ACM. doi:10.1145/1098918.1098942

Tanaka, S., Miki, M., Hiroyasu, T., & Yoshikata, M. (2009). An evolutional optimization algorithm to provide individual illuminance in workplaces. *Systems, Man and Cybernetics, 2009. SMC 2009. IEEE International Conference on* (pp. 941-947). Presented at the Systems, Man and Cybernetics, 2009. SMC 2009. IEEE International Conference on. doi:10.1109/ICSMC.2009.5346094

Tomishima, C., Miki, Mitsunori, Ashibe, M., Hiroyasu, Tomoyuki, & Yoshimi, M. (2010). Distributed Control of Illuminance and Color Temperature in Intelligent Lighting System. In L. Rutkowski, R. Scherer, R. Tadeusiewicz, L. Zadeh, & J. Zurada (Eds.), *Artifical Intelligence and Soft Computing*, Lecture Notes in Computer Science (Vol. 6114, pp. 411-419). Springer Berlin / Heidelberg. Retrieved from http://dx.doi.org/10.1007/978-3-642-13232-2_50

US Department of Energy. (2009, March). 2009 Buildings Energy Data Book. *http://buildingsdatabook.eren.doe.gov/.*

US Department of Energy. (n.d.). 2008 Commercial Energy End-Use Expenditure Splits, by Fuel Type. Retrieved May 5, 2011, from http://buildingsdatabook.eren.doe.gov/TableView.aspx?table=3.3.4

Wen, Y.-J., & Agogino, Alice M. (2008). Wireless networked lighting systems for optimizing energy savings and user satisfaction. *2008 IEEE Wireless Hive Networks Conference* (pp. 1-7). Presented at the 2008 IEEE Wireless Hive Networks Conference (WHNC), Austin, TX, USA. doi:10.1109/WHNC.2008.4629493

Wen, Y.-J., Granderson, J., & Agogino, A.M. (2006). Towards Embedded Wireless-Networked Intelligent Daylighting Systems for Commercial Buildings. *IEEE International Conference on Sensor Networks, Ubiquitous, and Trustworthy Computing - Vol 1 (SUTC'06)* (pp. 326-331). Presented at the IEEE International Conference on Sensor Networks, Ubiquitous, and Trustworthy Computing -Vol 1 (SUTC'06), Taichung, Taiwan. doi:10.1109/SUTC.2006.1636196

Wong, A. C. W., & So, A. T. P. (1997). Building automation in the 21st century. *Advances in Power System Control, Operation and Management, 1997. APSCOM-97. Fourth International Conference on (Conf. Publ. No. 450)* (Vol. 2, pp. 819-824 vol.2). Presented at the Advances in Power System Control, Operation and Management, 1997. APSCOM-97. Fourth International Conference on (Conf. Publ. No. 450).

Wong, J. K. W., Li, H., & Wang, S. W. (2005). Intelligent building research: a review. *Automation in Construction, 14*(1), 143-159. doi:10.1016/j.autcon.2004.06.001

Yeh, L.-W., Lu, C.-Y., Kou, C.-W., Tseng, Y.-C., & Yi, C.-W. (2010). Autonomous Light Control by Wireless Sensor and Actuator Networks. *IEEE Sensors Journal, 10*(6), 1029-1041. doi:10.1109/JSEN.2010.2042442

Zigbee Alliance. (2008, January). Zigbee Specification: ZigBee Document 053474r17.

Orientation and Tilt Dependence of a Fixed PV Array Energy Yield Based on Measurements of Solar Energy and Ground Albedo – a Case Study of Slovenia

Jože Rakovec[1,2], Klemen Zakšek[2,3], Kristijan Brecl[4],
Damijana Kastelec[5] and Marko Topič[4]
[1]Faculty of Mathematics and Physics, University of Ljubljana
[2]Centre of Excellence Space-SI
[3]Institute of Geophysics, University of Hamburg
[4]Faculty of Electrical Engineering, University of Ljubljana
[5]Biotechnical Faculty, University of Ljubljana
[1,2,4,5]Slovenia
[3]Germany

1. Introduction

In the last decade solar photovoltaic (PV) systems have become available as an alternative electrical energy source not only in remote locations but even in densely populated areas as their price decreases and their performance increases. The chapter discusses fixed PV array potential in Slovenia with great geographical and topographical variety, which is a reason that the climate, and also PV potential, changes rapidly already on short distances. The study is based on the meteorological measurements of solar irradiance, air temperature and albedo from the MODIS satellite data. Simulations for four meteorological stations were employed to determine combinations of azimuth and tilt angle for fixed PV arrays that would enable their maximum efficiency. As expected, large tilt with southern orientation is optimal during winter and almost flat installations are optimal during summer. The optimal PV gains are compared also with the results obtained by using the rule of a thumb tilt angle showing some significant differences in some cases.

2. Theoretical background

PV system users can define the orientation of their PV arrays: their azimuth angle (angle measured clockwise from North) and the tilt angle (the angle above the horizontal plane). Previous studies show that, if local weather and climatic conditions are not considered, the optimal fixed tilt angle of PV modules depends only on geographical latitude φ (and the optimal azimuth is always south in the northern hemisphere). Considering only direct solar irradiation, the optimal tilt angle during the year can be calculated as $\varphi - \delta_s$, where δ_s is the declination of the Sun. For example, for latitude $\varphi = 46°$ N the maximal direct irradiation on 21 March and 21 September is achieved at a tilt angle of 44°. On 21 June and on 21 December the tilt angle is changed by the declination of the sun (± 23.5°) to 20.5° and 67.5°.

Due to the diffuse light the optimal tilt angles differ from those in reality. Since modules are frequently incorporated into the architecture of some objects, often some "rule of thumb" is applied. By taking such an approach a certain "yearly optimum" is obtained – as suggested by Duffie and Beckman (1991) – the tilt angle should be 10°–15° more than the latitude during winter and 10°–15° less than the latitude during summer. The lower values are originally based on the classical report by Morse & Czarnecki (1958) from the mid-20th century. Their suggestion for the annually optimally fixed tilt angle is a value 0.9 times the latitude, which results in 40° for Slovenia. Other authors (Lewis, 1987; Heywood, 1971; Lunde, 1982; Garg, 1982) have concluded that the optimal tilt differs from the latitude in a range between ±8° and ±15°. An analytical equation to find the daily optimal tilt angle at any latitude has also been used (El-Kassaby & Hassab, 1994). For example, the average optimal tilt angle on Cyprus (latitude $\varphi = 35°N$) equals 48° in the winter months ($\varphi + 13°$) and 14° ($\varphi - 21°$) in the summer months (Ibrahim, 1995). The optimal tilt was estimated for Brunei Darussalam on the basis of maximising the global solar irradiation reaching the collector surface for each month and year (Yakup & Malik, 2001). The tilt optimised for winter in Poland equals 50°–65°, for summer 10°–25°, and the PV module does not necessarily have to be oriented directly to the south – a range in the azimuth angle of -60° to +60° from the South also provides good results (Chwieduk & Bogdanska, 2004). The optimal tilt angle in Turkey varies from 13°–61° from summer to winter (Kacira et al., 2004), while the monthly optimised tilt in Ireland can vary from 10° to 70° (Mondol et al., 2007).

The optimal tilt for the whole of Europe (PVGIS) shows that climate characteristics have a huge influence on the optimal tilt (Huld et al., 2008). In this contribution we particularly emphasise local weather and climatic conditions when computing the optimal orientation and tilt. As we will show in Section 3, these may differ considerably from the "maximum noon direct irradiation" as well as from the "rule of thumb" results.

2.1 Solar irradiance on a tilted plane

The most important parameter for computing the solar irradiance reaching the Earth's surface is cloud coverage. In clear-sky conditions the next most important factor is the optical path length as the transmissivity of the atmosphere exponentially depends on it, which implies the position of the Sun in the sky (its zenith angle ϑ_s and azimuth A_S that may be aggregated into unit vector $\vec{s}(\vartheta_s, \alpha_s)$ towards the Sun) changing over the course of a day and year. The **true solar time** (considering the geographical latitude and the equation of time) has been used for accurate computations and not the zonal time.

Actual irradiance on the tilted plane varies significantly with its orientation geometry (tilt τ – angle of inclination between the horizontal surface and the PV module's receiving plane, and orientation A – the azimuth angle between the North and the azimuthal component of the normal to the PV's plane; both may be aggregated into a unit normal vector of the plane $\vec{n}(\tau, A)$.

Solar irradiance is usually measured on a horizontal plane as global irradiance E_{gl}. The direct component $E_{gl, dir}$ and diffuse component $E_{gl, diff}$ of global irradiance must be considered separately because of their very different dependences on the tilted irradiance. When knowing the two components one can compute irradiance E_{tilt} on the tilted plane (in the meteorological community this is called quasi-global irradiance). The effect of the PV module's tilt and azimuth angle on the **direct part of solar tilted irradiance** $E_{tilt,dir}$ is described by a scalar product of the two unit vectors:

$$E_{tilt,dir} = \vec{s} \cdot \vec{n} \frac{E_{gl,dir}}{\cos \vartheta_s} \tag{1}$$

The effect of **diffuse tilted irradiance** ($E_{tilt,dif}$) can only be considered to be isotropic when there are no obstacles (e.g. mountains, buildings) on the horizon and the whole sky is covered by clouds of uniform brightness (Fig. 1). Many anisotropic models have therefore been developed: e.g. Brunger & Hooper (1993); a good overview is included in Kambezidis et al. (1994) but they mostly have an empirical background and thus their use is only suitable when a calibration of the model with measurements on the tilted surface is possible. Therefore, simplified isotropic models based on the parameter called the **sky-view factor** (*svf*) are mostly used. The sky-view factor is defined as the hemispherical fraction of unobstructed sky. There are several isotropic models of *svf* for inclined receivers. The 2D one: $svf = (1 + \cos \tau)/2$ (Mondol et al., 2007; Huld et al., 2008; Liu & Jordan, 1963), the improved one with a more realistic 3D consideration: $svf = (1 + \cos^2 \tau)/2$ (Badescu, 2002; Brunger & Hooper, 1993), as well as the 3D linear model by Tian et al. (2001):

$$svf = (\pi - \tau)/\pi. \tag{2}$$

Besides diffuse radiation from the sky, reflected (multiple scattered) radiation from the ground can also be important, especially for modules with a larger tilt and in areas of high ground reflectivity, like in a snow-covered landscape. Ground reflection is defined by the **ground-view factor** (*gvf*) that is a complementary parameter to the sky-view factor:

$$gvf = 1 - svf. \tag{3}$$

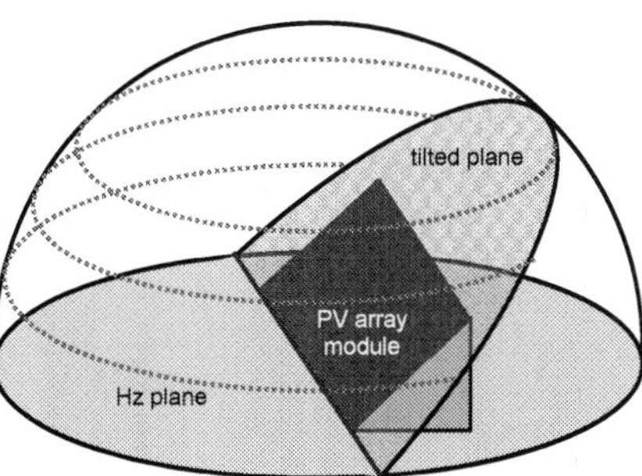

Fig. 1. a) 3D approach to estimating the sky-view factor – the visible sky is limited by the horizontal plane and tilted plane of the PV module, thus only that part of the hemisphere between these planes is visible.

For the diffuse irradiation coming from the ground is beside the geometry important also the ground reflectivity, characterised by the **albedo (reflectivity) of the surrounding surfaces** a_{gr}. A constant albedo of 0.2 (typical grassland) was used in most previous studies. Some other approaches as Gueymard (1993) also suggest a seasonal albedo model. Such an albedo changes over the year according to the latitude and land cover of the observed area or to anisotropic approaches (Arnfield, 1975). These models are mainly appropriate for areas where a direct reflection is possible.

Tilted solar irradiance E_{tilt} of a tilted plane is written by many authors as the sum of the abovementioned contributions. The diffuse component coming from the sky decreases with the tilt angle while at the same time the ground diffuse part of the irradiance increases:

$$E_{tilt} = E_{tilt,dir} + E_{tilt,dif} = E_{tilt,dir} = \vec{s} \cdot \vec{n} \frac{E_{gl,dir}}{\cos\vartheta_s} + E_{gl,dif} \cdot sfv + a_g \cdot (E_{gl,dir} + E_{gl,dif}) \cdot gfv. \tag{5}$$

Just recently the first two authors of this contribution have elaborated a more exact and conceptually proper approach based on the integration of isotropic radiance of sky L_{sky} that gives, when integrated over the whole hemisphere, the diffuse irradiance of horizontal receiving surface $E_{gl,dif} = \pi L_{sky}$. Integration over the hemisphere, for which part of it has the radiance of sky L_{sky} and the other part has a different radiance of ground L_{gr} results in irradiation of the tilted surface E_{tilt}. If the albedo a_{gr} and the coefficient k describing the contribution of the diffuse irradiation to the global irradiation ($E_{gl,dif} = k \cdot E_{gl}$) are also considered, an alternative, more accurate estimate of irradiation of the tilted receiving surface is obtained:

$$E_{tilt} = E_{tilt,dir} + E_{tilt,dif} = \vec{s} \cdot \vec{n} \frac{E_{gl,dir}}{\cos\vartheta_s} + E_{gl,dif} \frac{1}{2}\left(1 + \cos^2\tau + \frac{a_{gr}}{k}\sin^2\tau\right) \tag{5}$$

Expressions (4) and (5) differ only as regards diffuse irradiation; the difference depends on reflectivity a_{gr} and on contribution k of the diffuse irradiance to the global irradiance. For example, for $a_{gr} = 0.2$ and $k = 0.5$ the results differ by up to approximately ±6% for the diffuse irradiance, and up to approximately ±3% for the whole irradiance of tilted irradiance E_{tilt}. Here a more appropriate expression (5) was applied; more details about this are found in a submitted paper (Rakovec & Zakšek, n.d.).

2.2 Performance of PV modules and arrays

The energy conversion efficiency of a PV module or array as a group of electrically connected PV modules in the same plane is defined as the ratio between electrical power P_{PV} conducted away from the module, and the incidence power of the sun: $P_{PV}(t)/SE_{tilt}(t) = \eta$. Normally, their efficiency is defined under standard test conditions η_{STC} (STC – module temperature: $T_{STC} = 25\ °C$, irradiance: $E = 1000\ W/m^2$, spectrum: AM1.5; IEC 61836-TR/Ed.2:2007; IEC 60904; http://www.iec.ch). The output power of a PV module depends on several parameters, including the irradiance, incidence angle and PV cell temperature T as the most influential. Namely, the PV cell **efficiency also depends on its temperature** as in solar cells based on the p-n junction diode principle the efficiency decreases with increasing temperature due to the higher dark current (Green, 1982). The efficiency temperature dependence is normally expressed by a linear equation:

$$\eta\,(T) = \eta_{STC}[1+\gamma\,(T - T_{STC})]. \tag{6}$$

The value of γ is approximately -0.004/K for polycrystalline silicon cells and modules (Carlson et al., 2000).

Beside the irradiance, incidence angle and temperature dependence of the PV module, the output power of the PV system also depends on system losses: Joule losses in wirings of PV modules into PV arrays and inverter losses. These additional losses do not influence the tilt and azimuth dependence of the output energy since they only depend on the output power and on irradiance and not on time like the module's temperature. To obtain the system energy output from the PV module output energy we used a typical system performance factor of 85% in our study.

2.3 Thermal model of PV modules

How the temperature of the absorbing material of the receiving PV module increases depends on the energy exchanges between the absorber and its environment. Different assumptions can be made as regards the PV module energy balance equation. To explain only the basic energy exchange here we consider the PV module as a whole: cells with temperature T_c, the covering plate with its temperature T_p, eventually with the temperature on the surface (where it exchanges energy with the environment) also different from the one on the inner side of the plate are all considered to be one object with temperature T and with heat capacity c, having mass m and a receiving area S. Such a simplification neglects all the energy flows between the separate parts of the PV module, but on the other hand emphasises only the most important features of PV module energetics, without entering into particular details. In this paper we will also focus only on outdoor conditions. We also suppose, again to simplify the explanation, that all the surroundings have the same temperature as environmental air T_{env}. In principle, for both PV modules and solar thermal (ST) solar collectors the energy flows are the same (Petkovšek & Rakovec, 1983). The divergence of all these energy flows results in cooling (normally during the late afternoon and night), while convergence (i.e. negative divergence) results in a warming of the absorber (normally during morning and early afternoon hours). The result expressed as $(mc)dT/dt$ can be written as:

$$mc\frac{dT}{dt}(t) = P_s(t) + P_{conv}(t) + P_{cond}(t) + P_{IR}(t) + P_{lat}(t) - P_{PV}(t) =$$
$$= (1-a)SE_{tilt} - (K_{conv} + K_{cond})(T - T_{env}) + \sigma\varepsilon S(\varepsilon_{env}T_{env}^4 - T^4) + P_{lat} - P_{PV} \tag{7}$$

The terms in the equation are as follows: absorbed solar power $P_s = (1-a)SE_{tilt}$, the (turbulent) convective heat exchange between the absorber and its atmospheric environment $P_{conv} = -K_{conv}(T - T_{env})$, heat conduction between the absorber and the surrounding neighbouring parts of the module (e.g. supporting) $P_{cond} = -K_{cond}(T - T_{env})$, the infrared radiation energy exchange (in the "terrestrial" wavelengths interval, centred at about 10 µm) $P_{IR} = S\varepsilon(\varepsilon_{env}\sigma T_{env}^4 - \sigma T^4)$, eventually latent heat exchanges, due to condensation or evaporation at the module, due to precipitation falling upon it etc: P_{lat} and, of course, the flow of energy away from the absorber – the yield of the useful energy P_{PV}. For the meaning of some of the symbols, see the main text; the others are: a – albedo of the module for solar radiation, S – the area of the module, K_{conv} and K_{cond} are the heat exchange coefficients, σ is the Stefan-Boltzmann constant and ε and ε_{env} are the IR emissivities of the module and its environment, respectively. As regards the IR irradiation from above: for clear sky is IR emissivity ε_{env} of approximately 0.7, while for overcast sky it approaches one – the emissivity of the black body.

An analytical solution of equation (7) needs input data in analytical form; the two most important forms of environmental data are tilted irradiance E_{tilt} and environmental (air) temperature T_{env}. The climatological values exhibit an excellent similarity to the sinusoidal course and the same similarity is found for individual cases (see the example for Etilt in Fig. 3a) as shown in Fig. 2. But an equation in which some other coefficients also change in time differently from case to case can only be precisely integrated numerically for each of the governing conditions to give $T(t)$ and with that also $\eta[T(t)]$. The numerical approach is used to calculate PV characteristics – and $\eta[T(t)]$ – on the basis of the measured data.

For example, an increase in the module's temperature from the morning hours until noon $\Delta T \approx 47$ K (Fig. 3d) leads to a negative relative change in the module's conversion efficiency $\gamma \cdot \Delta T = \Delta \eta / \eta_{STC} \cong -0.19$, which is confirmed with measurements ($\eta = 10\%$ at noon in Fig. 3c compared to $\eta_{STC} = 12.3\%$).

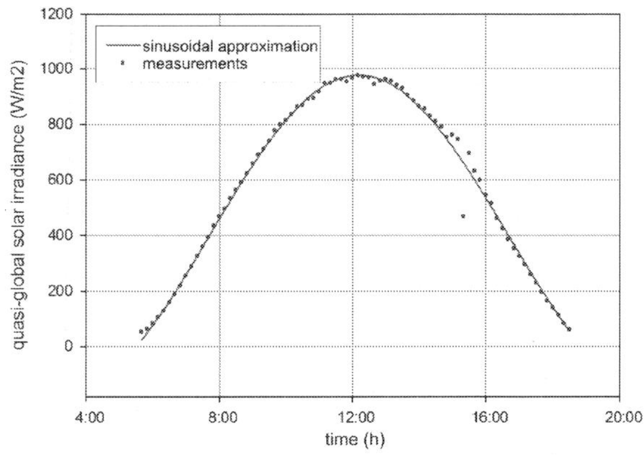

Fig. 2. Quasi-global solar irradiance fitted with a section of the sinusoidal function $E_{tilt0} + E_{tilt1} \sin \omega(t - t_0)$ (with $E_{tilt0} = 405$ W/m², $E_{tilt1} = 572$ W/m², $\omega = 8.46$ h⁻¹). The correlation coefficient between the data and the analytical function is 0.996.

2.4 Some experimental results

The Laboratory of Photovoltaics and Optoelectronics at the Faculty of Electrical Engineering of the University of Ljubljana (latitude: 46.07°N, longitude: 14.52°E) continuously monitors outdoor conditions of several variables and parameters relevant for PV (Kurnik et al., 2007; Kurnik et al., 2008), including E_{tilt}, P_{PV}, module temperature T and air temperature T_{air}.

One example for 20 July 2007 in Ljubljana is presented in Fig. 3. Based on these data the module efficiency was computed and is presented in Fig. 3c. Due to the higher reflection from the module by large incident angles, the efficiencies in early morning and late afternoon hours are quite low. Instead of being some 11 or 12% (as the module's temperature at that time is low!) the calculated values are even below 9%. Between 8:30 and 12:30, when solar rays are more perpendicular to the module (low reflection), the module

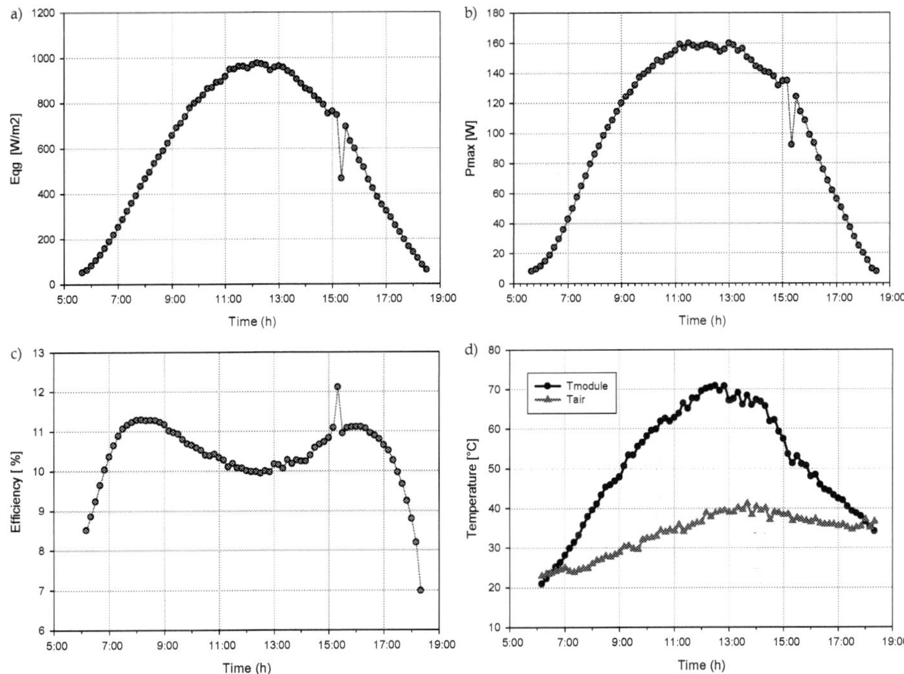

Fig. 3. a) Measured tilted irradiance E_{tilt} in the plane of the PV module (τ=30°, α=180°) oriented to the South on a clear day on 20 July 2007 in Ljubljana – with the Sun being occulted by a cloud at 15:20; b) measured power P_{PV} obtained of a typical polycrystalline module (S=1.634 m^2) with the same tilt and orientation; c) measured efficiency $\eta = P_{PV}/S/E_{tilt}$ (η_{STC} = 12.3%); and d) temperature of module T and of the surrounding air on roof T_{air} being higher than the one measured at the met station (Topič et al., 2007).

temperature increases from 45 °C to 71 °C and the η drops from 11.3% to 10.0%. The empirically estimated relative efficiency temperature coefficient is - 0.0044/K, which is close to the producer's specification of the temperature coefficient of the maximal output power γ = - 0.004/K.

3. Case study

The case study area presented in this chapter is Slovenia – a country on the south-east flank of the Alps between the Mediterranean and the Pannonian plain (approximately 13.5°-16.5°E and 45.5°-47.0°N). The country's great topographical variety significantly influences the climate characteristics, which results in annual solar radiation variations and influences the orientation of PV modules.

3.1 Data

The majority of pyranometers installed at meteorological stations in Slovenia have been functioning since 1993 or even later. The study was therefore done on just 10-year-long data sets (Kastelec et al., 2007) and not on a 30-year period, which is the climatologically established standard. Global solar irradiation was during 1994–2003 measured at 12 meteorological stations (on average, one per approximately 2,000 km²). Air temperature measurements were also taken from the same meteorological stations.

Map of ten-year average of annual global solar irradiation exposure was done by spatial interpolation of measured data on 12 locations and estimated data of global irradiation exposure on the basis of measured sunshine duration on 15 additional locations using Ångström's formula (Fig. 4).

Annual global radiance exposure changes significantly due to the country's great climatic variety even over short distances. No data across the Slovenian border was taken into account by spatial interpolation, so the accuracy of interpolated values is lower in the regions near the borders especially in the mountainous western and southern parts.

The diffuse part of the incoming solar energy was determined statistically by the Meteonorm 5.0 model package (Meteotest, 2003) at the remaining stations. The diffuse part of the incoming energy contributes a relatively smaller proportion to the global radiance exposure during summer (approximately 35–45%), and a relatively greater one during winter when there is even more diffuse than direct radiance exposure (up to 60%).

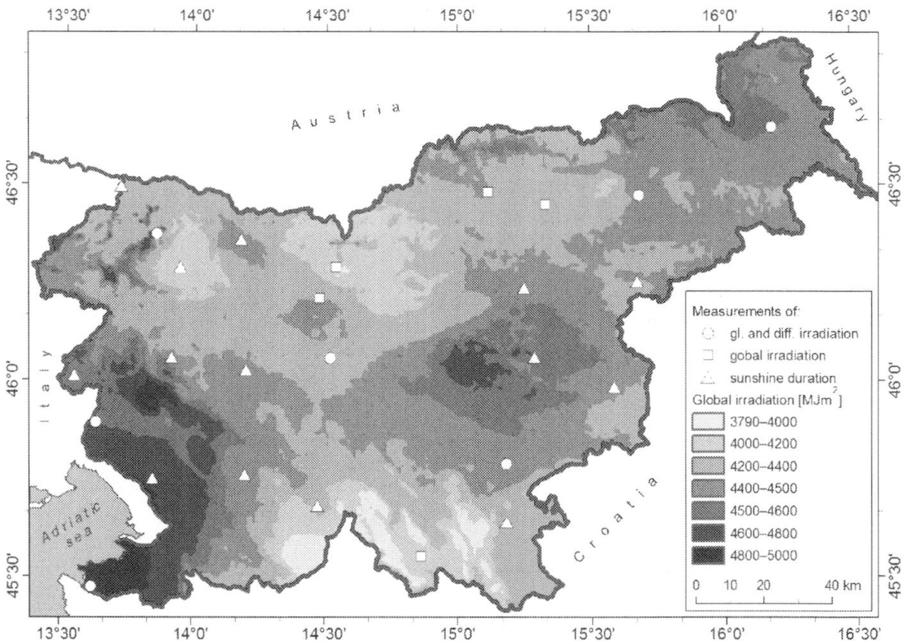

Fig. 4. Interpolated average annual global solar irradiation exposure has a heterogeneous spatial distribution in Slovenia (average for the 1994–2003 period; Kastelec et al., 2007).

The surface albedo was estimated by satellites. MODIS MOD43B3 albedo product (NASA, 2010) was used in the study, more specifically the shortwave (0.3–5.0μm) white sky broadband albedo. The MOD43B3 albedo product is prepared every 16 days in a one-kilometre spatial resolution. A reprocessed (V004) MOD43B3 albedo product is available from March 2000 till the present (thus not for the same time interval as used for global radiance exposure). Fig. 5 shows the annually averaged albedo over Slovenia for the 2000–2007 period.

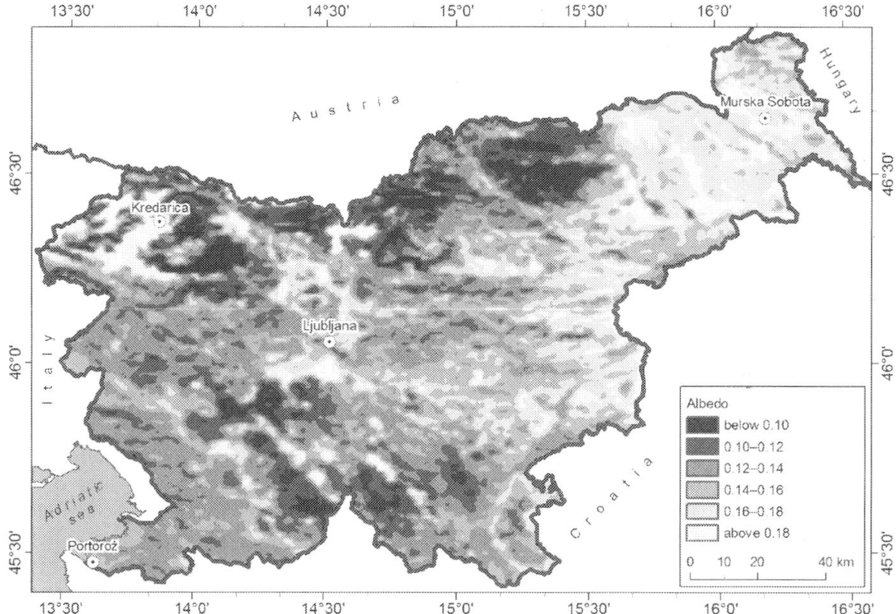

Fig. 5. Yearly averaged albedo (2000–2007) of the surface in Slovenia using MODIS images in a 1,000 m spatial resolution. Locations of four locations whose results are shown in the case study are also marked.

3.2 Computational simulation

We computed the energy output for each combination of a tilt and azimuth angle for all months and for the whole year. This gives us the optimum combination of both geometry parameters for each period. In the same way we get also the increase or decrease of the energy received on any orientation of a PV in the chosen period. Our results are the graphs showing this increase/decrease relative to tilt τ and azimuth angle α are the most important results of this study. We ran the simulation using the IDL language. It took several minutes for each computation using a relatively powerful personal computer.

Solar irradiance changes continuously over time in nature. Therefore, we decided to average the hourly measurements for 10-day periods. This resulted into 16-hourly averaged values (sunrise always after 4:00 and sunset always before 20:00) for each of the 36 periods. As meteorological measurements are performed at observing times according to UTC or to zonal time (CET) and not according to the true solar time, the distribution of the solar irradiance over the day is not symmetrical regarding the zonal noon. This can lead to errors of 20° by estimation of the optimal azimuth angle in March. The hourly data were thus fitted to a 5th order polynomial and then the irradiance and temperature values were estimated for each one hundredth of an hour. These values (at the end 16,000 for each of the 36 periods) were used in the simulation.

The MOD43B3 albedo product was averaged for the 2000–2007 period (this product was not available for earlier years) over each month. Then it was projected to the Slovenian national co-ordinate system into a regular grid of a 1,000 m spatial resolution. Due to cloud coverage, some albedo datasets contain data gaps; these were in our case study removed during temporal averaging.

4. Results

The results are presented for four locations in Slovenia (see their locations in Fig. 5). The graphs (Figs. 6–9) and Table 1 present the relative gain of energy (as a percentage) for the optimal combination of the inclination and orientation (marked by a cross) in comparison to energy on the horizontal surface. The abscise axes correspond to the azimuthal orientation (clockwise from the North) for azimuths from E to W (90° to 270°) and ordinate axes to the tilt (zero when the surface is horizontal and 90° for a vertical receiving surface). There are some differences among the four places, along with some common attributes. It is important to stress that optimal orientations and tilts are strongly affected by local weather and climatic conditions.

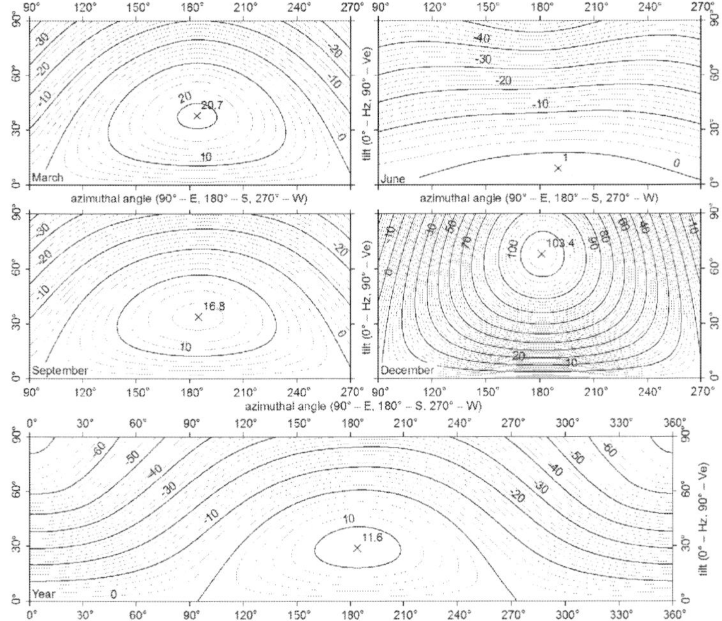

Fig. 6. Contour plots of the relative PV array energy yield regarding the horizontal surface as a function of a fixed orientation and tilt for March, June, September and December as well as the whole year for Portorož in the Mediterranean part of Slovenia.

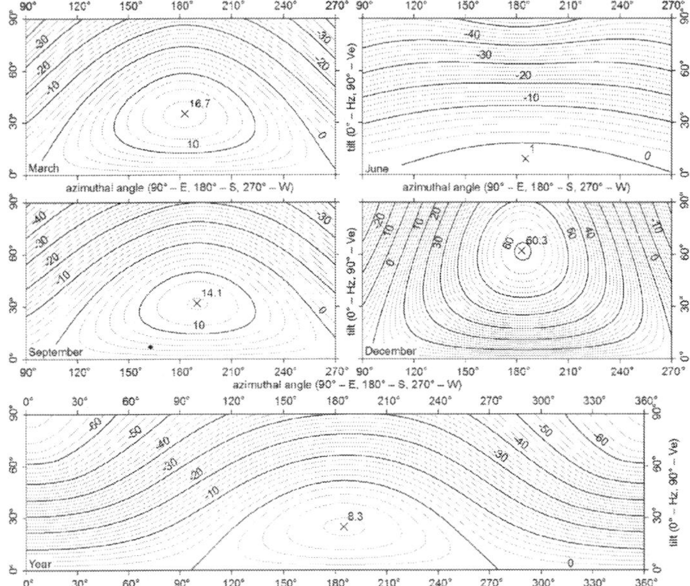

Fig. 7. As for Figure 6, but for Ljubljana in a basin in central Slovenia.

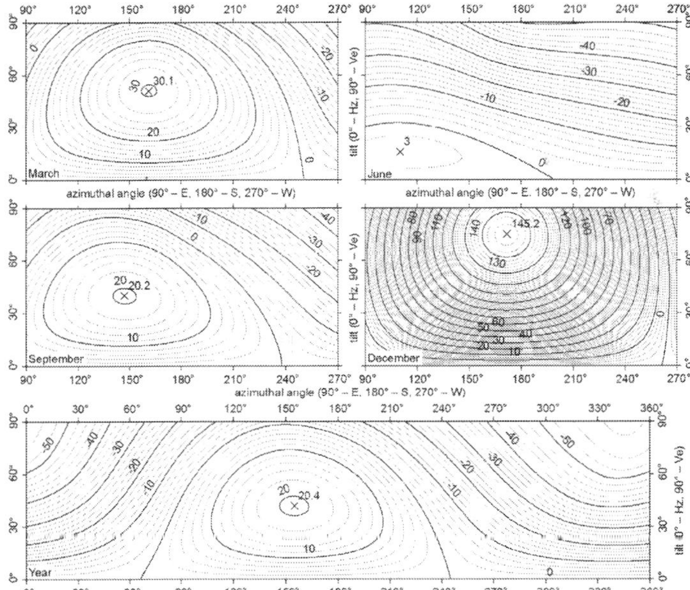

Fig. 8. As for Figure 6, but for Kredarica in high mountains.

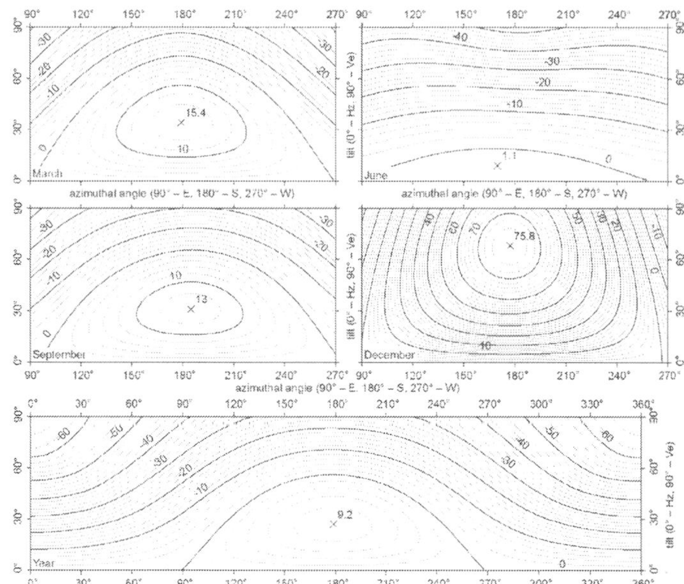

Fig. 9. As for Figure 6, but for Murska Sobota in the Pannonian part of Slovenia

Figures 6–9 and Table 1 show that the PV modules should be oriented more or less towards the South – but not exactly; in Portorož and Ljubljana the optimal orientation is around 5° from the South towards the West. The main reason for this is that the effect of morning fog or low cloudiness, making the irradiance asymmetrical around (true) noon, prevails over the effect of lower efficiency in early afternoon hours due to the higher temperature of the module. The situation at Kredarica in this respect is very specific due to the mountain wall of the top cone of Mt. Triglav to the West of the location. Since there is a lot of shadow in the afternoon, the modules should be considerably oriented towards SE (α=155°). In the Pannonian part of Slovenia, in the warm part of the year a considerable proportion of precipitation is caused by convective cloudiness – and the fact that convective clouds normally develop in the early afternoon is also reflected in radiance exposures in Murska Sobota – especially in June and September the orientation from the South more to the East is clearly expressed. Thus not only monthly but even the optimal fixed annual orientation and tilt perform slightly better than using "a rule of thumb", especially in places with a complex horizon (like at the mountainous Kredarica).

The tilt angles are more season dependent as azimuth angles. For example, in December the optimal orientation for clear sky conditions should be South (180°) and considering only direct irradiation the tilt should be from 67° to 70° (depending on the latitude). However, as there is often fog and low cloudiness on winter mornings, the tilt may change considerably. For example in Ljubljana located in a basin (where such phenomena are most frequent) the optimal tilt is only 62° and the orientation 183°. In contrast, in June it is best to have the module more or less horizontal. The reason for that (at first glance quite unexpected result) is the high solar elevation; in June the Sun rises north from East (at approximately ENE in Slovenia) and also sets north from West (at approximately WNW in Slovenia). So a PV module might be in the shadow (no direct insolation) during early morning hours and late in the afternoon. Further, a tilted surface also receives less diffuse irradiance.

	Global radiance exposures (kWh/m²)	Optimal orientation (°)	Tilt for maximum E_{dir} at solar noon (°)*	Optimal tilt (°)	Radiance exposures by optimal orientation and tilt (kWh/m²)	Increase by optimal tilt and orientation according to global radiance exposures (%)
Portorož on the Adriatic coast, φ = 45° 28´, λ = 13° 37´ h = 2 m a.s.l.						
March	110.2	184	45.5	38	133.0	20.7
June	202.2	190	22	9	204.1	1.0
Sept	126.4	185	45.5	34	147.6	16.8
Dec	33.6	181	67	68	68.4	103.4
year	1412.3	184		29	1573.1	11.4
Ljubljana in a basin in central Slovenia, φ = 46° 4´, λ = 14° 31´ h = 299 m a.s.l.						
March	97.4	183	46	35	113.8	16.7
June	178.5	185	22.5	9	180.4	1.1
Sept	108.7	190	46	32	124.1	14.1
Dec	23.8	183	69.5	62	38.2	60.3
year	1229.3	185		25	1329.3	8.1
Kredarica in high mountains, φ = 46° 23´, λ = 13° 51´ h = 2514 m a.s.l.						
March	121.3	161	46.5	51	157.9	30.2
June	155.9	110	23	16	160.6	3.0
Sept	107.7	147	46.5	40	129.5	20.2
Dec	44.3	172	70	75	108.7	145.3
year	1282.5	154		42	1538.5	20.0
Murska Sobota on Pannonian flatlands, φ = 46° 39´, λ = 16° 11´ h = 188 m a.s.l.						
March	99.7	179	46.5	34	115.1	15.4
June	184.9	170	23	9	186.9	1.1
Sept	109.8	185	46.5	31	124.1	13.0
Dec	26.8	177	70	68	47.1	75.8
year	1275.3	179		27	1390.0	9.0

* Tilts for maximum E_{dir} at solar noon are rounded to 0.5 of a degree

Table 1. Optimal azimuths and tilts according to months and the whole year and the resulting solar radiance exposures. The orientation for maximum E_{dir} at solar noon is 180° for all cases.

It is also interesting that taking the temperature dependence of the PV module on efficiency into account does not greatly influence the optimal orientation and tilt. A comparison with optima for the solar radiance exposures alone, i.e. the isolation of natural surfaces (Rakovec & Zakšek, 2008), only shows here and there some differences in optimal orientations and tilts. Most of the results are equal (within a degree or two). The main reason for this is evident in Figure 3 where one can notice that although the efficiencies are not exactly symmetrical around noon – they are slightly smaller in the afternoon than in the morning hours – the asymmetry does not influence the optimal orientation essentially – by more than a degree or so. The albedo of the surrounding landscape influences the gain mainly at greater tilts of the modules when they "see" a considerable proportion of the ground in their half space; with tilts of around 70° the proportion is roughly 40% ground and 60% sky. As high tilts are favourable in winter, and as it is possible that there is snow cover in winter, with a high albedo, the ground may be even brighter than the sky. Such details are not included in our "monthly average" albedo – except for mountainous locations where snow in winter is regular and hence captured by satellite data.

A comparison with some other studies for Slovenia, e.g. PVGIS (Huld et al., 2008) using a yearly averaged albedo and isotropic model proposed by (Liu & Jordan, 1963), shows that the optimal yearly tilts are significantly larger than in our study (PVGIS 35° for almost the whole of Slovenia versus ours e.g. 25° in Ljubljana; Table 1). If in addition the surface albedo is also overestimated (most models use 0.2, while the average yearly albedo equals e.g. 0.14 in Ljubljana) the results overestimate the actual gains by some 2–3%. The difference in gains could also be a consequence of an interpolation inaccuracy – PVGIS interpolated results for the whole of Europe and we estimated our results for chosen locations. The method considers the heterogeneity of the country by using solar irradiance characteristics that have different seasonal and daily courses in different parts of the country. These climatic differences accompanied by the albedo's heterogeneity therefore lead to different optimal azimuthal angles and tilts of the PV modules.

Location	H_{gl} (kWh/m²)	$maxH^*$ (kWh/m²)	H_{rt} (kWh/m²)	$H_{rt}/maxH$ (%)	H_{fix} (kWh/m²)	$H_{fix}/maxH^*$ (%)	W_{el} (kWh/m²)
Portorož	1412.3	1641.1	1560.8	95.1	1573.1	95.8	160.2
Ljubljana	1229.3	1367.0	1311.6	96.0	1329.3	97.2	134.6
Kredarica	1282.5	1655.7	1498.5	90.5	1538.5	92.9	158.0
Murska Sobota	1275.3	1442.4	1379.2	95.6	1390.0	96.4	141.0

* maxH – Maximum solar radiance exposure is determined by considering monthly optimal orientations and tilts; such a maximum could even be increased by changing the orientation and tilt daily – but this is not a realistic option; among other reasons also due to the changing weather from day to day.

Table 2. Average annual solar radiance exposures; H_{gl} – solar global radiance exposure, $maxH$ (see the note marked by *), H_{rt} – exposure by a fixed "rule of thumb" orientation 180° and tilt 35°, H_{fix} – by fixed optimal orientation and tilt, and W_{el} – electrical energy from a 215 Wp PV module with 85% system efficiency and with a fixed annual orientation and tilt at selected locations in Slovenia.

On the basis of our simulations, it may be concluded that the best solution would be to change the orientation and inclination of PV modules during the course of the year (monthly, if technically possible; Table 1). But also for a fixed annual orientation and tilt the optimal orientation and tilt perform somewhat better than using "a rule of thumb", especially in places with a complex horizon or specific climatic conditions – up to 3%.

5. Conclusion

To conclude, long-term measured meteorological values should be used to obtain reliable results on PV yield. Only then it is possible to dimension the PV system for yield optimization. We showed that the measured irradiation values are the most important parameter in photovoltaics. If only measurements of global irradiation are available, the diffuse part of irradiation can be simulated. Temperature measurements have merely a small effect on optimal orientation of PV system. Accurate albedo values are also irrelevant for the system orientation during the summer as albedo is usually low and optimal tilt angles are small. However in regions, where the albedo changes significantly during the year, is its accuracy important especially during winter, as the ground covered by snow is often even brighter than the sky.

6. Acknowledgment

This work was funded by the Slovenian Research Agency under the P2/0197 programme and by the Slovenian Environmental Agency under contract 2523-04-300351. The Centre of Excellence for Space Sciences and Technologies SPACE-SI is an operation partly financed by the European Union, European Regional Development Fund and Republic of Slovenia, Ministry of Higher Education, Science and Technology.

7. References

Arnfield, A.J. (1975). A Note on the Diurnal, Latitudinal and Seasonal Variation of the Surface Reflection Coefficient. *Journal of Applied Meteorology*, Vol. 14, No. 8, pp. 1603-1608, ISSN 0021-8952

Badescu, V. (2002). 3D isotropic approximation for solar diffuse irradiance on tilted surfaces. *Renewable Energy*, Vol. 26, No. 2, pp. 221-233, ISSN 0960-1481

Brunger, A.P. & Hooper, F.C. (1993). Anisotropic sky radiance model based on narrow field of view measurements of shortwave radiance. *Solar Energy*, Vol. 51, No. 1, pp. 53-64, ISSN 0038-092X

Carlson, D.E., Ling, G. & Ganguly, G. (2000). Temperature dependence of amorphous silicon solar cell PV parameters. *IEEE Phot Spec C*, Vol. 28, pp. 707-712

Chwieduk, D. & Bogdanska, B. (2004). Some recommendations for inclinations and orientations of building elements under solar radiation in Polish conditions. *Renewable Energy*, Vol. 29, No. 9, pp. 1569-1581, ISSN 0960-1481

Duffie, J.A. & Beckman, W.A. (1991). *Solar Engineering of Thermal Processes*, 2nd ed., Wiley-Interscience, ISBN 0471510564, New York (USA)

El-Kassaby, M.M. & Hassab, M.H. (1994). Investigation of a variable tilt angle Australian type solar collector. *Renewable Energy*, Vol. 4, No. 3, pp. 327-332, ISSN 0960-1481.

Garg, H.P.C. (1982). *Treatise on solar energy: Volume 1: Fundamentals of solar energy.*, John Wiley & Sons, ISBN 047110180X, New York (USA)

Green, M. (1982). *Solar cells : operating principles, technology, and system applications*, Prentice-Hall, ISBN 9780138222703, Englewood Cliffs NJ

Gueymard, Ch. (1993). Mathematically integrable parameterization of clear-sky beam and global irradiances and its use in daily irradiation applications. *Solar Energy*, Vol. 50, No. 5, pp. 385-397, ISSN 0038-092X

Heywood, H. (1971). Operational experience with solar water heating. *J Inst Heat Vent Energy*, Vol. 39, pp. 63-69

Huld, T., Šúri, M. & Dunlop, E.D. (2008). Comparison of potential solar electricity output from fixed-inclined and two-axis tracking photovoltaic modules in Europe. *Progress in Photovoltaics: Research and Applications*, Vol. 16, No. 1, pp. 47-59, ISSN 1099-159X

Ibrahim, D. (1995). Optimum tilt angle for solar collectors used in Cyprus. *Renewable Energy*, Vol. 6, No. 7, pp. 813-819, ISSN 0960-1481

Kacira, M., Simsek, M., Babur, Y. & Demirkol, S. (2004). Determining optimum tilt angles and orientations of photovoltaic panels in Sanliurfa, Turkey. *Renewable Energy*, Vol. 29, No. 8, pp. 1265-1275, ISSN 0960-1481

Kambezidis, H.D., Psiloglou, B.E. & Gueymard, C. (1994). Measurements and models for total solar irradiance on inclined surface in Athens, Greece. *Solar Energy*, Vol. 53, No. 2, pp. 177-185, ISSN 0038-092X

Kastelec, D., Rakovec, J. & Zakšek, K. (2007). *Sončna energija v Sloveniji (Solar energy in Slovenia)*, Založba ZRC, ISBN 978-961-254-002-9, Ljubljana (Slovenia)

Kurnik, J., Brecl, K., Jankovec, M. & Topič, M. (2007). First measurement results of effective efficiency of different PV modules under field conditions. In: *Proceedings of the international conference*, 22nd European Photovoltaic Solar Energy Conference. Milan, Italy

Kurnik, J., Jankovec, M., Brecl, K. & Topič, M. (2008). Development of outdoor photovoltaic module monitoring system. *Informacije MIDEM*, p. 80

Lewis, G. (1987). Optimum tilt of a solar collector. *Solar and Wind Energy*, Vol. 4, No. 3, pp. 407-410

Liu, B.Y.H. & Jordan, R.C. (1963). The long-term average performance of flat-plate solar energy collectors. *Solar Energy*, Vol. 7, pp. 53-74, ISSN 0038-092X

Lunde, P.J. (1982). *Solar Thermal Engineering: Space Heating and Hot Water Systems: Solutions Manual*, Wiley, ISBN 0471891770, New York (USA)

Meteotest (April 2011). Meteonorm, In: *Meteotest*. Accessed April 5, 2011. Available from: http://www.meteotest.ch/en/footernavi/solar_energy/meteonorm

Mondol, J.D., Yohanis, Y.G. & Norton, B. (2007). The impact of array inclination and orientation on the performance of a grid-connected photovoltaic system. *Renewable Energy*, Vol. 32, No. 1, pp. 118-140, ISSN 0960-1481

Morse, R.N. & Czarnecki, J.T. (1958). Flat heat solar absorbers: The effect on incident radiation of inclination and orientation. In: Rep E.E. 6 Commonwealth CSIRO. Melbourne, Australia

NASA (May 2010). MCD43B3, In: *LP DAAC*. Accessed April 5, 2011. Available from https://lpdaac.usgs.gov/lpdaac/products/modis_products_table/albedo/16_day _13_global_1km/mcd43b3

Petkovšek, Z. & Rakovec, J. (1983). The influence of meteorological parameters on flat-plate solar energy collector. *Archives for Meteorology, Geophysics, and Bioclimatology Series B*, Vol. 33, No. 1-2, pp. 19-30, ISSN 0066-6424

Rakovec, J. & Zakšek, K. (2008). Sončni obsevi različno nagnjenih in različno orientiranih površin (Solar radiance exposures of differently tilted and differently oriented surfaces). *EGES*, Vol. 12, No. 2, pp. 93-95

Rakovec, J. & Zakšek, K. On the proper analytical expression for the sky-view factor and the diffuse irradiation of a slope for the isotropic sky. *Submitted to Renewable energy*, ISSN 0960-1481

Tian, Y.Q., Davies-Colley, R.J., Gong, P. & Thorrold, B.W. (2001). Estimating solar radiation on slopes of arbitrary aspect. *Agricultural and Forest Meteorology*, Vol. 109, No. 1, pp. 67-74, ISSN 0168-1923

Topič, M., Brecl, K. & Sites, J. (2007). Effective efficiency of PV modules under field conditions. *Progress in Photovoltaics: Research and Applications*, Vol. 15, No. 1, pp. 19-26, ISSN 1099-159X

Yakup, M.A. bin H.M. & Malik, A.Q. (2001). Optimum tilt angle and orientation for solar collector in Brunei Darussalam. *Renewable Energy*, Vol. 24, No. 2, pp. 223-234, ISSN 0960-1481

8

Optimal Design of Cooling Water Systems

Eusiel Rubio-Castro, José María Ponce-Ortega
and Medardo Serna-González
Chemical Engineering Department, Universidad Michoacana de San Nicolás de Hidalgo
México

1. Introduction

Re-circulating cooling water systems are generally used to remove waste heat from hot process streams in conditions above the ambient temperature in many types of industries such as chemical and petrochemical, electric power generating stations, refrigeration and air conditioning plants, pulp and paper mills, and steel mills. Typical re-circulating cooling water systems are constituted by a mechanical draft wet-cooling tower that provides the cooling water that is used in a set of heat exchangers operated in parallel as can be seen in Figure 1. The economic optimization of re-circulating cooling water systems includes the simultaneous selection of the optimal design variables of the cooling tower and each heat exchanger in the cooling network, as well as the optimal structure of the cooling water network. The question is then how to reach this goal. Earlier work on cooling water systems has concentrated on the optimization of stand-alone components, with special attention given on the individual heat exchangers of the cooling water network. Other publications have dealt with the problem of designing minimum-cost cooling towers for a given heat load that must be dissipated (see Söylemez 2001, 2004; Serna-González et al., 2010). Most of the methodologies previously reported have concentrated their attention in the optimal synthesis of cooling water networks (see Kim and Smith, 2001; Feng et al., 2005; Ponce-Ortega et al., 2007). All previous formulations simplified the network configurations because they consider the installation of only one cooling tower; however, the industrial practice shows that sometimes it is preferable to use a set of cooling towers connected in series, parallel, and series-parallel arrangements to improve the performance of the cooling towers reducing the operational cost and, hence, to decrease the overall total annual cost for the cooling water system. In addition, previous methodologies do not have considered several arrangements for the cooling water that can improve the performance in the coolers and reduce their capital costs. Another limitation for the previously reported methodologies is that they are based on the use of simplified formulations for the design of cooling towers.

This chapter presents an optimization model for the simultaneous synthesis and detailed design of re-circulating cooling water systems based on the superstructure of Figure 2. The model considers all the potential configuration of practical interest and the results show the significant savings that can be obtained when it is applied.

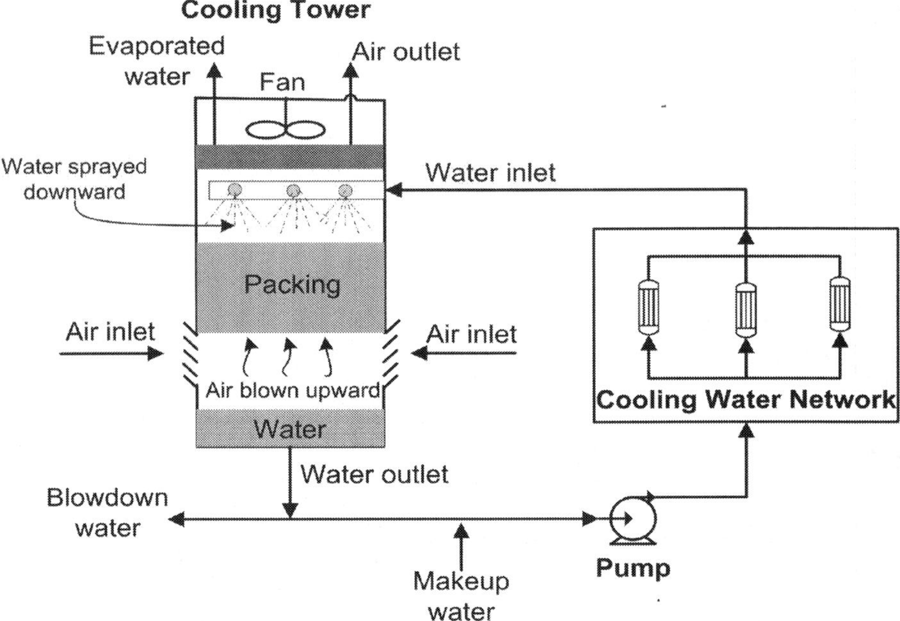

Fig. 1. Typical cooling water system.

2. Model formulation

This section presents the relationships for the proposed model, which is based in the superstructure of Figure 2. In the next equations, the set NEF represents the cooling medium streams leaving the cooler network, *ST* the stages of the cooling network, *HP* the hot process streams and *NCT* represents the cooling towers. The subscripts *av, b, cu, d, dis, ev, f, fi, fr, i, j, k, e, l, m, in, n, nct, out, p, pl, r, s, t* and *WB* are used to denote average, blowdown, cooling medium, drift, end of the cooling tower network, evaporation, fan, fill, cross-sectional, hot process stream, cold process stream, stage in the cooling network, type of packing, constants to calculate the heat and mass transfer characteristics for a particular type of packing, constants for the loss coefficient correlation for a particular type of packing, inlet, temperature increment index, cooling tower, outlet, pump, parallel arrangement, makeup, series arrangement, total and wet-bulb, respectively. The superscript *max* is an upper limit and *min* is a lower limit. In addition, the scalars *NOK* is the total number of stages in the cooling network, *NCP* is the total number of cooling medium streams at the hot end of the cooling network and *LCT* is the last cooling tower in the cooling tower network.

The heat of each hot process stream (QHP_i) is calculated by the multiplication between the heat capacity flowrate of each hot process stream (FCP_i) and the difference of the inlet and outlet temperatures of each stream ($THIN_i$, $THOUT_i$), All terms of the above equation

$$(THIN_i - THOUT_i)FCP_i = QHP_i \tag{1}$$

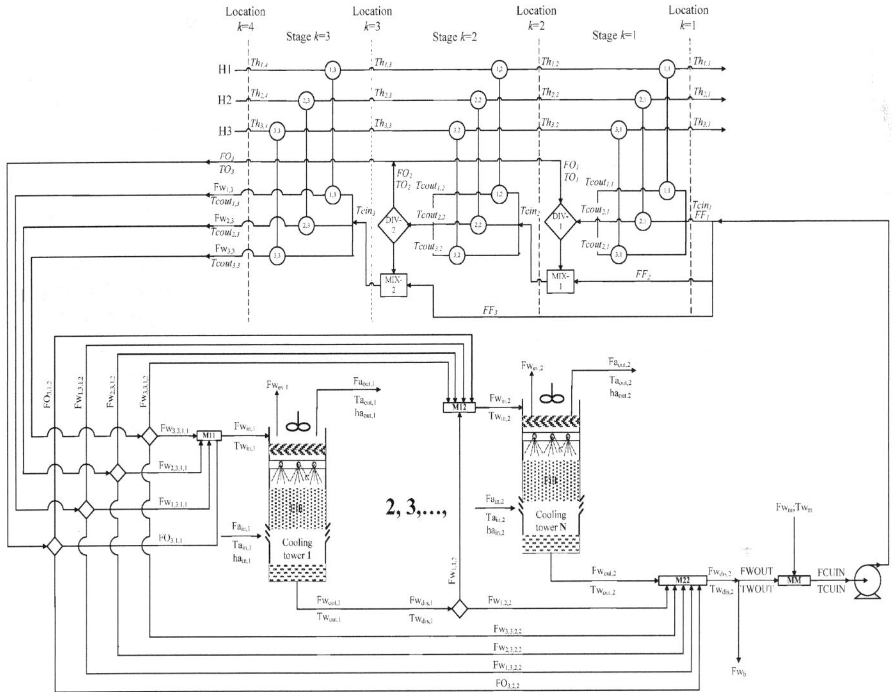

Fig. 2. Proposed superstructure.

$$QHP_i = \sum_{k \in ST} q_{i,k}, \quad i \in HP \tag{2}$$

are parameters with known values. The heat absorbed by the cooling medium in the matches is equal to the transferred heat by the hot process streams, where $q_{i,k}$ is the heat exchanged in each match. In addition to the above balance, the model includes balances for the splitters and mixers on each stage for the cooling medium.

A heat balance for each match of the superstructure is required to determine the intermediate temperatures of the hot process stream and the cooling medium as well as the cooling medium flowrate at each match. For problems with NH hot process stream, the number of stages in the superstructure NOK must be equal to NH, to allow arrangements completely in series. In this case, there are $NOK+1$ temperatures for each hot process stream, because the outlet temperature at one stage is equal to the inlet temperature in the next stage. It is required to identify the inlet and outlet temperatures at each stage for the cooling medium. Stage $k=1$ represents the lowest level of temperature. The heat balance for each match is given as follows:

$$\left(Th_{i,k+1} - Th_{i,k}\right)FCP_i - q_{i,k}, \quad k \in ST, i \in HP \tag{3}$$

$$\left[\left(Tcout_{i,k} - Tcin_k\right)F_{i,k}\right]CP_{cu} = q_{i,k}, \quad k \in ST, i \in HP \tag{4}$$

where $F_{i,k}$ is the flowrate for the cooling medium in each match. In addition, $Th_{i,k}$ is the temperature of each hot stream in each match, $Tcin_k$ is the inlet temperature for the cooling medium in each stage, $Tcout_{i,k}$ is the outlet temperature for the cooling medium in each match and CP_{cu} is the heat capacity for the cooling medium. In previous equations, CP_{cu} is a parameter known prior to the optimization process.

Mass and energy balances are required to calculate the inlet flow rate and temperature for the cooling water to each stage. For the kth stage of the cooler network, the mass balance in the mixer is given by the sum of the water flow rates that are required by the matches in that stage ($\sum_{i\in HP} F_{i,k}$) and the supply cold-water flow rate to the hotter adjacent stage $k+1$ (FF_{k+1}) minus the sum of the bypass water flow rate (FO_k) running from the splitter of that stage to the cooler network outlet and the flow rates of cooling water streams that are required by the matches of the hotter adjacent stage $k+1$ ($\sum_{i\in HP} F_{i,k+1}$),

$$\sum_{i\in HP} F_{i,k} + FF_{k+1} = FO_k + \sum_{i\in HP} F_{i,k+1}, \quad k \in ST-1 \tag{5}$$

To calculate the temperature of the bypass cooling-water stream of each stage (TO_k), the following heat balance in the splitters is required,

$$\left(\sum_{i\in HP} F_{i,k}\right) TO_k = \sum_{i\in HP} \left(F_{i,k} Tcout_{i,k}\right), \quad k \in ST-1 \tag{6}$$

Note that using the heat balance given in equation (4) is possible to know the inlet cooling medium temperature ($Tcin_{k\neq1}$) to each stage, except to the first one (i.e., for $k = 1$). And for the first stage, the cooling medium temperature is given by the cooling tower,

$$\left(\sum_{i\in HP} F_{i,k}\right) TO_k + FF_{k+1} TCU_{in} = FO_k TO_k + \sum_{i\in HP} \left(F_{i,k+1} Tcin_{k+1}\right), \quad k \in ST-1 \tag{7}$$

The above set of equations is necessary for $NOK-1$ stages.

The following mass balance must be included for the first stage to determine the flowrate of the cold water provided by the cooling tower network to the cooler network, considering that only cold water is used in the first stage.

$$FF_k = \sum_{i\in HP} F_{i,k}, \quad k = 1 \tag{8}$$

The inlet temperatures of the hot process streams define the last location for the superstructure. In other words, the inlet temperature of the hot process stream i is the temperature of such stream in the hot end of the cooling network ($Th_{i,NOK+1}$),

$$THIN_i = Th_{i,NOK+1}, \quad i \in HP \tag{9}$$

The outlet temperatures of hot process streams give the first location for the superstructure. Therefore, the outlet temperature of the hot process stream i is the temperature of such stream in the cold end of the cooling network ($Th_{i,1}$),

$$THOUT_i = Th_{i,1}, \quad i \in HP \tag{10}$$

In addition, the inlet temperature for the cooling medium ($TCUIN$) represents the inlet temperature at the first stage ($Tcin_1$), considering that the outlet temperature in each match is restricted by an upper limit ($\Omega^{max}_{Tcout_{i,k}}$) to avoid operational problems.

$$TCUIN = Tcin_1 \tag{11}$$

$$Tcout_{i,k} < \Omega^{max}_{Tcout_{i,k}}, \quad i \in HP, k \in ST \tag{12}$$

To ensure a monotonically decrement for the temperatures through the stages of the superstructure, the next constraints are included. It is necessary to specify that the temperature of each hot process stream in the stage k must be lower or equal than the temperature of each hot process stream in the stage $k+1$,

$$Th_{i,k} \leq Th_{i,k+1}, \quad k \in ST, i \in HP \tag{13}$$

The inlet cooling medium temperature to stage k must be lower or equal than the outlet cooling medium temperature in the match i, k,

$$Tcin_k \leq Tcout_{i,k}, \quad k \in ST, i \in HP \tag{14}$$

The temperature inlet cooling medium to the cooling network must be less or equal than the inlet cooling medium temperature in the stage k,

$$TCU_{in} \leq Tcin_k, \quad k \in ST, \forall k > 1 \tag{15}$$

Finally, the outlet cooling medium temperature in the match i,k should be lower or equal than the outlet cooling medium temperature in the match $i,k+1$,

$$Tcout_{i,k} \leq Tcout_{i,k+1}, \quad k \in ST, i \in HP \tag{16}$$

Logic constraints and binary variables are used to determine the existence of the heat exchangers between the hot process stream i in the stage k with the cooling medium. These constraints are stated as follow,

$$q_{i,k} - \Omega^{max}_{q_{i,k}} z^1_{i,k} \leq 0, \quad i \in HP, k \in ST \tag{17}$$

Here $\Omega^{max}_{q_{i,k}}$ is an upper limit equals to the heat content of the hot process stream i and $z^1_{i,k}$ is a binary variable used to determine the existence of the heat exchangers.

Because the area requirements for each match ($A_{i,k}$) are included in the objective function, the temperature differences should be calculated. The model uses a pair of variables for the temperature difference in the cold side ($dtfri_{i,k}$) and the hot side ($dtcal_{i,k}$) of each match. In addition, binary variables are used to ensure positive temperature differences and greater than a given value of ΔT_{MIN} when a match exists.

$$dtcal_{i,k} \leq Th_{i,k+1} - Tcout_{i,k} + \Gamma_i\left(1 - z^1_{i,k}\right), \quad k \in ST, i \in HP \tag{18}$$

$$dtfri_{i,k} \leq Th_{i,k} - Tcin_k + \Gamma_i\left(1 - z_{i,k}^1\right), \quad k \in ST, i \in HP \tag{19}$$

$$dtcal_{i,k} \geq \Delta T_{MIN}, \quad k \in ST, i \in HP \tag{20}$$

$$dtfri_{i,k} \geq \Delta T_{MIN}, \quad k \in ST, i \in HP \tag{21}$$

where Γ_i is an upper limit for the temperature difference for the hot process stream i. The value of Γ_i is a constant known previous to the optimization and it is given by,

$$\Gamma_i = \max\left(0, THIN_i - TCUIN, -THOUT_i - TCU_{in}, THIN_i - Tcout_{max}, THOUT_i - Tcout_{max}\right) \tag{22}$$

Equations (18) and (19) are written as inequalities because the heat exchanger costs decrease when the temperature differences increase. Note that the use of binary variables allows the feasibility because if the match does not exist, the parameter Γ_i ensures that these restrictions are met. When a heat exchanger for the hot process stream i exists at the stage k, the binary variable $z_{i,k}^1$ is equal to one, then the constraint is applied and the temperature differences are properly calculated.

The flowrates (Fw_j) and temperatures (Tw_j) of the cooling water streams that are directed to the splitters at the inlet of the cooling tower network are determined in the last stage of the cooler network. It is important to note that a problem with NH hot process streams will have $NH+1$ cooling water streams leaving the cooler network, because in addition to the flowrates of the cooling water for each match in the last stage it is generated an overall cooling water stream that results from combining the bypass cooling-water streams of the previous stages. Therefore, the value for the set of cooling water streams directed to the cooling tower network is $NEF=NH+1$.

$$Fw_j = F_{i,k}, \quad k = NOK; j \neq NCP; j = i \tag{23}$$

$$Fw_j = \sum_{k \in ST} FO_k, \quad j = NCP \tag{24}$$

$$Tw_j = Tcout_{i,k}, \quad k = NOK; j \neq NCP; j = i \tag{25}$$

$$Tw_j Fw_j = \sum_{k \in ST} TO_k FO_k, \quad j = NCP \tag{26}$$

The outlet cooling water stream flowrate from the cooling network can be sent to each tower of the cooling tower network ($Fw_{1,j,net}$) and/or directed to the end of the cooling tower network ($Fw_{2,j}$),

$$Fw_j = \sum_{nct \in NCT} Fw_{1,j,nct} + Fw_{2,j}, \quad j \in NEF \tag{27}$$

the inlet water flowrate ($Fw_{in,net}$) and temperature ($Tw_{in,net}$) to the cooling towers are generated by the mix of the portions of cooling medium streams sent to the cooling towers and the flowrate from the cooling tower ($FTT_{net-1,net}$),

$$Fw_{in,nct} = \sum_{j \in NEF} Fw_{1,j,nct} + FTT_{nct-1,nct}, \quad nct \in NCT \tag{28}$$

$$Tw_{in,nct}Fw_{in,nct} = \sum_{j \in NEF} Tw_j Fw_{1,j,nct} + Tw_{out,nct-1}FTT_{nct-1,nct}, \quad nct \in NCT \tag{29}$$

There is a loss of water in the cooling towers by evaporation ($Fw_{ev,nct}$) and the drift of water by the air flowrate ($Fw_{d,nct}$). The water evaporated is obtained from the following relationship:

$$Fw_{ev,nct} = Fa_{nct}\left(w_{out,nct} - w_{in,nct}\right), \quad nct \in NCT \tag{30}$$

while the drift loss of water is 0.2 percent of the inlet water flowrate to the cooling tower (Kemmer, 1988):

$$Fw_{d,nct} = 0.002Fw_{in,nct}, \quad nct \in NCT \tag{31}$$

Thus, for each cooling tower, the outlet water flowrate is given by the following expression:

$$Fw_{out,nct} = Fw_{in,nct} - Fw_{ev,nct} - Fw_{d,nct} \tag{32}$$

and the outlet cooling tower flowrate ($Fw_{out,nct}$) can be split and sent to the next cooling tower and/or to the end of the cooling tower network (Fws_{nct}),

$$Fw_{out,nct} = FTT_{nct,nct+1} + Fws_{nct}, \quad nct \in NCT \tag{33}$$

The flowrate (Fw_{dis}) and temperature (Tw_{dis}) for the end of the cooling tower network are the sum of the bypassed flowrates of the cooling medium streams and the outlet cooling tower flowrates,

$$Fw_{dis} = \sum_{j \in NEF} Fw_{2,j} + \sum_{nct \in NCT} Fws_{nct} \tag{34}$$

$$Tw_{dis}Fw_{dis} = \sum_{j \in NEF} Tw_j Fw_{2,j} + \sum_{nct1 \in NCT} Tw_{out,nct}Fws_{nct} \tag{35}$$

To avoid salts deposition, usually a little blowdown flowrate (Fw_b) is applied over the water flowrate treated in the cooling tower network, which can be determined by,

$$Fw_b = \frac{Fw_r}{N_{CYCLES}} - \sum_{nct \in NCT} Fw_{d,nct} \tag{36}$$

Note that the last term is the total drift loss of water by the air in the cooling tower network. Also, Fw_r is the makeup flowrate and N_{CYCLES} is the number of concentration cycles. Then, the outlet flowrate of the cooling tower network ($Fwctn$) is equal to the flowrate at the end of cooling network minus the blowdown,

$$Fwctn = Fw_{dis} - Fw_b \tag{37}$$

but the cooling medium temperature in the outlet of cooling network (T_{wctn}) is,

$$Twctn = Tw_{dis} \tag{38}$$

To maintain the cooling medium flowrate constant in the cooling system, it is necessary a makeup flowrate to replace the lost water by evaporation, drift and blowdown,

$$Fw_r = \sum_{nct \in NCT} Fw_{ev,nct} + \sum_{nct \in NCT} Fw_{d,nct} + Fw_b \tag{39}$$

Note that the total water evaporated and drift loss of water in the cooling tower network are considered. The flowrate required by the cooling network (FCU_{in}) is determined as follow:

$$FCU_{in} = Fwctn + Fw_r \tag{40}$$

and the inlet cooling medium temperature to the cooling network is obtained from,

$$TCU_{in}FCU_{in} = TwctnFwctn + Tw_rFw_r \tag{41}$$

To avoid mathematical problems, the recycle between cooling towers is not considered; therefore, it is necessary to specify that the recycle in the same cooling tower and from a cooling tower of the stage nct to the cooling tower of stage $nct-1$ is zero,

$$FTT_{nct1,nct} = 0, \quad nct, nct1 \in NCT; nct1 \le nct \tag{42}$$

The following relationships are used to model the design equations for the cooling towers to satisfy the cooling requirements for the cooling network. First, the following disjunction is used to determine the existence of a cooling tower and to apply the corresponding design equations,

$$\begin{bmatrix} z_{nct}^2 \\ \Psi_{nct} \le \Psi_{nct}^{max} \\ \Psi_{nct} \ge \Psi_{nct}^{min} \end{bmatrix} \vee \begin{bmatrix} -z_{nct}^2 \\ \Psi_{nct} = 0 \end{bmatrix}, \quad nct \in NCT$$

Here Z_{NCT}^2 is a Boolean variable used to determine the existence of the cooling towers, Ψ_{nct}^{max} is an upper limit for the variables, Ψ_{nct}^{min} is a lower limit for the variables, Ψ_{nct} is any design variable of the cooling tower like inlet flowrate, mass air flowrate, Merkel number, and others. For example, when inlet flowrate to the cooling tower is used, previous disjunction for the inlet flowrate to the cooling tower is reformulated as follows:

$$Fw_{in,nct} - \Omega_{Fw_{in,nct}}^{max} z_{nct}^1 \le 0, \quad nct \in NCT \tag{43}$$

$$Fw_{in,nct} - \Omega_{Fw_{in,nct}}^{min} z_{nct}^1 \ge 0, \quad nct \in NCT \tag{44}$$

where $\Omega_{Fw_{in,nct}}^{max}$ and $\Omega_{Fw_{in,nct}}^{min}$ are upper and lower limits for the inlet flowrate to the cooling tower, respectively. Notice that this reformulation is applied to each design variable of the cooling towers. The detailed thermal-hydraulic design of cooling towers is modeled with Merkel's method (Merkel, 1926). The required Merkel's number in each cooling tower, Me_{nct}, is calculated using the four-point Chebyshev integration technique (Mohiudding and Kant, 1996),

$$Me_{nct} = 0.25CP_{cu}\left(Tw_{in,nct} - Tw_{out,nct}\right)\sum_{n=1}^{4} 1/\Delta h_{n,nct}; \quad nct \in NCT \tag{45}$$

where n is the temperature-increment index. For each temperature increment, the local enthalpy difference ($\Delta h_{n,nct}$) is calculated as follows

$$\Delta h_{n,nct} = hsa_{n,nct} - ha_{n,nct}, \quad n = 1,...,4; nct \in NCT \tag{46}$$

and the algebraic equations to calculate the enthalpy of bulk air-water vapor mixture and the water temperature corresponding to each Chebyshev point are given by,

$$ha_{n,nct} = ha_{in,nct} + \frac{CP_{cu}Fw_{in,nct}}{Fa_{nct}}\left(Tw_{n,nct} - Tw_{out,nct}\right), \quad n = 1,...,4; nct \in NCT \tag{47}$$

$$Tw_{n,nct} = Tw_{out,nct} + TCH_n\left(Tw_{in,nct} - Tw_{out,nct}\right), \quad n = 1,...,4; nct \in NCT \tag{48}$$

where TCH_n is a constant that represents the Chebyshev points (TCH_1=0.1, TCH_2=0.4, TCH_3=0.6 and TCH_4=0.9). The heat and mass transfer characteristics for a particular type of packing are given by the available Merkel number correlation developed by Kloppers and Kröger (2005):

$$Me_{nct} = c_{1,nct}\left(\frac{Fw_{in,nct}}{A_{fr,nct}}\right)^{c_{2,nct}}\left(\frac{Fa_{nct}}{A_{fr,nct}}\right)^{c_{3,nct}}\left(L_{fi,nct}\right)^{1+c_{4,nct}}\left(Tw_{in,nct}\right)^{c_{5,nct}}, \quad nct \in NCT \tag{49}$$

To calculate the available Merkel number, the following disjunction is used through the Boolean variable Y_{nct}^e:

$$\begin{bmatrix} Y_{nct}^1 \\ (\text{splash fill}) \\ c_{l,nct} = c_{l,nct}^1, \quad l=1,...,5 \end{bmatrix} \vee \begin{bmatrix} Y_{nct}^2 \\ (\text{trickle fill}) \\ c_{l,nct} = c_{l,nct}^2, \quad l=1,...,5 \end{bmatrix} \vee \begin{bmatrix} Y_{nct}^3 \\ (\text{film fill}) \\ c_{l,nct} = c_{l,nct}^3, \quad l=1,...,5 \end{bmatrix}, \quad nct \in NCT$$

Notice that only when the cooling tower ntc exists, its design variables are calculated and only one fill type must be selected; therefore, the sum of the binary variables referred to the different fill types must be equal to the binary variable that determines the existence of the cooling towers. Then, this disjunction can be described with the convex hull reformulation (Vicchietti et al., 2003) by the following set of algebraic equations:

$$y_{nct}^1 + y_{nct}^2 + y_{nct}^3 = z_{nct}^2, \quad nct \in NCT \tag{50}$$

$$c_{l,nct} = c_{l,nct}^1 + c_{l,nct}^2 + c_{l,nct}^3, \quad l=1,...,5; nct \in NCT \tag{51}$$

$$c_{l,nct}^e = b_l^e\, y_{nct}^e, \quad e=1,...,3.; l=1,...,5; nct \in NCT \tag{52}$$

Values for the coefficients b_l^e for the splash, trickle, and film type of fills are given in Table 1 (Kloppers and Kröger, 2005); these values can be used to determine the fill performance. For each type of packing, the loss coefficient correlation can be expressed in the following form (Kloppers and Kröger, 2003):

$$K_{fi,nct} = \left[d_{1,nct}\left(\frac{Fw_{in,nct}}{A_{fr,nct}}\right)^{d_{2,nct}}\left(\frac{Fa_{nct}}{A_{fr,nct}}\right)^{d_{3,nct}} + d_{4,nct}\left(\frac{Fw_{in,nct}}{A_{fr,nct}}\right)^{d_{5,nct}}\left(\frac{Fa_{nct}}{A_{fr,nct}}\right)^{d_{6,nct}} \right]L_{fi,nct}, \quad nct \in NCTa \quad (53)$$

The corresponding disjunction is given by,

$$\begin{bmatrix} Y^1_{nct} \\ (\text{splash fill}) \\ d_{m,nct} = d^1_{m,nct}, m = 1,...,6 \end{bmatrix} \vee \begin{bmatrix} Y^2_{nct} \\ (\text{trickle fill}) \\ d_{m,nct} = d^2_{m,nct}, m = 1,...,6 \end{bmatrix} \vee \begin{bmatrix} Y^3_{nct} \\ (\text{film fill}) \\ d_{m,nct} = d^3_{m,nct}, m = 1,...,6 \end{bmatrix}, \quad nct \in NCT$$

Using the convex hull reformulation (Vicchietti et al., 2003), previous disjunction is modeled as follows:

$$d_{m,nct} = d^1_{m,nct} + d^2_{m,nct} + d^3_{m,nct}, \quad m = 1,...,6; nct \in NCT \quad (54)$$

$$d^e_{m,nct} = c^e_m\, y^e_{nct}, \quad e = 1,...,3; m = 1,...,6; nct \in NCT \quad (55)$$

	b^e_l		
1	e=1 (splash fill)	e=2 (trickle fill)	e=3 (film fill)
	0.249013	1.930306	1.019766
2	-0.464089	-0.568230	-0.432896
3	0.653578	0.641400	0.782744
4	0	-0.352377	-0.292870
5	0	-0.178670	0

Table 1. Constants for transfer coefficients

Values for the coefficients c^e_m for the three fills are given in Table 2 (Kloppers and Kröger, 2003). These values were obtained experimentally and they can be used in the model presented in this chapter. The total pressure drop of the air stream is given by (Serna-González et al., 2010),

$$\Delta P_{t,nct} = 0.8335\frac{Fav^2_{av,nct}}{\rho_{av,nct}A^2_{fr,nct}}\left(K_{fi,nct}L_{fi,nct} + 6.5\right), \quad nct \in NCT \quad (56)$$

where $Fav_{m,nct}$ is the arithmetic mean air-vapor flowrate through the fill in each cooling tower,

$$Fav_{av,nct} = \frac{Fav_{in,nct} + Fav_{out,nct}}{2}; \quad nct \in NCT \quad (57)$$

and $\rho_{av,nct}$ is the harmonic mean density of the moist air through the fill calculated as:

$$\rho_{av,nct} = 1/\left(1/\rho_{in,nct} + 1/\rho_{out,nct}\right), \quad nct \in NCT \tag{58}$$

m	c_m^e		
	e=1 (splash fill)	e=2 (trickle fill)	e=3 (film fill)
1	3.179688	7.047319	3.897830
2	1.083916	0.812454	0.777271
3	-1.965418	-1.143846	-2.114727
4	0.639088	2.677231	15.327472
5	0.684936	0.294827	0.215975
6	0.642767	1.018498	0.079696

Table 2. Constants for loss coefficients

The air-vapor flow at the fill inlet and outlet $Fav_{in,nct}$ and $Fav_{out,nct}$ are calculated as follows:

$$Fav_{in,nct} = Fa_{nct} + w_{in,nct}Fa_{nct}, \quad nct \in NCT \tag{59}$$

$$Fav_{out,nct} = Fa_{nct} + w_{out,nct}Fa_{nct}, \quad nct \in NCT \tag{60}$$

where $w_{in,nct}$ is the humidity (mass fraction) of the inlet air, and $w_{out,nct}$ is the humidity of the outlet air. The required power for the cooling tower fan is given by:

$$PC_{f,nct} = \frac{Fav_{in,nct}\Delta P_{t,nct}}{\rho_{in,nct}\eta_{f,nct}}; \quad nct \in NCT \tag{61}$$

where $\eta_{f,nct}$ is the fan efficiency. The power consumption for the water pump may be expressed as (Leeper, 1981):

$$PC_p = \left(\frac{g}{gc}\right)\left[\frac{FCU_{in}\left(L_{fi,t} + 3.048\right)}{\eta_p}\right] \tag{62}$$

where η_p is the pump efficiency. As can be seen in the equation (62), the power consumption for the water pump depends on the total fill height ($L_{fi,t}$), which depends on the arrangement of the cooling tower network (i.e., parallel ($L_{fi,t,pl}$) or series ($L_{fi,t,s}$));

$$L_{fi,t} = L_{fi,t,pl} + L_{fi,t,s} \tag{63}$$

If the arrangement is in parallel, the total fill height is equal to the fill height of the tallest cooling tower, but if the arrangement is in series, the total fill height is the sum of the cooling towers used in the cooling tower network. This decision can be represented by the next disjunction,

$$\begin{bmatrix} z^3_{nct,nct} \\ FTT_{nct,nct} \leq FTT^{max}_{nct,nct} \\ FTT_{nct,nct} \geq FTT^{min}_{nct,nct} \end{bmatrix} \vee \begin{bmatrix} -z^3_{nct,nct} \\ FTT_{nct,nct} = 0 \end{bmatrix}$$

This last disjunction determines the existence of flowrates between cooling towers. Following disjunction is used to activate the arrangement in series,

$$\begin{bmatrix} z^4_s \\ \sum_{nct}\sum_{nct} z^3_{nct,nct} \geq \Phi^{min}_{\sum_{nct}\sum_{nct} z^3_{nct,nct}} \\ L_{fi,t,s} = \sum_{nct \in NCT} L_{fi,s,nct} \\ \Phi^{min}_{\sum_{nct}\sum_{nct} z^3_{nct,nct}} = 1 \end{bmatrix} \vee \begin{bmatrix} -z^4_s \\ \sum_{nct}\sum_{nct} z^3_{nct,nct} = 0 \\ L_{fi,t,s} = 0 \end{bmatrix}$$

here $\Phi^{min}_{\sum_{nct}\sum_{nct} z^3_{nct,nct}}$ is the minimum number of interconnections between cooling towers when a

series arrangement is used. The reformulation for this disjunction is the following:

$$\sum_{nct}\sum_{nct} z^{3,1}_{nct,nct} \geq \Phi^{min}_{Z_{nct,nct}} z^{4,1}_s \tag{64}$$

$$L_{fi,nct,s} \leq \Omega^{max}_{L_{fi}} z^4_s \tag{65}$$

$$L_{fi,t,s} = \sum_{nct \in NCT} L_{fi,nct,s} \tag{66}$$

If a series arrangement does not exist, then a parallel arrangement is used. In this case, the total fill height is calculated using the next disjunction based on the Boolean variable $Z^{5,nct}_p$, which shows all possible combination to select the biggest fill height from the total possible cooling towers that can be used in the cooling tower network:

$$\begin{bmatrix} Z^{5,1}_{pl} \\ L_{fi,nct=1,pl} \geq L_{fi,nct \neq 1,pl} \\ L_{fi,t,pl} = L_{fi,nct=1,p} \end{bmatrix} \vee \begin{bmatrix} Z^{5,2}_{pl} \\ L_{fi,nct=2,pl} \geq L_{fi,nct \neq 2,pl} \\ L_{fi,t,pl} = L_{fi,nct=2,p} \end{bmatrix} \cdots \vee \begin{bmatrix} Z^{5,LCT}_{pl} \\ L_{fi,nct=LCT,pl} \geq L_{fi,nct \neq LCT,pl} \\ L_{fi,t,pl} = L_{fi,nct=LCT,pl} \end{bmatrix}$$

The reformulation for the disjunction is:

$$z^{5,1}_{pl} + z^{5,2}_{pl} + ... + z^{5,LCT}_{pl} = \left(1 - z^4_s\right) \tag{67}$$

Notice that when z^4_s is activated, then any binary variable $z^{5,nct}_{pl}$ can be activated, but if z^4_s is not activated, only one binary variable $z^{5,nct}_{pl}$ must be activated, and it must represent the tallest fill. The rest of the reformulation is:

$$L_{fi,nct=1,pl} = L^1_{fi,nct=1,pl} + L^2_{fi,nct=1,pl} + ... + L^{LCT}_{fi,nct=1,pl} \tag{68}$$

$$L_{fi,nct=2,pl} = L^1_{fi,nct=2,pl} + L^2_{fi,nct=2,pl} + \dots + L^{LCT}_{fi,nct=2,pl} \tag{69}$$

$$L_{fi,nct=LCT,pl} = L^1_{fi,nct=LCT,pl} + L^2_{fi,nct=LCT,pl} + \dots + L^{LCT}_{fi,nct=LCT,pl} \tag{70}$$

$$L_{fi,t,pl} = L^1_{fi,t,pl} + L^2_{fi,t,pl} + \dots + L^{LCT}_{fi,t,pl} \tag{71}$$

$$
\begin{aligned}
L^1_{fi,nct=1,pl} &\geq L^1_{fi,nct\neq1,pl} \\
L^2_{fi,nct=2,pl} &\geq L^2_{fi,nct\neq2,pl} \\
&\;\;\vdots \\
L^{LCT}_{fi,nct=LCT,p} &\geq L^{LCT}_{fi,nct\neq LCT,pl}
\end{aligned}
\tag{72}
$$

$$
\begin{aligned}
L^1_{fi,t,pl} &= L^1_{fi,nct=1,pl} \\
L^2_{fi,t,pl} &= L^2_{fi,nct=2,pl} \\
&\;\;\vdots \\
L^{LCT}_{fi,t,pl} &= L^{LCT}_{fi,nct=NCT,pl}
\end{aligned}
\tag{73}
$$

$$
\begin{aligned}
L^1_{fi,nct=1,pl} &\leq \Omega^{\max}_{L_{fi,nct}} z^{5,1}_{pl} \\
L^2_{fi,nct=1,pl} &\leq \Omega^{\max}_{L_{fi,nct}} z^{5,2}_{pl} \\
&\;\;\vdots \\
L^{LCT}_{fi,nct=1,pl} &\leq \Omega^{\max}_{L_{fi,nct}} z^{5,LCT}_{pl}
\end{aligned}
\tag{74}
$$

$$
\begin{aligned}
L^1_{fi,nct=2,pl} &\leq \Omega^{\max}_{L_{fi,nct}} z^{5,1}_{pl} \\
L^2_{fi,nct=2,pl} &\leq \Omega^{\max}_{L_{fi,nct}} z^{5,2}_{pl} \\
&\;\;\vdots \\
L^{NCT}_{fi,nct=2,pl} &\leq \Omega^{\max}_{L_{fi,nct}} z^{5,LCT}_{pl}
\end{aligned}
\tag{75}
$$

$$
\begin{aligned}
L^1_{fi,nct=LCT,pl} &\leq \Omega^{\max}_{L_{fi,nct}} z^{5,1}_{pl} \\
L^2_{fi,nct=LCT,p} &\leq \Omega^{\max}_{L_{fi,nct}} z^{5,2}_{pl} \\
&\;\;\vdots \\
L^{LCT}_{fi,nct=LCT,pl} &\leq \Omega^{\max}_{L_{fi,nct}} z^{5,LCT}_{pl}
\end{aligned}
\tag{76}
$$

Finally, an additional equation is necessary to specify the fill height of each cooling tower depending of the type of arrangement,

$$L_{fi,nct} = L_{fi,nct,pl} + L_{fi,nct,s}, \quad nct \in NCT \tag{77}$$

According to the thermodynamic, the outlet water temperature in the cooling tower must be lower than the lowest outlet process stream of the cooling network and greater than the inlet wet bulb temperature; and the inlet water temperature in the cooling tower must be lower than the hottest inlet process stream in the cooling network. Additionally, to avoid the fouling of the pipes, 50°C usually are specified as the maximum limit for the inlet water temperature to the cooling tower (Serna-González et al., 2010),

$$Tw_{out,nct} \geq TWB_{in,nct} + 2.8, \quad nct \in NCT \tag{78}$$

$$Tw_{out,nct} \leq TMPO - \Delta T_{MIN}, \quad nct \in NTC \tag{79}$$

$$Tw_{in,nct} \leq TMPI - \Delta T_{MIN}, \quad nct \in NTC \tag{80}$$

$$Tw_{in,nct} \leq 50°C, \quad nct \in NTC \tag{81}$$

here $TMPO$ is the inlet temperature of the coldest hot process streams, $TMPI$ is the inlet temperature of the hottest hot process stream. The final set of temperature feasibility constraints arises from the fact that the water stream must be cooled and the air stream heated in the cooling towers,

$$Tw_{in,nct} > Tw_{out,nct}, \quad nct \in NTC \tag{82}$$

$$TA_{out,nct} > TA_{in,nct}, \quad nct \in NTC \tag{83}$$

The local driving force (hsa_{nct}-ha_{nct}) must satisfy the following condition at any point in the cooling tower (Serna-González et al., 2010),

$$hsa_{n,nct} - ha_{n,nct} > 0 \quad n = 1,...,4; \; nct \in NTC \tag{84}$$

The maximum and minimum water and air loads in the cooling tower are determined by the range of test data used to develop the correlations for the loss and overall mass transfer coefficients for the fills. The constraints are (Kloppers and Kröger, 2003, 2005),

$$2.90 \leq Fw_{in,nct}/A_{fr,nct} \leq 5.96, \quad nct \in NTC \tag{85}$$

$$1.20 \leq Fa_{nct}/A_{fr,nct} \leq 4.25, \quad nct \in NTC \, . \tag{86}$$

Although a cooling tower can be designed to operate at any feasible $Fw_{in,nct}/Fa_{nct}$ ratio, Singham (1983) suggests the following limits:

$$0.5 \leq Fw_{in,nct}/Fa_{nct} \leq 2.5, \quad nct \in NTC \tag{87}$$

The flowrates of the streams leaving the splitters and the water flowrate to the cooling tower have the following limits:

$$0 \leq Fw_{1,j,nct} \leq Fw_j, \quad j \in NEF; nct \in NCT \tag{88}$$

$$0 \leq Fw_{2,j} \leq Fw_j \qquad j \in NEF \tag{89}$$

The objective function is to minimize the total annual cost of cooling systems (*TACS*) that consists in the total annual cost of cooling network (*TACNC*), the total annual cost of cooling towers (*TACTC*) and the pumping cost (*PWC*),

$$TACS = TACNC + TACTC + PWC \tag{90}$$

$$PWC = H_Y ce PC_p \tag{91}$$

where H_Y is the yearly operating time and ce is the unitary cost of electricity. The total annual cost for the cooling network is formed by the annualized capital cost of heat exchangers (*CAPCNC*) and the cooling medium cost (*OPCNC*).

$$TACNC = CAPCNC + OPCNC \tag{92}$$

where the capital cooling network cost is obtained from the following expression,

$$CAPCNC = K_F \left[\sum_{i \in HP} \sum_{k \in ST} CFHE_i z_{i,k} + \sum_{i \in HP} \sum_{k \in ST} CAHE_i A_{i,k}^\beta \right] \tag{93}$$

Here $CFHE_i$ is the fixed cost for the heat exchanger i, $CAHE_i$ is the cost coefficient for the area of heat exchanger i, K_F is the annualization factor, and β is the exponent for the capital cost function. The area for each match is calculated as follows,

$$A_{i,k} = q_{i,k} / \left(U_i \Delta TML_{i,k} + \delta \right) \tag{94}$$

$$U_i = 1 / \left(1/h_i + 1/h_{cu} \right) \tag{95}$$

where U_i is the overall heat-transfer coefficient, h_i and h_{cu} are the film heat transfer coefficients for hot process streams and cooling medium, respectively. $\Delta TML_{i,k}$ is the mean logarithmic temperature difference in each match and δ is a small parameter (i.e., $1x10^{-6}$) used to avoid divisions by zero. The Chen (1987) approximation is used to estimate $\Delta TML_{i,k}$,

$$\Delta TML_{i,k} = \left[\left(dtcal_{i,k} \right) \left(dtfri_{i,k} \right) \left(\left(dtcal_{i,k} + dtfri_{i,k} \right) / 2 \right) \right]^{1/3} \tag{96}$$

In addition, the operational cost for the cooling network is generated by the makeup flowrate used to replace the lost of water in the cooling towers network,

$$OPCNC = CUw H_Y Fw_r \tag{97}$$

where CUw is the unitary cost for the cooling medium. The total annual cost of cooling towers network involves the investment cost for the cooling towers (*CAPTNC*) as well as the operational cost (*OPTNC*) by the air fan power of the cooling towers. The investment cost for the cooling towers is represented by a nonlinear fixed charge expression of the form (Kintner-Meyer and Emery, 1995):

$$CAPTNC = K_F \sum_{nct \in NCT} \left[C_{CTF} z_{nct}^2 + CCTV_{nct} A_{fr,nct} L_{fi,nct} + C_{CTMA} Fa_{nct} \right] \tag{98}$$

where C_{CTF} is the fixed charge associated with the cooling towers, $CCTV_{net}$ is the incremental cooling towers cost based on the tower fill volume, and C_{CTMA} is the incremental cooling towers cost based on air mass flowrate. The cost coefficient $CCTV_{net}$ depends on the type of packing. To implement the discrete choice for the type of packing, the Boolean variable Y_{nct}^e is used as part of the following disjunction,

$$\begin{bmatrix} Y_{nct}^1 \\ (\text{splash fill}) \\ CCTV_{nct} = CCTV_{nct}^1 \end{bmatrix} \vee \begin{bmatrix} Y_{nct}^2 \\ (\text{trickle fill}) \\ CCTV_{nct} = CCTV_{nct}^2 \end{bmatrix} \vee \begin{bmatrix} Y_{nct}^3 \\ (\text{film fill}) \\ CCTV_{nct} = CCTV_{nct}^3 \end{bmatrix}$$

This disjunction is algebraically reformulated as:

$$CCTV_{nct} = CCTV_{nct}^1 + CCTV_{nct}^2 + CCTV_{nct}^3, \quad nct \in NCT \tag{99}$$

$$CCTV_{nct}^e = a^e y_{nct}^e, \quad e = 1,...,3, nct \in NCT \tag{100}$$

where the parameters a^e are 2,006.6, 1,812.25 and 1,606.15 for the splash, trickle, and film types of fill, respectively. Note that the investment cost expression properly reflects the influence of the type of packing, the air mass flowrate (Fa_{net}) and basic geometric parameters, such as height ($L_{fi,nct}$) and area ($A_{fi,nct}$) for each tower packing. The electricity cost needed to operate the air fan and the water pump of the cooling tower is calculated using the following expression:

$$OPTNC = H_Y ce \sum_{nct=1} PC_{f,nct} \tag{101}$$

This section shows the physical properties that appear in the proposed model, and the property correlations used are the following. For the enthalpy of the air entering the tower (Serna-González et al., 2010):

$$ha_{in} = -6.4 + 0.86582 * TWB_{in} + 15.7154 \exp\left(0.0544 * TWB_{in}\right) \tag{102}$$

For the enthalpy of saturated air-water vapor mixtures (Serna-González et al., 2010):

$$hsa_i = -6.3889 + 0.86582 * Tw_i + 15.7154 \exp\left(0.054398 * Tw_i\right), \quad i = 1,...4 \tag{103}$$

For the mass-fraction humidity of the air stream at the tower inlet (Kröger, 2004):

$$w_{in} = \left(\frac{2501.6 - 2.3263\left(TWB_{in}\right)}{2501.6 + 1.8577\left(TA_{in}\right) - 4.184\left(TWB_{in}\right)} \right) \left(\frac{0.62509\left(PV_{WB,in}\right)}{P_t - 1.005\left(PV_{WB,in}\right)} \right) -$$
$$\left(\frac{1.00416\left(TA_{in} - TWB_{in}\right)}{2501.6 + 1.8577\left(TA_{in}\right) - 4.184\left(TWB_{in}\right)} \right) \tag{104}$$

where $PV_{WB,in}$ is calculated from Equation (115) and evaluated at $T = TWB_{in}$. For the mass-fraction humidity of the saturated air stream at the cooling tower exit (Kröger, 2004):

$$w_{out} = \frac{0.62509 PV_{out}}{P_t - 1.005 PV_{out}} \qquad (105)$$

where PV_{out} is the vapor pressure estimated with Equation (115) evaluated at $T = TA_{out}$, and P_t is the total pressure in Pa. Equation (115) was proposed by Hyland and Wexler (1983) and is valid in the range of temperature of 273.15 K to 473.15 K,

$$\ln(PV) = \sum_{n=-1}^{3} c_n T^n + 6.5459673 \ln(T) \qquad (106)$$

PV is the vapor pressure in Pa, T is the absolute temperature in Kelvin, and the constants have the following values: $c_{-1} = 5.8002206 \times 10^3$, $c_0 = 1.3914993$, $c_1 = -4.8640239 \times 10^{-3}$, $c_2 = 4.1764768 \times 10^{-5}$ and $c_3 = -1.4452093 \times 10^{-7}$. For the outlet air temperature, Serna-González et al. (2010) proposed:

$$hsa_{out} + 6.38887667 - 0.86581791 * TA_{out} - 15.7153617 \exp\left(0.05439778 * TA_{out}\right) = 0 \qquad (107)$$

For the density of the air-water mixture (Serna-González et al., 2010):

$$\rho = \frac{P_t}{287.08\ T}\left[1 - \frac{w}{w + 0.62198}\right]\left[1 + w\right] \qquad (108)$$

where P_t and T are expressed in Pa and K, respectively. The density of the inlet and outlet air are calculated from the last equation evaluated in $T = TA_{in}$ and $T = TA_{out}$ for $w = w_{in}$ and $w = w_{out}$, respectively.

3. Results

Two examples are used to show the application of the proposed model. The first example involves three hot process streams and the second example involves five hot process streams. The data of these examples are presented in Table 3. In addition, the value of parameters ce, H_Y, K_F, N_{CYCLES}, η_f, η_p ,P_t, C_{CTF}, C_{CTMA}, CU_w, CP_{cu}, β, $CFHE$, $CAHE$ are 0.076 $US/kWh, 8000 hr/year, 0.2983 year^{-1}, 4, 0.75, 0.6, 101325 Pa, 31185 $US, 1097.5 $US/(kg dry air/s), 1.5449x10^{-5} $US/kg water, 4.193 kJ/kg°C, 1, 1000$US, 700$US/m^2, respectively. For the Example 1, fresh water at 10 °C is available, while the fresh water is at 15°C for the Example 2.

For the Example 1, the optimal configuration given in Figure 3 shows a parallel arrangement for the cooling water network. Notice that one exchanger for each hot process stream is required. In addition, only one cooling tower was selected; consequently, the cooling tower network has a centralized system for cooling the hot process streams. The selected packing is the film type, and the lost water is 13.35 kg/s due to the evaporation lost (75%), and the drift and blowdown water (4.89% and 20.11%), while a 70.35% of the total power consumption is used by the fan and the rest is used by the pump (29.64%). The two above terms represent the total operation cost of the cooling system; therefore, both the evaporated water and the power fan are the main components for the cost in this example. Notice that the water flowrate in the cooling network is 326.508 kg/s, but the reposition water only is 13.25 kg/s, which represents a save of freshwater of 95.94% respect to the case when is not used a cooling tower for thermal treatment of the cooling medium. The total annual cost is 468,719.906$US/year. The contribution to total annual cost for the cost of cooling network is 66%, while for cooling tower network and the pump are 31% and 2.96%, respectively. These results are given in the Table 4.

Respect to the Example 2, Figure 4 presents the optimal configuration, which shows a parallel arrangement to the cooling water network, while the cooling towers network is formed by a distributed system composed by two cooling towers to treat the effluents from the cooling network and to meet the cooling requirements. The selected fill is the film type, and the lost water by evaporation, drift and blowdown represent a 74.99%, 3.94% and 21.07% of the total water lost, respectively. Respect the total power consumption in the cooling system, the fan demands a 65.37% and the pump use a 34.62% of the total cost. The economical results are given in the Table 4. The optimal cooling system shows costs for the cooling network, cooling tower and water pump equal to 61.21%, 36.34% and 2.44%, respectively, of the total annual cost. In addition, for the case that only one cooling tower is selected, the total annual cost is 143,4326.66$US/year, which is 7% more expensive than the optimal configuration. The savings obtained are because the distributed system is able to find a better relationship between the capital cost and the operation cost, which depends of the range, inlet water flowrate and inlet air flowrate to the cooling tower network; therefore,

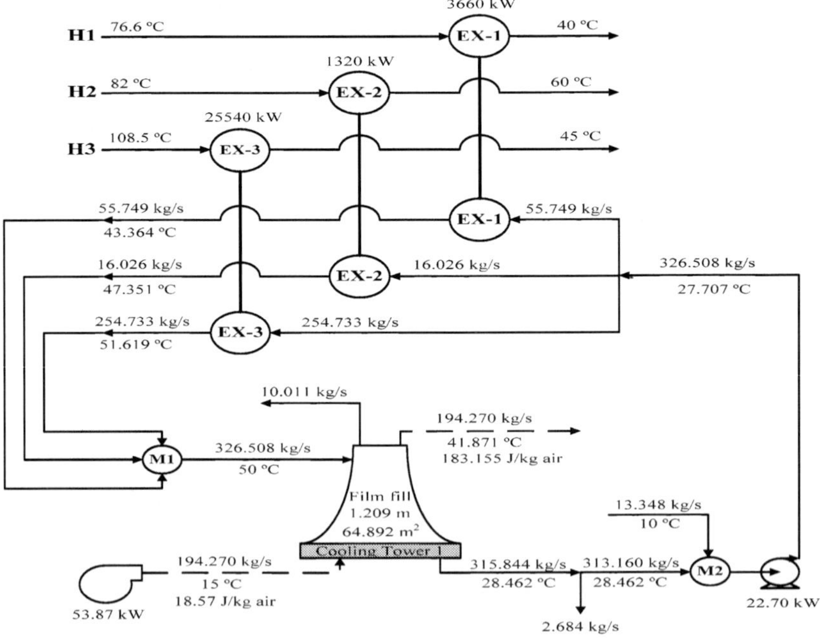

Fig. 3. Optimal configuration for the Example 1

in the distributed systems there are more options. In this case, the use of freshwater by the cooling network is reduced by 94.92% with the use of the cooling towers. Other advantage of use a distributed system is that depending of the problem data just one cooling tower could not meet with the operational and/or thermodynamic constraints and could be necessary to use more than one cooling tower.

Example 1					
Streams	THIN (°C)	THOUT (°C)	FCP (kW/°C)	Q (kW)	h (kW/m²°C)
1	40	76.6	100	3660	1.089
2	60	82	60	1320	0.845
3	45	108.5	400	25540	0.903

Example 2					
Streams	THIN (°C)	THOUT (°C)	FCP (kW/°C)	Q (kW)	h (kW/m²°C)
1	80	60	500	10000	1.089
2	75	28	100	4700	0.845
3	120	40	450	36000	0.903
4	90	45	300	13500	1.025
5	110	40	250	17500	0.75

Table 3. Data for examples

	Example 1	Example 2
TACS (US$/year)	468,719.906	1,334,977.470
TACNC (US$/year)	309,507.229	817,192.890
TACTC (US$/year)	145,336.898	485,196.940
OPCNC (US$/year)	6,131.013	16,588.250
OPTNC (US$/year)	32,958.635	61,598.140
PWC (US$/year)	13,875.780	32,587.640
CAPCNC (US$/year)	303,376.216	800,604.640
CAPTNC (US$/year)	112,378.262	181,000.330

Table 4. Results for examples

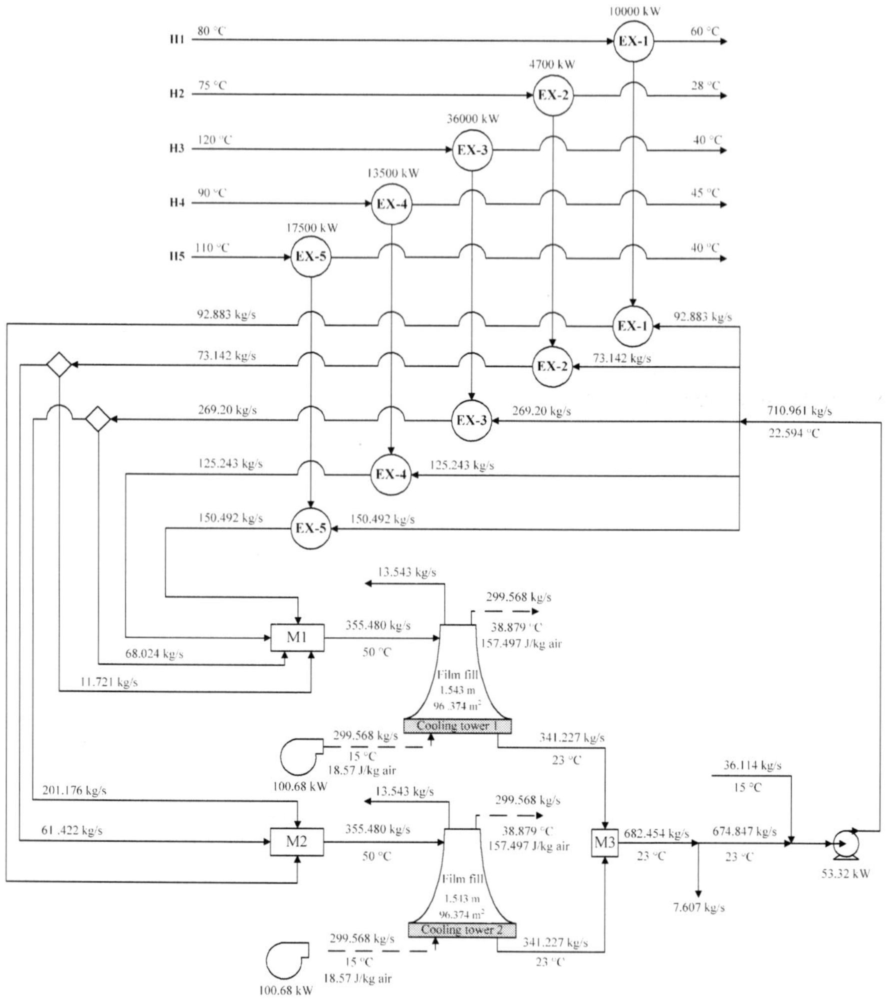

Fig. 4. Optimal configuration for the Example 2

4. Conclusion

This chapter presents a new model for the detailed optimal design of re-circulating cooling water systems. The proposed formulation gives the system configuration with the minimum total annual cost. The model is based on a superstructure that considers simultaneously series and parallel arrangements for the cooling water network and cooling tower network, in which the cooling medium can be thermally treated using a distributed system. Significant savings were obtained with the distributed cooling systems for the interconnection between cooling water network and cooling towers. Evaporation represents the main component for the lost of water (70-75%); while the drift and blowdown represent the 3-5% and 20-25%, respectively. The fan power consumption usually represents the 65-70% of the total power consumption in the cooling system; and the pump represents around the 30-35%. For re-circulating cooling water systems the costs of cooling network, cooling tower network and the water pump represent the 60-70%, 30-40% and 2-5% of the total cooling system cost, respectively. When re-circulating cooling water systems are used, the use of freshwater in the cooling network is significantly reduced (i.e., 95%).

5. References

Chen, J.J. (1987). Letter to the Editors: Comments on improvement on a replacement for the logarithmic mean. *Chemical Engineering Science*, Vol.42, pp. 2488-2489.

Feng, X.; Shen, R.J. & Wang, B. (2005). Recirculating cooling-water network with an intermediate cooling-water main. *Energy & Fuels*, Vol.19, No.4, pp. 1723-1728.

Hyland, R.W. & Wexler, A. (1983). Formulation for the thermodynamic properties of the saturated phases of H_2O from 173.15 K and 473.15 K. *ASHRAE Transactions*, Vol.89,No.2A, pp. 500-519.

Kemmer, F.N. (1988). *The NALCO Water Handbook*, McGraw-Hill, New York, USA.

Kim, J.K. & Smith, R. (2001). Cooling water system design. *Chemical Engineering Science*, Vol.56, No.12, pp. 3641-3658.

Kintner-Meyer, M. & Emery, A.F. (1995). Cost-optimal design for cooling towers. *ASHRAE Journal*, Vol.37, No.4, pp. 46-55.

Kloppers, J.C. & Kröger, D.G. (2003). Loss coefficient correlation for wet-cooling tower fills, *Applied Thermal Engineering*, Vol.23, No.17, pp. 2201-2211.

Kloppers, J.C. & Kröger, D.G. (2005). Refinement of the transfer characteristic correlation of wet-cooling tower fills. *Heat Transfer Engineering*, Vol.26, No.4, pp. 35-41.

Kröger, D.G. (2004). *Air-Cooled Heat Exchangers and Cooling Towers*, PenWell Corp., Tulsa, USA.

Leeper, S.A. (1981). Wet cooling towers: rule-of-thumb design and simulation. *Report, U.S. Department of Energy*, 1981.

Merkel, F. (1926).Verdunstungskuhlung, *VDI Zeitchriff Deustscher Ingenieure*, Vol.70, pp. 123-128.

Mohiudding, A.K.M. & Kant, K. (1996). Knowledge base for the systematic design of wet cooling towers. Part I: Selection and tower characteristics. *International Journal of Refrigeration*, Vol.19, No.1, pp. 43-51.

Ponce-Ortega, J.M.; Serna-González, M. & Jiménez-Gutiérrez, A. (2007). MINLP synthesis of optimal cooling networks. *Chemical Engineering Science*, Vol.62, No.21, pp. 5728-5735.

Serna-González, M.; Ponce-Ortega, J.M. & Jiménez-Gutiérrez, A. (2010). MINLP optimization of mechanical draft counter flow wet-cooling towers. *Chemical Engineering Research and Design*, Vol.88, No.5-6, pp. 614-625.

Singham, J.R. (1983). *Heat Exchanger Design Handbook*, Hemisphere Publishing Corporation, New York, USA.

Söylemez, M.S. (2001). On the optimum sizing of cooling towers. *Energy Conversion and Management*, Vol.42, No.7, pp. 783-789.

Söylemez, M.S. (2004). On the optimum performance of forced draft counter flow cooling towers. *Energy Conversion and Management*, Vol.45, No.15-16, pp. 2335-2341.

Vicchietti, A., Lee, S. & Grossmann, I.E. (2003). Modeling of discrete/continuous optimization problems characterization and formulation of disjunctions and their relaxations. *Computers and Chemical Engineering*, Vol.27, No.3, pp. 433-448.

A New Supercapacitor Design Methodology for Light Transportation Systems Saving

Diego Iannuzzi and Davide Lauria
University of Naples Federico II, Electrical Engineering Department
Italy

1. Introduction

Light transportation systems are not new proposals since their utilization started at dawn of electric energy spreading at industrial level. The light transportation systems class includes tramways, urban and subway metro-systems as well as trolley buses. These systems are intrinsically characterized by low investment costs and high quality service in terms of environmental impact and energy efficiency. The vehicle technology, which is continuously improving, allows lighter and economical solutions.

These attractive characteristics has generated a renewed interest of the researchers and transportation companies in trying to obtain better performances of these systems, being foreseeable their remarkable spreading in the next future. However, even though the perspectives appear to be particularly bright, the increase of the electrical power demand associated to the higher number of vehicles circulating at the same time, require to investigate new solutions for optimizing the whole transportation system performance and particularly the energy consumption, also by exploiting the possibility tendered by the advent of technology innovation. The improvement of the energy efficiency is a crucial issue, both in planning stage and during operating conditions, which cannot be deferred. It has to be highlighted that also for already existing light transportation systems, the energy saving can be pursued by integrating new technological components or apparatuses as energy storage systems (Chymera et al,2006), which allow contemporaneously to obtain high energy saving and to reduce the of load peaks requested to the supply system. More specifically, the storage devices employment, which may be on board (Steiner et al,2007), or located at both the substations or along the track (Barrero et al,2008), are attractive means for obtaining contemporaneously energy saving, energy efficiency, pantograph voltage stabilization and peak regularization (Hase et al,2003). Storage devices may play also a fundamental role in enhancing the dynamic response of the overall light transit system, if they are used in combination with properly controlled power converters. All these previously mentioned benefits surely are convincing arguments for persuading to upgrade the existing light transportation systems, all the more that capital costs exhibit very reduced payback periods.

The high energy densities make the supercapacitors attractive means for real time energy optimization, voltage regulation and high reduction of peak powers requested to feeding substations during the acceleration and braking phases. Many solutions have been suggested in the relevant literature, both oriented to the employment of distributed supercapacitors stationary stations along contact line sections (Konishi,2004) and to the use of onboard storage devices (Iannuzzi,2008). The supercapacitors exhibit energy densities (6 Wh/kg) lower than those of batteries and flywheels but higher power densities (6 kW/kg), with discharge times ranging from ten of seconds to minutes (Conway,1999). This characteristic suggests their utilization for supplying power peaks, for energy recovery and for compensating quickly voltage drop.

The design procedure and management strategy of these innovative systems are often defined on the basis of specific case studies and realized prototype (Hase et al,2002). High difficulties are related to modeling aspects, since the time varying nature of the light transportation system has to be properly performed, this affecting dramatically the identification of a rational procedure for choosing the fundamental characteristics of the storage device.

In the paper, the supercapacitor design problem for light transportation systems saving is handled in terms of isoperimetric problem. Some analytical results can be obtained only with respect to simple case studies, even if they are very interesting because their analyses permit to capture the relationships between fundamental storage device parameters and the transportation ones. For more complex cases, it results quite impossible to have analytical closed solutions. In any case, the design problem can be addressed to a general constrained multiobjective optimization problem which without restrictions is able to handle all the interest cases for deriving the energy management strategies. The optimization procedure results particularly useful for sensitivity analyses, which could be requested also for identifying the optimal allocation and configuration, taking properly into account the timetable. The paper is organized as follows. First of all the fundamental characteristics of supercapacitor devices are described in section II. Some preliminary consideration with respect to optimization methodologies are summarized in section III. Hence the light transportation systems modeling, indispensable for applying the optimization procedure, is derived in section IV, with reference both to the case of the application of stationary storage systems and to the on-board one. The choice of the objective function of the constrained optimization problem, over a prefixed time horizon, is deeply investigated. A numerical application is reported in section V for a case study with two trains along double track dc electrified subway networks, both for stationary and on-board application. The numerical results demonstrate the feasibility and the validity of the proposed systemic design methodology.

The authors will try to tempt in future works to extend the proposed procedure, conceived for the planning stage, to real time control strategy.

2. Electrical energy storage system based on supercapacitors device

The Energy Storage System includes the storage unit, i.e. the modules of supercapacitors, a DC-DC, switching power converter whose control system acts in order to exchange regulated power flows between the storage device and the electrical network. The storage unit is realized connecting together several modules of supercapacitors in series and/or in parallel in order to attain the values of voltage and current required for the specific application.

2.1 Supercapacitors devices

Various papers discuss the physical construction of the double-layer capacitor (DCL). The DLC consists of activated carbon particles that act as polarizable electrodes. These particles, strongly packed, are immersed in an electrolytic solution, forming a double-layer charge distribution along the contact surface between carbon and electrolyte. The physics of the double-layer charge distribution is discussed in (Kitahara et al.,1984) and (R. Morrison, 1990). Three major aspects of the physics of the double-layer charge distribution affect the structure of the equivalent circuit model, as summarized in the following. Firstly, by taking into account the electrochemistry of the interface between two materials in different phases, the double-layer charge distribution of differential sections of the interface is modeled as RC circuit. The resistive element represents the resistivity of the materials constituting the double-layer charge distribution. The capacitive element represents the capacitance between the two materials. As far as the second aspect is concerned, based on the *theory of the interfacial tension in the double-layer*, the capacitance of the double-layer charge distribution depends on the potential difference across the material.

DLC's measurements highlight the same non linear relationship between capacitance and terminal voltage in the device. Furthermore, the measurements put in evidence that, in the interest voltage range of the device, the DLC capacitance varies linearly as function of the capacitor terminal voltage.

By taking into account on the physical aspects and on the basis of the both previously mentioned considerations and the requirement of a practical engineering model, the equivalent circuit can be obtained by employing:

1.   RC circuits, by keeping the number of RC elements as low as possible for practical reasons;
2.   a non linear capacitance to be included only in one RC element;
3.   a parallel leakage resistor.

In order to avoid an arbitrary modeling, a proper choice of the RC circuits number of the equivalent circuit model, depending on the time span of the transient response, is required.

Extensive experiences resulting from measurements have oriented to propose a circuit model exhibiting three RC branches characterized by different time constants, covering the interest time horizon. This choice corresponds to the least number for obtaining a satisfactory degree of accuracy over a time horizon nearly equal to 30 min. The different time constants allow to capture the significant dynamics of the supercapacitor device. The first branch, including the voltage-dependent capacitor (in F/V), dominates the initial time behavior of the DLC, in the time window of seconds order. The second branch, named delayed branch, refers to the slower dynamics in the time window of minutes order. Finally, the third one or long-term branch determines the behavior of time windows longer than 10 min.

For taking into account the voltage dependence of the capacitance, the first branch is modeled as a voltage-dependent differential capacitor. The differential capacitor consists of a fixed capacitance and a voltage-dependent capacitor. A leakage resistor, inserted in parallel to the terminals, is added for representing the self discharge property. The proposed equivalent circuit is shown in Fig. 1.

It has to be highlighted that, however, most of the ultra-capacitor models presented in the literature consider a non-linear (voltage dependent) transmission line or finite ladder RC network (F. Belhachemi et al., 2000), (N. Rizoug et al.,2006). For simplicity of the analysis, the transmission line effect is neglected, and a first order nonlinear model is used (R. Faranda,2007). The internal equivalent resistance R_i is a constant and frequency independent resistance. The ultra-capacitor total capacitance is a voltage-controlled capacitance:

$$C(V_{ci})=C_0+k\,V_{ci};$$

where C_0 is the initial linear capacitance representing electrostatic capacitance and K is a proper coefficient that takes into account the effects of the diffused layer of the supercapacitor *(R. Kotz, et al., 2000)*.

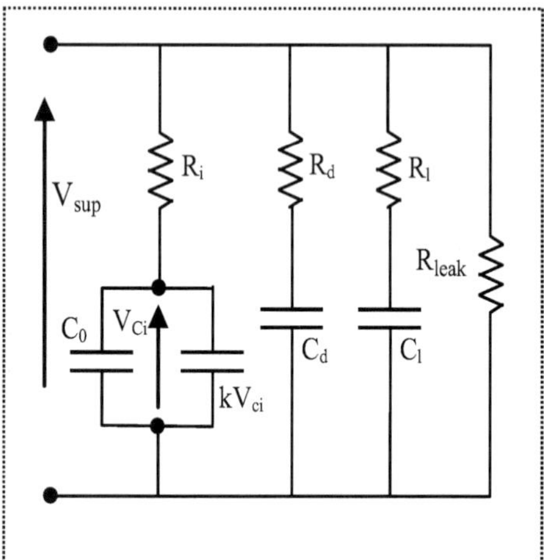

Fig. 1. SC equivalent circuit suggested in (Luis Zubieta et al.,2000)

2.2 Dc-Dc power converter and control systems

The switching power converter, for interfacing the storage system and the electrical network is boost type with bidirectional power flow. The bidirectional property allows the discharge and the recharge of supercapacitors. The converter is connected to the contact line and it is able to regulate the voltage at its output terminals, since input and output behave like voltage sources. The converter may be both current-controlled and voltage-controlled. So the duty-cycle may be evaluated on the basis of the reference output current or voltage of the converter itself.

On the basis of type of control adopted (voltage or current mode control) the whole system of supercapacitors and converter can be modeled as an ideal voltage or current source. In the case of current source, the control system consists of a supercapacitors side current control on the basis of actual value of the supercapacitors current and state of charge of supercapacitors. The set-point of supercapacitors current depends on the energy strategy adopted. For example, in the case of on-board application, the set-point for the supercapacitors charge and discharge is calculated on the base of kinetic energy of the train, thanks to the knowledge of actual value of train speed. In Figg. 2, and 3 simple schematic current and voltage mode control are depicted.

3. Light transportation system and modelling

3.1 Physical system

A light transportation system characterized by double line track, depicted in Fig.4, is investigated. The system represents a large class of actual systems and the analyses performed can be easily generalized to more complex transportation systems.

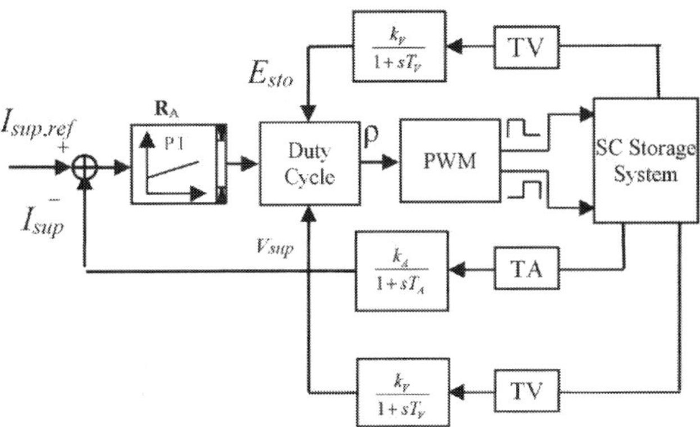

Fig. 2. Supercapacitor current-mode control diagram block

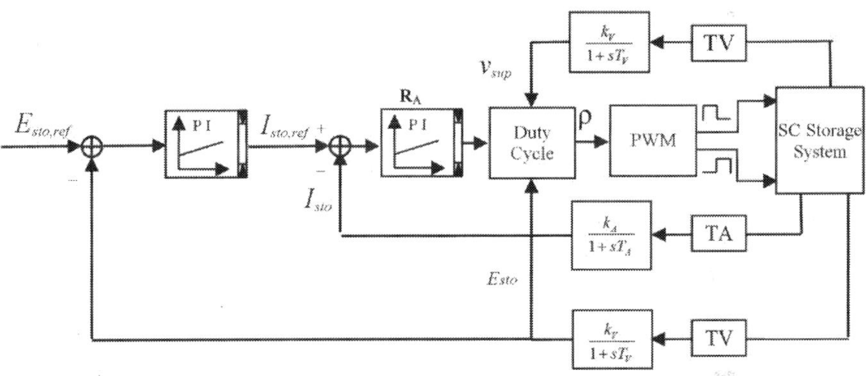

Fig. 3. Storage voltage-mode control diagram block

The overhead contact line consists of one main wire having a section of 120 mm² for each direction of the vehicles, as depicted in Fig. 4. In Tab. I the main parameters of the test system are listed. For simplicity, the simulated track in the paper refers to a branch of 1.5 km with two regular stops and two trains traveling in different direction. The system during operating conditions may be affected by high pantograph voltage drop consequent to the train peak powers, this strongly depending on driving cycles and their diplacement, load dynamic behaviours and network characteristics. The optimal design of storage devices based upon supercapacitors is deeply investigated in the following at the aim of obtaining contemporaneously energy saving, energy efficiency, pantograph voltage stabilization and peak regularization.

Two case studies will be considerd considered: the first one refers to a storage system based upon supercapacitors (SC) employment, located at the end of line and the second with supercapacitors installed onboard.

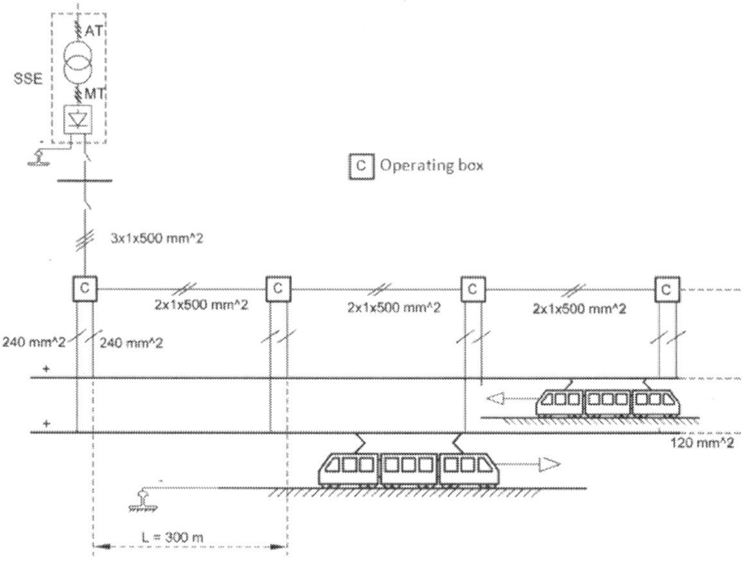

Fig. 4. The light transportation system under study

Parameter	Unit	Quantity
Track Length	[km]	1.5
Contact Wire Resistance (Copper 150 mm²)	[Ω/km]	0.125
Rail resistance	[Ω/km]	0.016
Substation internal Resistance	[mΩ]	20
Rated Voltage	[V]	750
N° Substations	-	1
N° Trains	-	2
Average Train acceleration/deceleration	[m/s²]	0.7/0.9
Maximum Train Power	[kW]	800
Maximum Braking Power	[kW]	400
Train Mass	[T]	60

Table I. Light Transportation system Parameters

3.2 Electrical network modeling with stationary ESS

The equivalent circuit of the traction system and the energy storage system located on end of the line are shown in Fig. 5 in which the subscripts odd and even refer respectively to the traction system parameters (contact wire resistance, track resistance, train currents and pantograph voltages) of both the odd and even tracks. In particular contact wire resistances will vary as a function of trains positions with respect to the feeding substations. The railway electrical system can be considered, broadly speaking, as a distribution system.

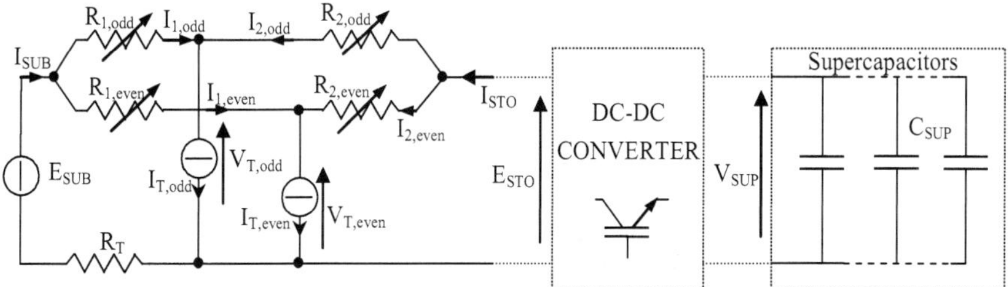

Fig. 5. Equivalent electrical circuit with wayside energy storage system.

In the model, the traction loads are modeled as current sources, I_{Ti} , whose values depend on the powers required by the trains with reference to the track diagram and on the pantograph voltages through the relation at the at k-th time step:

$$I_{Ti}^{(k)} = \frac{P_{Ti}^{(k)}}{V_{Ti}^{(k)}} \qquad k=1, 2,\ldots\ldots,K \tag{1}$$

where K corresponds to the final state.

The discrete mathematical model is expressed in terms of non linear system where the power trains and the substation voltage, at generic instant (k), are known quantities. The unknown quantities are represented by the trains voltage, substation current and storage current and voltage.

$$
\begin{bmatrix} I_{SUB}^{(k)} \\ I_{STO}^{(k)} \\ \dfrac{P_{T,odd}^{(k)}}{V_{T,odd}^{(k)}} \\ \dfrac{P_{T,even}^{(k)}}{V_{T,even}^{(k)}} \end{bmatrix} =
\begin{bmatrix}
\left(\dfrac{1}{R_{1,odd}^{(k)}}+\dfrac{1}{R_{1,even}^{(k)}}\right) & 0 & -\dfrac{1}{R_{1,odd}^{(k)}} & -\dfrac{1}{R_{1,even}^{(k)}} \\
0 & \left(\dfrac{1}{R_{2,odd}^{(k)}}+\dfrac{1}{R_{2,even}^{(k)}}\right) & -\dfrac{1}{R_{2,odd}^{(k)}} & -\dfrac{1}{R_{2,even}^{(k)}} \\
-\dfrac{1}{R_{1,odd}^{(k)}} & -\dfrac{1}{R_{2,odd}^{(k)}} & \left(\dfrac{1}{R_{1,odd}^{(k)}}+\dfrac{1}{R_{2,odd}^{(k)}}\right) & 0 \\
-\dfrac{1}{R_{1,odd}^{(k)}} & -\dfrac{1}{R_{2,odd}^{(k)}} & 0 & \left(\dfrac{1}{R_{1,odd}^{(k)}}+\dfrac{1}{R_{2,odd}^{(k)}}\right)
\end{bmatrix}
\begin{bmatrix} E_{SUB}^{(k)} \\ E_{STO}^{(k)} \\ V_{T,odd}^{(k)} \\ V_{T,even}^{(k)} \end{bmatrix}, \tag{2}
$$

k = 1, 2, ... , K.

4.3 Electrical network modeling with ESS on board

In 2nd case, the equivalent circuit of the traction system and the energy storage systems located on board are shown in Fig.6. The currents absorbed at trains pantograph is sum of the actual trains current and storage currents.

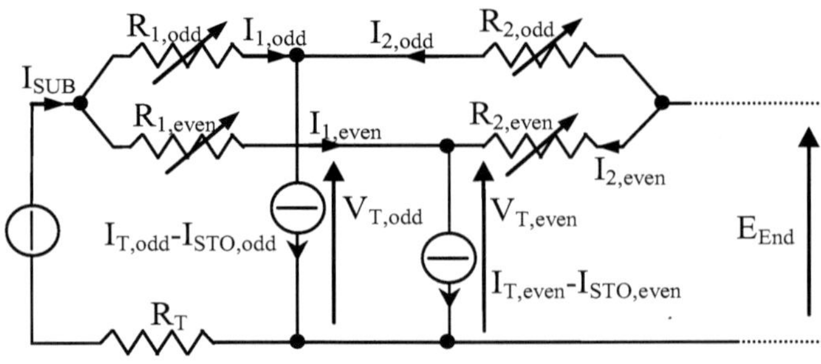

Fig. 6. Equivalent circuit with energy storage systems on board

The following mathematical model holds:

$$
\begin{bmatrix} I_{SUB}^{(k)} \\ 0 \\ \dfrac{P_{T,odd}^{(k)}}{V_{T,odd}^{(k)}} - I_{sto,odd}^{(k)} \\ \dfrac{P_{T,even}^{(k)}}{V_{T,even}^{(k)}} - I_{sto,even}^{(k)} \end{bmatrix} = \begin{bmatrix} \left(\dfrac{1}{R_{1,odd}^{(k)}} + \dfrac{1}{R_{1,even}^{(k)}}\right) & 0 & -\dfrac{1}{R_{1,odd}^{(k)}} & -\dfrac{1}{R_{1,even}^{(k)}} \\ 0 & \left(\dfrac{1}{R_{2,odd}^{(k)}} + \dfrac{1}{R_{2,even}^{(k)}}\right) & -\dfrac{1}{R_{2,odd}^{(k)}} & -\dfrac{1}{R_{2,even}^{(k)}} \\ -\dfrac{1}{R_{1,odd}^{(k)}} & -\dfrac{1}{R_{2,odd}^{(k)}} & \left(\dfrac{1}{R_{1,odd}^{(k)}} + \dfrac{1}{R_{2,odd}^{(k)}}\right) & 0 \\ -\dfrac{1}{R_{1,odd}^{(k)}} & -\dfrac{1}{R_{2,odd}^{(k)}} & 0 & \left(\dfrac{1}{R_{1,odd}^{(k)}} + \dfrac{1}{R_{2,odd}^{(k)}}\right) \end{bmatrix}
$$

$$
\cdot \begin{bmatrix} E_{SUB}^{(k)} \\ E_{End}^{(k)} \\ V_{T,odd}^{(k)} \\ V_{T,even}^{(k)} \end{bmatrix} , \quad for \ \ k = 1, 2, \dots , K. \tag{3}
$$

where the power trains, the substation voltage, at generic instant (k), are known quantities. The unknown quantities are represented by the train voltages, storage currents and end line voltage.

Finally, both systems are completed taking into account the relation between converter and supercapacitors device. In fact, with respect to the boost converter laws, the quasi stationary modeling becomes:

$$
\begin{cases} V_{sup,j}^{(k+1)} - V_{sup,j}^{(k)} + \dfrac{I_{sup,j}^{(k)}}{C_{sup,j}} \Delta t = 0 & k = 1,2,\dots,K\text{-}1; \\[2ex] I_{sup,j}^{(k)} = \dfrac{V_{sup,j}^{(k)} - \sqrt{V_{sup,j}^{(k)2} - 4R_{C,j}E_{sto,j}^{(k)}I_{sto,j}^{(k)}}}{2R_{C,j}} & k = 1,2,\dots,K; \end{cases} \tag{4}
$$

with $j = \{odd, even\}$.

In the 1st case (stationary), the storage system is the same for both tracks (*odd* and *even*); otherwise in the 2nd case the storage systems are different and the terminal voltage on dc side of power converters $E_{sto,j}$ are the same of terminal voltages at trains pantograph $V_{T,j}$.

The above relationship can be easily deduced by the converter power balance. Hence, by neglecting the fast transients, the electrical systems can be described as a sequence of stationary states whose input data are the substation voltages and the train powers for each current position.

4. Optimal design

Some preliminary concepts are briefly summarized in order to better understand the design optimization procedure based upon the formulation of an isoperimetric problem.

A rational way to face with this kind of problem is to make the recourse to classical calculus of variations. Substantially, the objective is to search the functions of extrema of a functional, subject to known side-conditions. In the following, the Euler-Lagrange formalism of the calculus of variations is adopted (Pierre 1986).

Let us consider the problem of identifying the real curve x*(t) which yields the minimum or maximum of the functional:

$$J = \int_{t_a}^{t_b} f(x,\dot{x},t)\,dt,$$

where t_a, t_b, $x(t_a) = c_a$ and $x(t_b) = c_b$ are assigned. Provided that the real-valued function $f(x,\dot{x},t)$ is of class C_2 with respect to all of its argument, in short, a necessary condition is the well-known Euler-Lagrange equation:

$$\frac{d}{dt}(f_{\dot{x}}) - f_x = 0,$$

If a constraint equation of the following kind is imposed:

$$h = \int_{t_a}^{t_b} g(x,\dot{x},t)\,dt,$$

where h is a constant and g a known real-valued function, this equation is usually called isoperimetric condition. The solution x*(t) which yields the minimum or maximum of the functional, while satisfying the isoperimetric constraint, is the one obtained by assuming that x*(t) is a first-variational curve resulting in the minimum or maximum of the functional:

$$J_1 = \int_{t_a}^{t_b} f(x,\dot{x},t) + \lambda\,g(x,\dot{x},t)\,dt,$$

where λ is the Lagrange multiplier.

On the other hand, it is quite impossible to obtain analytical closed solutions for very large and complex systems, especially if the side-conditions are posed in the form of inequalities. However, after a discretization procedure, the optimization problem can be formulated as a nonlinear programming problem, as performed in (Battistelli et al. 2009) at the aim of determining the optimal size of supercapacitor storage systems for transportation systems. In mathematical terms, the constrained optimization problem can be summarized as:

$$\min \phi\left[\mathbf{x}, \mathbf{u}, \mathbf{m}\right]$$
$$\theta\left(\mathbf{x}, \mathbf{u}, \mathbf{m}\right)=0,$$
$$\psi\left(\mathbf{x}, \mathbf{u}, \mathbf{m}\right)\leq 0.$$

where $\mathbf{x}$ is the state variables vector, $\mathbf{u}$ the control variables vector, $\mathbf{m}$ the parameters vector , ϕ is the objective function to minimize and θ, ψ refer to equality and inequality constraints respectively.

The optimal sizing of the energy storage device has to be effected guaranteeing contemporaneously the voltage profile regularization at both train pantographes, the substation current minimization and the supercapacitor size reduction. In the case of a single stationary storage device, this can be pursued by selecting the following objective function ϕ to be minimized:

$$\phi = \int_0^T \left[w_1\left(V_{T,even} - V_{ref}\right)^2 + w_2\left(V_{T,odd} - V_{ref}\right)^2 + w_3 I_{SUB}^2 + w_4 I_{sup}^2 \right] dt \tag{5}$$

where w_1, w_2, w_3 and w_4 are suitable coefficients which are able to weight the previously mentioned requirements, V_{ref} being the rated line voltage. In an analogous way, the proper objective function for on board arrangement can be determined.

The energy storage conservativeness on the whole time cycle can be described by the following isoperimetric condition:

$$\int_0^T V_{sup} I_{sup} dt = 0 \tag{6}$$

The isoperimetric problem is completed by the equality constraints which have been described in 4.3 which substantially take into account the electrical network relationships and the electrical modeling of components.

In (D. Iannuzzi et al., 2011) the authors have provided an analytical solution to this problem for a simple case study, on the assumption that the input of the design procedure are the currents rather than the traction powers, this permitting to obtain a closed form to the optimization problem. In this paper the discretized version of the optimization problem is arranged, providing in this way a numerical solution. The sequential quadratic programming method, which belongs to the class of iterative methods, is employed which solves at each step a quadratic programming problem.

5. Numerical application

In order to verify the validity of the proposed procedure a realistic case with respect to actual operation, a 1.5 km double track line, 750 V nominal voltage, is investigated. A 120 seconds operation has been foreseen with two regular stops. The trains, equipped with regenerative braking, depending on the load dynamic behavior, absorb or generate the corresponding electrical powers. The simulation data are reported in Table I.

The driving cycle used for simulation is based on the observations of the real route measurements. It follows the theoretical directives of accelerating up to 75 km/h with an acceleration of 1 m/s², whenever it is possible. The electrical power required by the vehicle has been deduced by measurement at the pantograph during the travel on a typical track. The data have been post-processed and interpolated. The speed and electrical power cycles are shown in Fig. 7. It is assumed that the two trains are timely shifted of 20 s. Substation no load voltage is assumed to be constant and equal to V_0 = 750 V. The storage system has been located at the end of the line in the first case and then they are located on board.

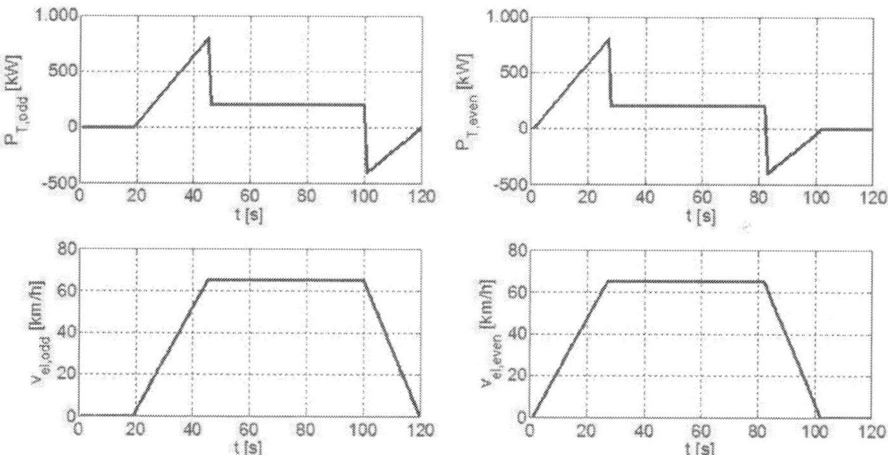

Fig. 7. Traction cycles of the two trains in terms of electrical power at pantograph and vehicles speed

At this purpose, it has to be highlighted that the traction powers has to be regarded as an input data in the optimization procedure, the most convenient vehicle displacement being not investigated.

In order to compare the effectiveness of the storage devices, the reference case, characterized by the absence of storage device, has been simulated. In Fig.8 the total feeding substation current, the odd and even pantograph voltages are depicted. In particular during the acceleration the substation current reaches a peak value of 1.5 kA, the line drop voltage on both tracks can be observed. The odd voltage at pantograph reaches a minimum value of 600 V with a decreasing of 20% of rated value (750V). On contrary, during the breaking time the train electrical powers became negative with consequence inversion of the substation current and increasing of line voltage. In particular the substation current reaches a negative peak of 500 A and an increasing of line voltage referred to even track equal to 7% of rated value.

Successively, two cases are examined for which the proposed optimization procedure is applied. The first one refers to the on-board solution.

The following constraints are imposed:

$$\begin{cases} I_{SUB}^{(k)} \geq 0\,[A], \\ 600\,[V] \leq V_{T,odd}^{(k)} \leq 850\,[V], \\ 600\,[V] \leq V_{T,even}^{(k)} \leq 850\,[V], \qquad k=1,2,\ldots\ldots,K. \\ 550\,[V] \leq E_{SUB}^{(k)} \leq 900\,[V], \\ 300\,[V] \leq V_{sup}^{(k)} \leq 500\,[V], \end{cases}$$

The optimization procedure is performed, by choosing the following weight coefficients: w_1, w_2. The supercapacitor value has been evaluated by imposing a constraint in terms of weight. More specifically the weight of the storage device has been constrained to be less than 2% of the train one.

By following this choice the supercapacitor equivalent capacitance has been resulted equal to 57 [F] for each train. In the Fig.9 the total feeding substation current, the odd and even pantograph voltages.

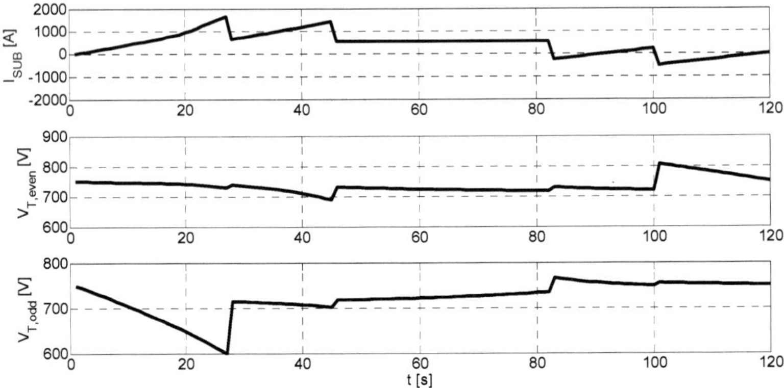

Fig. 8. Substation current and terminal voltages at trains pantograph in the case of absence of energy storage devices.

In this case the substation current diagram is quite flat and it is unidirectional reaching the peak value at 600 A, in fact it can be observed a drop voltages at pantograph about the 6-7% of rated value. This is due to effect of the presence of two supercapacitors devices located on board. The supercapacitors voltages and the storage currents are reported in Fig.10.

The supercapacitors devices, located on trains *odd* and *even*, supply the train during the acceleration giving a peak currents of about 750 A and 900 A respectively. In fact the supercapacitors voltages at its terminal decrease up to 300 V during the acceleration. On the contrary, the electrical energy recovery can be observed during the braking time when the supercapacitors voltages increase up to their rated values (500 V). So it is quite immediate to capture the actions of the two storage systems. The energy saving with respect to the base case is equal to 15,4%.

As far as the second case is concerned, the storage subsystem is placed at the end of a single-side supplied line.

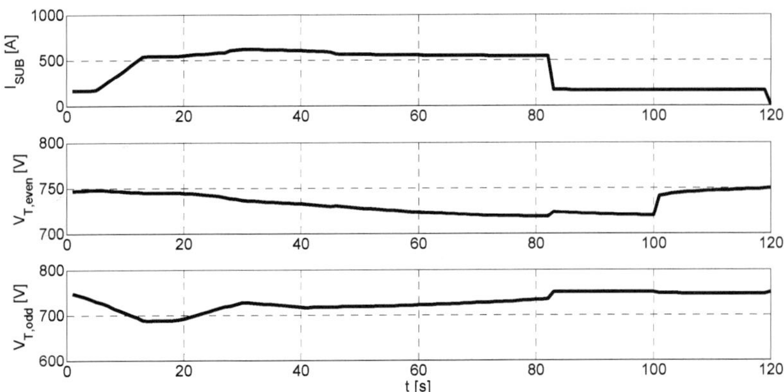

Fig. 9. Substation current and terminal voltages at trains pantograph with the energy storage devices on board.

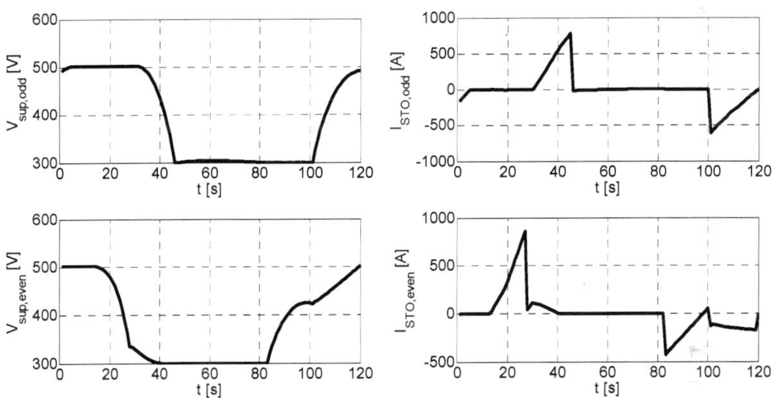

Fig. 10. Supercapacitors voltages and storage currents for each train

The case corresponding to the weight choice $w_1 = w_2 = w_3 = 1$ is reported. This choice was motivated for emphasizing the systemic role played by the storage device which modulates continuously the electric power, in order to contribute both at voltage profile regularization and substation current minimization. Also in this case, it can be observed the quite flat profile of the substation current and the reduced value of the pantographs voltage drop. By following this choice the supercapacitor equivalent capacitance has been resulted equal to 188 [F]. In the Fig.11 the total feeding substation current, the odd and even pantograph voltages.

The supercapacitors voltage and the storage current are reported in Fig.12. It can be observed that in the case of storage device located at the end of line the supercapacitors current profile is very similar to substation current shown in the fig.8. This shows the compensation action of supercapacitors during the different operation conditions of the electrical line.

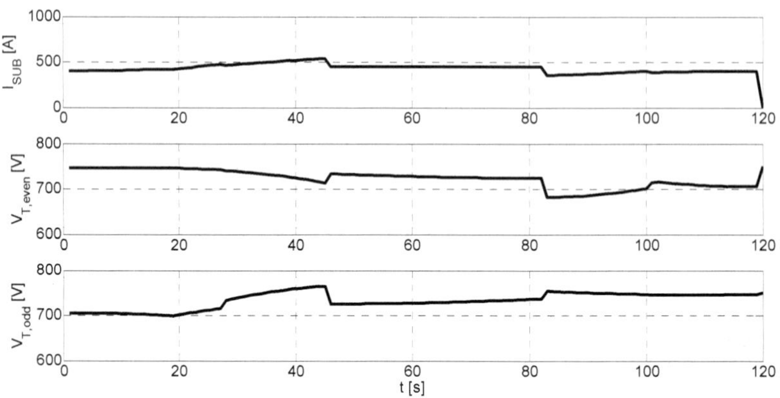

Fig. 11. Substation current and terminal voltages at trains pantograph in the case of energy storage devices located at end of line.

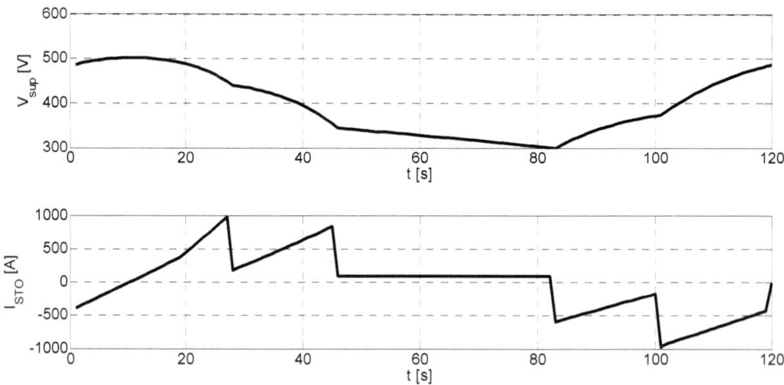

Fig. 12. Supercapacitor voltage and storage current

The energy saving with respect to the base case is equal to *11,6%*.

6. Conclusion

In the paper a new Supercapacitor Design Methodology for Light Transportation Systems Saving has been described. The supercapacitor design has been directed towards the energy efficiency improvement, voltage regulation and high reduction of peak powers requested to feeding substations during the acceleration and braking phases.

More specifically, the supercapacitor design problem for light transportation systems energy saving has been handled in terms of isoperimetric problem. Starting from this point, the problem has been tailored as a constrained multiobjective optimization problem which without restrictions has been proven able to face with all the interest cases.   The optimization procedure has been tested both for both stationary supercapacitors and for on-board arrangement.   The procedure output are the supercapacitor storage size and the supercapacitor reference voltage which can be employed as reference time trajectory to track during operating conditions. A numerical application has been performed for a case study with two trains along double track dc electrified subway networks, both for stationary and on-board configurations. The obtained numerical results allow to confirm the feasibility and the goodness of the proposed optimal design technique.

7. References

Chymera, M., Renfrew A. C., Barnes, M., 2006. *Energy storage devices in railway systems. Seminar on innovation in the railways: evolution or revolution?*, Austin Court, Birmingham, UK .

Rufer, A., Hotellier D., Barrade, P., 2004. *A Supercapacitor-based energy storage substation for voltage compensation in weak transportation networks.* IEEE Trans. on Power Delivery, 19 (2), pp. 629-636.

Barrero, R., Tackoen, X., Van Mierlo, J., 2008. *Improving Energy efficiency in Public Transport: Stationary supercapacitor based energy storage systems for a metro-network.* Proceedings of Vehicle Power and Propulsion Conf. VPPC'08, Harbin, China, Sept. 2008, pp.1-8.

Hase, S., Konishi, T., Okui, A., Nakamichi, Y., Nara,H., Uemura, T., 2002. Fundamental *study on Energy Storage Systems for dc Electric Railway Systems.* Proceedings of Power Conversion Conf. PCC Osaka 2002 , Osaka, Japan, 2002, pp.1456-1459.

Konishi, T., Hase, S., Nakamichi, Y., 2004-5. *Energy Storage System for DC Electrified Railway Using EDLC.* Quarterly Report of RTRI, 45 (2), pp.53-58.

Hase, S., Konishi, T., Okui, A., Nakamichi, Y., Nara,H., Uemura, T., 2003. *Application of Electric Double-layer Capacitors for Energy Storage on Electric Railway.* IEEJ Trans. on Ind. Applicat., 123 (5), pp.517-524.

Iannuzzi, D., 2008. *Improvement of the energy recovery of traction electrical drives using supercapacitors.* 13th Int. Power Electronics and Motion Control Conf. EPE-PEMC, Poznan, Poland, 1-3 Sept., 2008.

Steiner, M., Klohr, M., Pagiela, S.: *"Energy Storage System with UltraCaps on Board of Railway Vehicles"*, Proc.of the 12th European Conf. on Power Electronics and Applications, Aalborg, Denmark , 2-5 Sept., 2007.

Zubieta, L., Bonert, R., 1998. *Characterization of double-layer capacitors (DLCs) for power electronics applications.* IEEE Conf. Ind. Appl., St. Louis, MO, USA, 12-15 Oct., 1998, pp. 1149-1154.

Conway, B. E., 1999. *Electrochemical Supercapacitors: Scientific Fundamentals and Technological Applications.* Plenum Publishers Press, New York.

Battistelli, L., Ciccarelli, F., Lauria, D., Proto, D., 2009. *Optimal design of DC electrified railway Stationary Storage Systems.* 2nd ICCEP Conference, Capri, Italy, June 9-11, 2009, pp. 739-745.

Luis Zubieta, Richard Bonert, 2000. *Characterization of Double-Layer Capacitors for Power Electronics Applications* ,IEEE Transaction on Ind. Applications, Vol. 36, No. 1

Kitahara and A. Watanabe, 1984. *Electrical Phenomena at Interfaces: Fundamentals, measurements and Applications*. New York: Marcel Dekker.

R. Morrison, 1990. *The Chemical Physics of Surfaces*. New York: Plenum.

F. Belhachemi, S. Raiel, and B. Davat, 2000. *"A physical based model of power electric double-layer supercapacitors,"* in Proc. Ind. Appl. Conf., vol. 5, pp. 3069–3076.

R. Farande,M. Gallina, and D. T. Son, 2007 *"A new simplified model of doublelayer capacitors,"* in Proc. ICCEP, pp. 706–710.

N. Rizoug, P. Bartholomeüs, and P. Le Moigne, 2006. *"Modelling of supercapacitors with a characterization during cycling,"* in Proc. PCIM.

R. Kotz and M. Carlen, 2000. *"Principles and applications of electrochemical capacitors,"* Electrochim. Acta, vol. 45, no. 15, pp. 2483–2498.

Pierre AD (1986) *Optimization theory with applications*. Dover Publications, INC., New York.

Iannuzzi D., Lauria D., Tricoli P, Submitted 2011. *Optimal Design of Stationary Supercapacitors Storage Device for Light Electrical Transportation Systems*, Engineering Optimization Journal (Under Review)

10

Management of Locomotive Tractive Energy Resources

Lionginas Liudvinavičius and Leonas Povilas Lingaitis
Vilnius Gediminas Technical University, Faculty of Transport Engineering
Department of railway transport
Lithuania

1. Introduction

The paper addresses some basic theoretical and engineering problems of electrodynamic braking, presenting methods of braking force regulation and using of regenerative braking returning energy (energy saving systems) and diesel engine or any form of hybrid traction vehicles systems, circuit diagrams, electrical parameters curves. Environmental awareness plus reduced operating costs are now major considerations in procuring advanced rail vehicles for considerations in procuring advanced rail vehicles. It is needed to reduce electric demand, to use new energy savings and power supply optimization, hybrid traction vehicles systems, which are using regenerative braking energy. Electric braking is effective on the all speed. Air brake cannot be used. When a vehicle brakes, energy is released to date, most of this energy is being wasted in air. The challenging alternative is to store the braking energy on the train and use it during acceleration of operation of the vehicle. Presenting energy savings power systems, which are using regenerative braking-returning energy and diesel engine or any form of hybrid traction vehicles systems, light vehicles catenary free operation, circuit diagrams, electrical parameters curves (Liudvinavičius L. *New locomotive energy management systems,* 2010; Sen P. C., Principles of Electric Machines, 1996).

2. New elements-supercapacitors of energy accumulation

Companies of electronics created capacitors of big capacity, which are called in different countries as ultra condenser, pseudo condenser, supercapacitors, ultracapacitors. In English literature besides is found the name *Electric Double Layer Capacitors.* The characteristics of

Fig. 1. High-performance double layer technology capacitor (ultra capacitor) picture

supercapacitors are very high. Single module capasities are 3000F, at the tension 2,7V and even more powerful. (P. Barrade, *Series connexion...*, 2001). All this has given an impuls to the various scientific researches. Structure of the supercapacitor is given in Fig.1

Comparative characteristics of the supercapacitors and accumulators are given in the table below:

Performance	Accumulator	Supercapacitor
Energy (Wh/kg)	10 – 100	1 – 10
Number of cycles	1000	> 500 000
Specific power (W/kg)	< 1000	< 10 000

Table 1. Characteristics of accumulator and supercapacitor.

The charge – discharge time of conventional accumulative batteries is very long, because chemical reaction depends on time. The charge – discharge time of supercapacitors (J. D. Boyes..., *Technologies for energy...*, 2000). is only few seconds. In addition, their period of duty is incomparably longer. The authors performed first experiments on purpose to evaluate their technical characteristics in 1997. The diesel engines are used for creating of primary energy, which power is up to 6000kW. JSC *Lithuanian Railways* uses diesel engines, which power is up to 4000 hp. Using conventional systems of starting, from alkaline or acid accumulators, starting of such engines is very complicated because it requires powerful batteries of accumulators. During cold season the starting of such power diesel engines is particularly complicated. If in two or three attempts of starting the diesel engine fails, it is necessary to change the locomotive in line. If starting of diesel engine is not successful, main systems of diesel engine freeze, causing considerable material damage. Starting of high-power diesel engines also is a very complicated in ships. In this case, the consequences even worse than in the railway. The locomotives TEP-60 and TEP-70, which power of diesel engines is up to 4000 hp are used for pulling coaches. The locomotives TEP-60 and TEP-70 are with electrical drive. Conventional 110V X 550Ah accumulative batteries, weight of 3400 kg, are used for starting of diesel engines. The experts of Vilnius Gediminas Technical University and Vilnius locomotive depot have been researching how to extend the life of battery, reduce their weight, improve the conditions of diesel engine starting up. In Russia the supercapacitors were bought, for which evaluation of technical abilities the authors suggested to use them for starting up of the most powerful diesel engine of *Lithuanian Railways*, the locomotive TEP-60 with DC/DC current system. The supercapacitor assembled in a block (in Figure SCB), combining the separate elements sequentially, for the possibility to connect the capacitor to direct current (DC) of 110V voltage network, and parallelly, the total capacity must be enlarged (in Farads). For a fast discharge (charge) cycle of the capacitor, which is calculated by $T = RC$, the authors suggested to charge the supercapacitors from conventional charging equipment of accumulators, existing in locomotive. Fig.2 shows the first (preparatory) phase of diesel engine starting up: the charge of the supercapacitor (R. G. V. Hermann, *High performance...,2001*).

Charged supercapacitors to connect parallelly to accumulative battery (conventional battery of 110V X 550Ah) of much smaller capacity.

The structure of the locomotive TEP-60 electric drive Traction generator is used to start the diesel engine, i.e. is running as a conventional starter. In Fig. 3 the diagram is given, where the generator G, during the starting up is running in mode of direct current (DC) engine. The scheme of starting up of the diesel locomotive TEP-60 diesel engine is given in Fig.4.

Closing the chain of the contactor K, the starting up of the diesel engine is running, feeding from accumulative battery (of 110V X 550Ah) of much smaller capacity and parallelly connected supercapacitors.

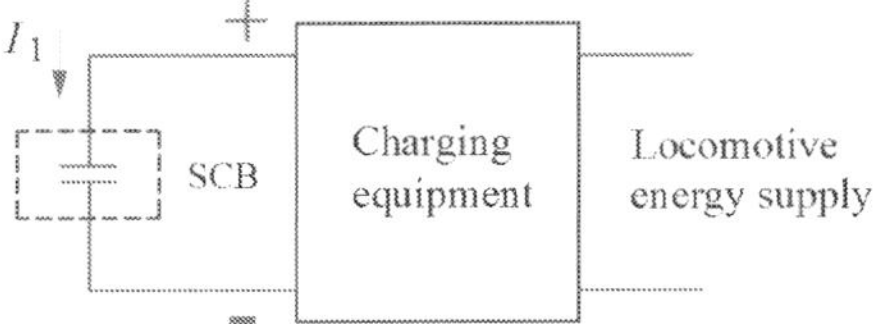

Fig. 2. The charge of the supercapacitors from the energy source of locomotive

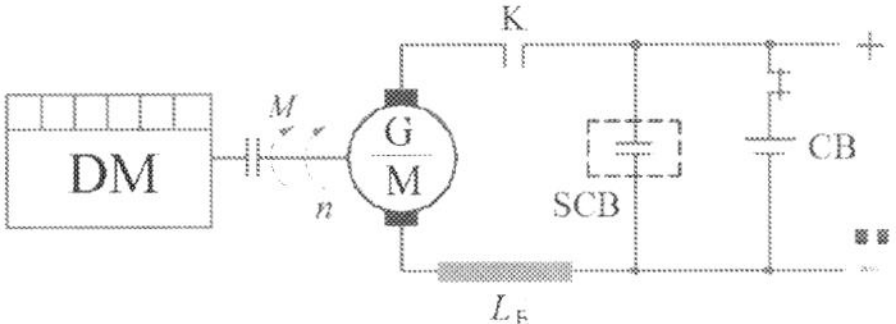

Fig. 3. The scheme of starting up of the diesel locomotive TEP-60 diesel engine:

DM- diesel engine; G/M- DC electric machine (generator or motor mode G/M) CB-conventional battery; SCB-block of supercapacitors; L_E- series existitation winding

3. The results of the research on new energy accumulation elements – using of the supercapacitors in starting up of diesel engines

In Fig.4 the diagram of locomotive TEP-60 diesel engines' starter's running of current accumulators in chain is given, where the diesel engine is starting up from conventional batteries (CB), whose parameters are 110V x 110V 550Ah, without SCB and the diagram 2 of current run, when the diesel engine is started using accumulative batteries of smaller capacity (110V x 160 Ah) and the block in parallel connected supercapacitors. Using the

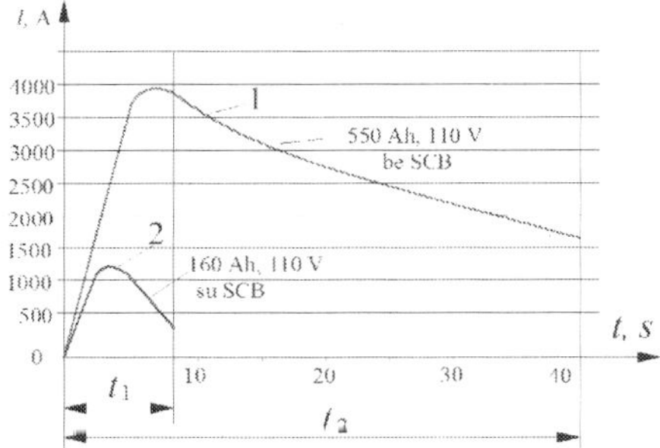

Fig. 4. The diagrams of starting up of the TEP-60 diesel engines starter in chain of current accumulators: 1- baterry current without SCB, when traction generator operates in a starter mode; 2- baterry current with SCB, when traction generator operates in a starter mode

conventional system of current starting up in chain of accumulators is up to 3700A. Using the conventional system of current starting up in chain of accumulators suggested by the authors is up to 1200A. The time of Diesel engine starting up, using the conventional system is 40-50 seconds, and using a complementary system is 7-10 seconds.

4. Locomotive energy saving systems

At this period of time locomotives new energy (3) saving technologies include: 1-optimized desing vehicle; 2-energy management control system; 3-energy storage system; 4- low energy climate system; 5-clean diesel motor power pack; 6- new technologies traction motor. Energy saving up to 8-15% using aeroefficient otimized train, up to 10-15% using energy management control system, up to 25-30% using energy management control system, up to 25-30% (Liudvinavičius..., *The aspect of vector...*, 2009) using energy storage system, up to 25-30% using low energy climate system. Clean diesel motor power pack reduced particle emission 70-80%. New technologies traction motor increased energy effiency 2- 4% at reduced volume and weight. New technogies can create energy savings up to 50%. Fig. 5 shows the possibilities of new energy saving technologies.

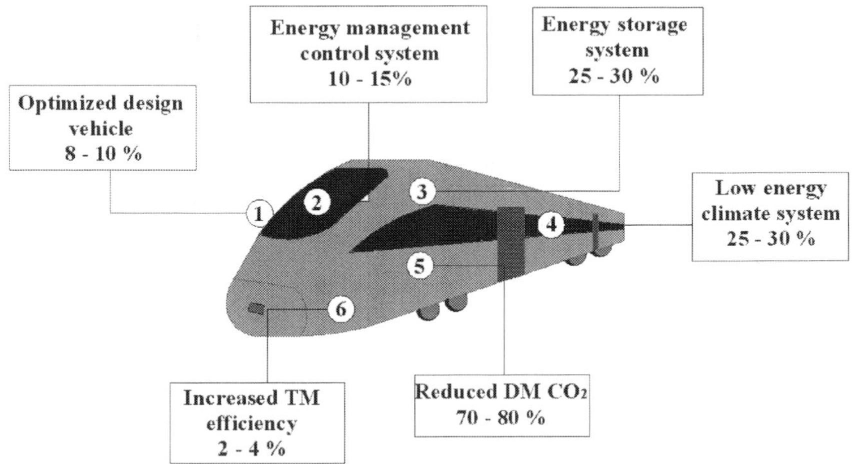

Fig. 5. Diagram of locomotive energy saving structure

5. Possibilities of new locomotives regenerative braking

Locomotive electric braking system may be divided into dynamic, and regenerative. Thus, the dynamic braking energy is converted into heat and dissipated from the system. In other words, electric energy generated is the typically wasted. In a typical prior art AC locomotive, however, the dynamic braking grids are connected to the DC traction bus because each traction motor is normally conected to the bus by the way of autonomous inverter. Fig. 6 shows that conventional structures electric locomotive AC traction energy transformed into heat through the braking resistor–R_b (Liudvinavičius..., *Electrodynamic braking...*, 2007).

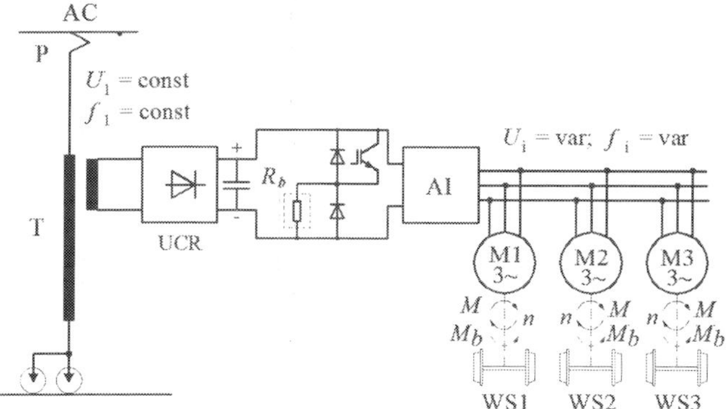

Fig. 6. A circuit diagram of AC/AC conventional electric locomotive dynamic braking: UCR-uncontrolled rectifier; AI– autonomous inverter; R_b –braking resistor; M1, M2, M3–one bogie asynchronous traction motors; WS,...,WS3-wheel-sets

Regenerative braking is more energy effective because power given to catenary power system is either used by another electric train or returned to power system. Thus, the conditions for the motor being idle to exceed point n_0 of torque-speed characteristic $n = f(M)$, which is required in regenerative braking, cannot be satisfied (see Fig 8). Locomotive traction motor regenerative braking energy is possiblly returned in to energy supply system then AC traction motor's speed is above no -load speed n_0. The traction motor goes to the generator mode, while electromagnetic moment, becomes a braking moment, and the power produced by generator is given to the catenary (energy power supply system).

6. Methods of new asynchronous traction motors speed control

The most modern kind of speed control of three-phase induction motors is the control by changing frequency f_1. (Lingaitis L.P. ..., Electric drives..., 2006; Strekopytov V...., Electric drives..., 2003). It ensures a wide control of range of the speed and causes only little additional losses.

Relative slip expressed by the formula:

$$s = \frac{n_1 - n_2}{n_1};$$

(1)

Where:

n_1 – the speed of the rotary field; n_2 –speed of the rotor (rotor speed on load)

f_1-. main frequency is: $f_1 = \dfrac{pn_1}{60}$, f_2 -frequency of the rotor voltage $f_2 = \dfrac{pn_2}{60}$ (there p-is number of pole pairs). Then:

$$s = \frac{f_1 - f_2}{f_1}.$$

(2)

Asynchronous motor's rotor speed:

$$n_2 = n_1 (1-s) = \frac{60 f_1}{p}(1-s) ; \tag{3}$$

may be adjusted in the following ways: by adjusting supply voltage U_1; by adjusting main frequency f_1; by varying the number of pole pairs-p, speed of the rotor's rotating field can be discretely changed; by adjusting slip s (not using slip energy), the nature of the speed-torque characteristic can be changed; by adjusting slip s (using a part of slip energy- cascade speed control circuits of asynchronous motors). Asynchronous motors with squirrel-cage rotors and their parameters expressed by the formula:

$$M = \frac{p_1 m_1 U_1^2 \frac{r_1}{s}}{2\pi f_1 \left(r_2' + \frac{r_1}{s} \right) + \left(x_1 + x_2' \right)} ; \tag{4}$$

Where p_1 and m_1 – are numbers of the stator's poles and phases; r_1 and x_1 – denote resistance and inductive impedance of stator; r_2 and x_2 – denote resistance and inductive impedance of rotor reduced in accordance with the stator's parameters; U_1 –is supply voltage of the stators windings. Optimal mode of operation of asynchronous motors with squirrel – cage rotors (Lingaitis L. P. ..., *Electric drives of traction rolling stocks with AC motors*, 2006):

$$\frac{U_1}{U_1'} = \frac{f_1}{f_1'} \sqrt{\frac{M_1}{M_1'}} . \tag{5}$$

Hence, an optimal mode of operation of asynchronous motors with squirrel –cage rotors is defined by the relationship between their three parameters - amplitude of voltage U_1, frequency f_1 and the developed torque M_1. A mode of operation of a locomotive can be described by the locomotive speed V and traction or braking force F_k of wheel - set. It was found that: $V = 0{,}188 \dfrac{D}{\mu} \dfrac{60 f_1}{p}(1-s)$ or $V = 0{,}188 \dfrac{D}{\mu} \dfrac{60 f_1}{p} = C_1 f_1 ,$ and $F_k = \dfrac{2M}{D} \mu \eta_p = C_2 M$

(here: D – is diameter of the locomotive wheel-set; μ – is gear ratio; η_p –is gear efficiency)

On the basis of the formula (8), we can determine mode control of locomotives with asynchronous motors:

$$\frac{U_1}{U_1'} = \frac{V_1}{V_1'} \sqrt{\frac{M_1}{M_1'}} \ \text{or} \ \frac{U_1}{U_1'} = \frac{V_1}{V_1'} \sqrt{\frac{F_k}{F_k'}} . \tag{6}$$

In this case, speed V_1 and traction or braking force F_1 correspond frequency f_1, and supply U_1, or V_1' and F_k' – traction or braking force in presence of frequency f_1' and voltage U_1'

When the supply voltage increases, the characteristics move the area of higher speed (Fig 7, line 2). By changing simultaneously the supply of voltage U_1 and its frequency f_1, depending on mode of regulation, any flat characteristics can be obtained.

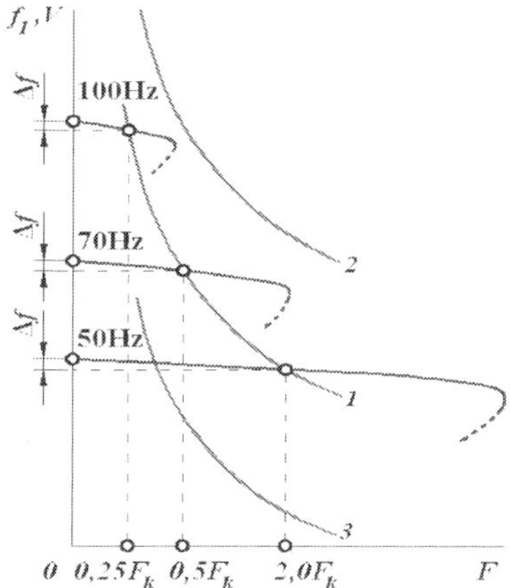

Fig. 7. Torque-speed characteristic of induction traction motor's traction modes by changing main frequency f_1–f_i– parameters

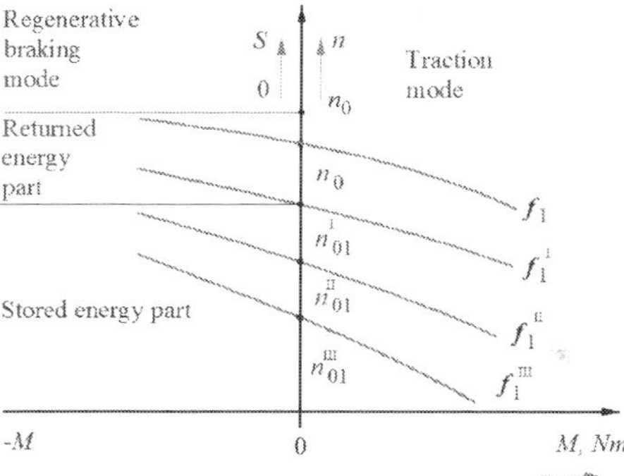

Fig. 8. Torque-speed characteristic of induction traction motor's regenerative braking and traction modes by changing main frequency f_1–f_i– parameters: n_{o1} – n_{oi} is AC traction motor's no–load speed

The frequency controlled squirrel-cage induction motor can be easily showed down by reducing the supply frequency.

Traction motor's no-load speed n_o is possible by changing the frequency f_1 and to receive more regenerative braking characteristics and regenerative braking energy returned to network supply or charging storage battery. Fig. 9 shows AC traction motors new possibilities of traction and regenerative braking modes operating. The energy management structure suggested by the authors in Fig.9 will allow the full use of regenerative braking capabilities: in a high-speed range to return energy for the energy system, in a low-speed range - to accumulate the energy in a battery of energy accumulating for further use. The characteristics given in Fig.8 illustrate these findings.

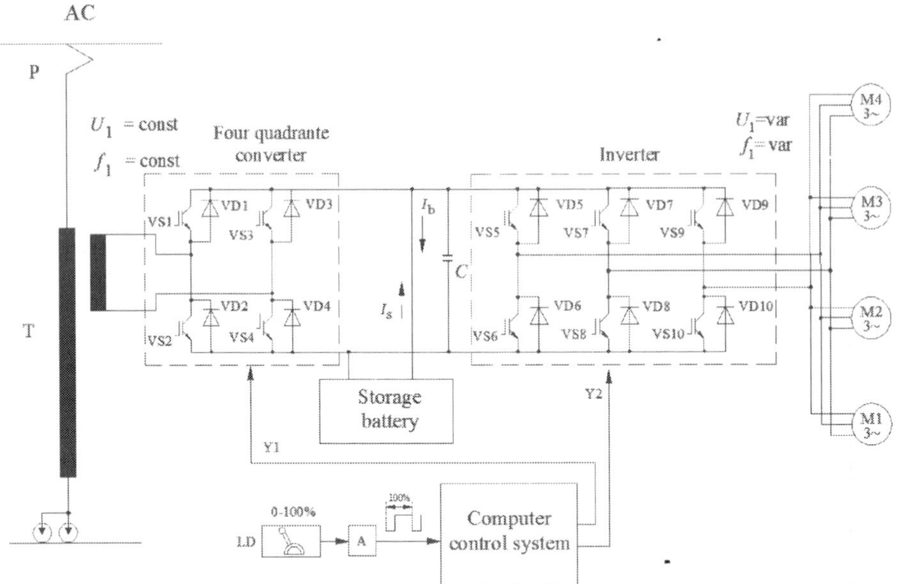

Fig. 9. A circuit diagram of AC/AC current system electric locomotive regenerative braking energy computer control system: M1-M4 – AC traction motors; LD- locomotive driver; A- analogic – digital converter; T- traction transformer; P- pantograph; VS1-VS10-IGBT transistors; VD1-VD10-diodes; Y1-four quadrant drive control signals; Y2- inverter drive control signals; I_b-braiking current; I_s- stored current; WS1-WS4- wheel-sets

Authors suggested to install storage battery into AC/AC current system conventional electric locomotive. Fig.9. shows principle of the braking energy management system used in AC/AC electric locomotive, when a part of regenerative braking energy is returned into energy supply system and part of energy is stored in storage battery.

7. Hybrid traction propulsion systems

Hybrid traction technology. Energy-saving propulsion system using storage-battery technology. As the train uses its traction motors the authors suggest to apply a hybrid propulsion system combining an engine generator with storage batteries (A. Rufer ..., *A supercapacitor-based energy storage...*, 2002.). A hybrid energy locomotive system having an energy storage and regeneration system. The system uses a series-hybrid configuration, designed to allow immediate system conversion (by replacing conventional diesel-powered train the engine generator with a fuel-cell unit, in pursuance locomotive modernisation and ect.). We offer to use a hybrid traction technology. Conventional diesel locomotives powered with electical transmision can not use regenerative braking energy. Any recovered energy can be used for traction.

This is expected to give fuel savings of approximately 20%-25% compared with conventional diesel-powered trains. An engine cutout control is also employed to reduce noise and fuel consumption while trains are stopping at stations.

Hybrid system configuration. This system uses a series-hybrid configuration (see Fig.10) that first converts the engine output into electrical power and then uses only motors for propulsion. Storage batteries are located on the intermediate DC section of the main converter. The charging and discharging of the storage batteries are controlled by using output adjustments of the converter and inverter. Charging and discharging processes of storage batteries are controlled by the converter and inverter output for management system.

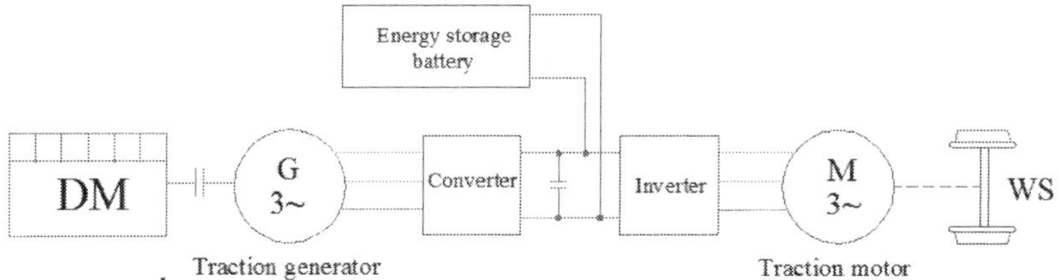

Fig. 10. A circuit diagram of Hybrid Traction System configuration with AC traction motors: DM-diesel engine; G-synchronous traction generator; M-induction traction motor.

A hybrid energy locomotive system can be used in AC or DC electric traction motors. When using DC traction motors, output of the alternator is typically rectified to provide appropriate DC power. When using AC traction motors, the alternator output is typically rectified to DC and traction inverter is shifted to three–phase AC before being supplied to traction motors.

Regenerative braking mode. The traction motors act as generators and recovered energy is used to charge the batteries. Storage battery operation of charging mode. Control of regenerative braking energy is carried out by using a bidirectional semiconductor converter of DC-AC energy which operates at an rectifier mode. In the braking mode, the control is directed from the energy source (AC – asynchronous traction motor) to energy storage batteries. (Takashi Kaneko.,et al.,. *Easy Maintenance…, 2004*).

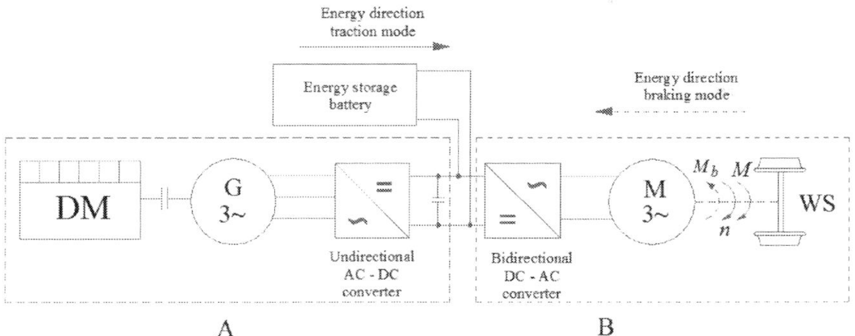

Fig. 11. AC/AC power structure diesel-electric locomotive complementary energy management system in regenerative braking and traction mode: DM-diesel engine; G-synchronous traction generator; M- AC traction motor. M_b- electromagnetic moment in the braking mode; n-speed of the rotor; A-energy generation part in the traction mode; B-energy generation part in the braking mode.

Departure and acceleration mode. The traction motors act as generators and while at a station, the engine can be stopped For and necessary hotel power can be provided from the battery. Upon departure, the train accelerates using the recovered energy only. Storage battery operation of discharging mode.

8. Main requirements for hybrid traction system

The technical trends in train traction systems are shown in Fig.11, 12, 13. In line with those trends, it is possible to develop rolling stock electrical-system with the following features to meet the demand for reduced maintenance, energy savings, environmental friendliness, and compact and light weight structures (Strekopytov, V. ..., Electric drives..., 2003; Yamaguchi, J. , *Automotive Engineering, 2006*).

Propulsion equipment requirements.

Hybrid traction system is used for high power variable voltage variable digital frequency VVVF traction converter. The authors offer to implement the conventional converter-inverter systems and auxiliary power converters which were applied vector control [5] techniques and new semiconductor elements high-voltage IGBT (insulated gate bipolar transistors) Commutation frequency of IGBT transistors is $f_k > 20000$ Hz. (one element parameters: 3300V, 1500A)

Traction motor speed control requirements. The authors suggest using AC traction motor speed drive sensor less vector control methods. AC traction motor speed sensor less vector control eliminating the speed sensor of traction motors creates space for increasing their power and improving their maintenance.

All-speed range electric brake control. Hybrid traction system needs to use more accurate traction motors speed estimation technology which enables the combination of sensor less speed vector control (Liudvinavičius, Lionginas ..., *The aspect of vector...*, 2009) with all-speed range electric brake control.

Storage battery system. Storage battery system can be of different types, such as: C-with capacitors; CB- conventional battery; C-CB- capacitors; CB- conventional battery.

Storage battery energy management system. Storage battery management technology enables storing brake energy in diesel-powered trains, expanding regenerative brake energy into high-speed region in electric trains, and stably supplying DC (direct current) power to the auxiliary power converter. Using hybrid traction system electric train inverter control technology makes regenerative braking possible and regenerated energy, temporarily stored in the batteries, can be used as auxiliary power for acceleration. Fig.12. shows the Hybrid Traction System auxiliary power inverter schematic circuit diagram.

The authors suggest using stored energy for starting the diesel engine. Fig 13 shows block circuit diagram of the Hybrid Traction System equipment: the inverter control technology enables using regenerated energy as auxiliary power for acceleration, i.e. the Figure shows technical possibilities of using stored energy for starting the diesel engine.

Using this system, it is possible to start a 2000- 6000 kW diesel engine in a short time. Traction motors regenerative braking energy charges ultra-capacitors block and latter ultra-capacitors block energy is used for starting the diesel engine. The authors propose to use the main inverter of Hybrid Traction System which operates on pulse- width modulation (PWM) principle function.

Function principle of PWM inverter. The function of an inverter is to transpose the DC voltage of the intermediate circuit into symmetric tree- phase voltage of variable frequency and amplitude. The necessary pulse diagram and the generated main voltage U_{UV} (at the terminals) are shown in Fig. 15. However, this voltage is non-sine-shaped. The effective value of the assumed sine-shaped supply voltage must be proportional to the frequency by changing the width of the single pulses in relation to the period duration. This kind of voltage control is called pulse-width modulation (PWM)

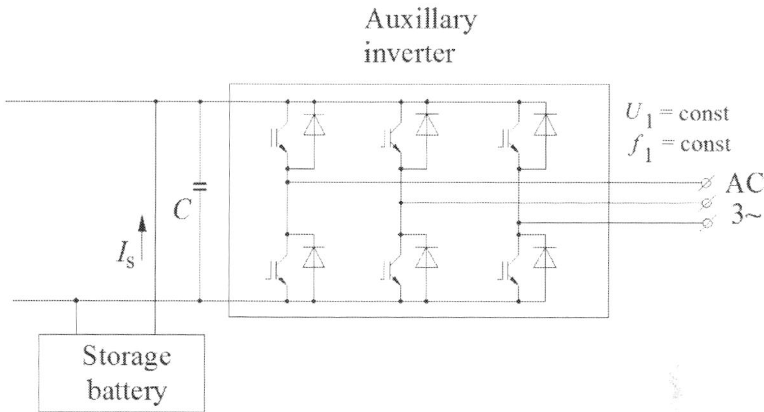

Fig. 12. Principal circuit diagram of Hybrid Traction System auxiliary power inverter

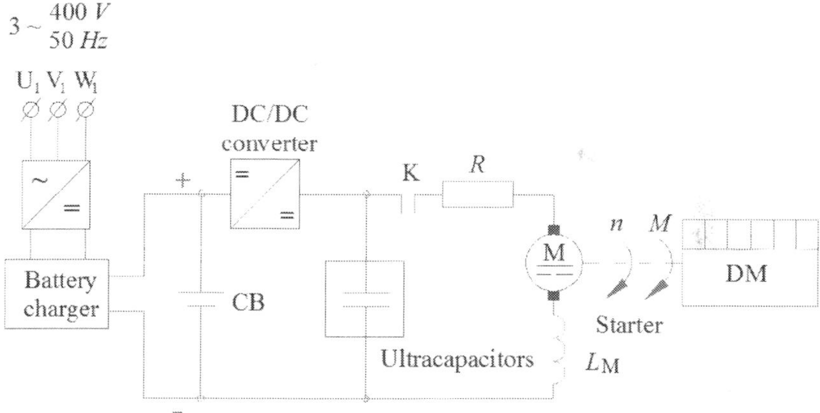

Fig. 13. Block circuit diagram of the Hybrid Traction System diesel engine start operation mode: DM- diesel engine; CB- conventional battery; K-contactor;

Fig 14 shows the example of a pulsed voltage block with five pulses per half-wave and the resulting main voltage U_{UV}. In this process, the voltage pulses of the main voltage become wider towards the middle of a half-wave as they first approach a sine-shaped course of the main voltage. For this reason, this kind of drive is called sine-weighted pulse-width modulation.

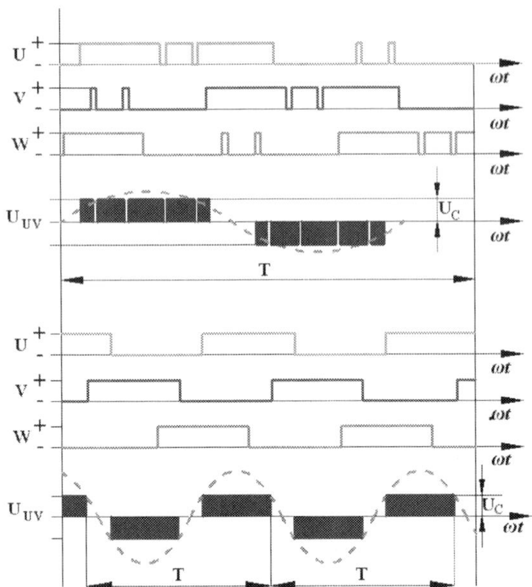

Fig. 14. Pulse diagram for 5 pulses and inverter per main wave

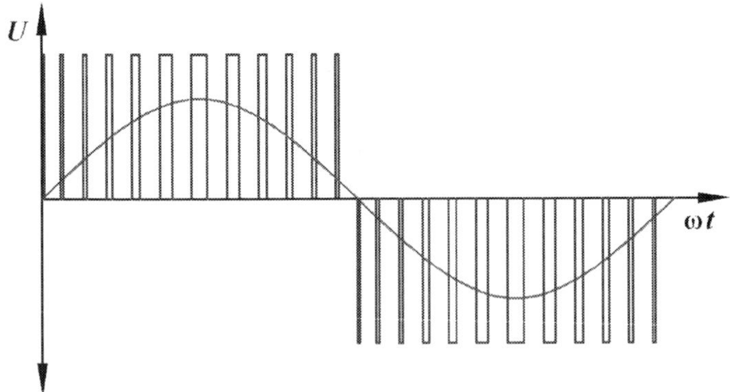

Fig. 15. The main voltage U_{UV}. pulse resulting diagram of the principle generation of a sine-weighted PWM

Fig. 15 shows the generation of pulse sequence for valve control and the resulting main voltage U_{UV}. Pulse frequency of the converter is determined by the frequency of delta voltage. The higher the switching frequency is, the better the sine weighting of the converter output voltage and the smaller the harmonic portion of the output currents are. A reduction of the harmonic portion leads to smaller oscillating torque and losses of the motor. Thus, the switching frequency should be as high as possible.

The authors suggest using an externally supplied energy system with energy tender.

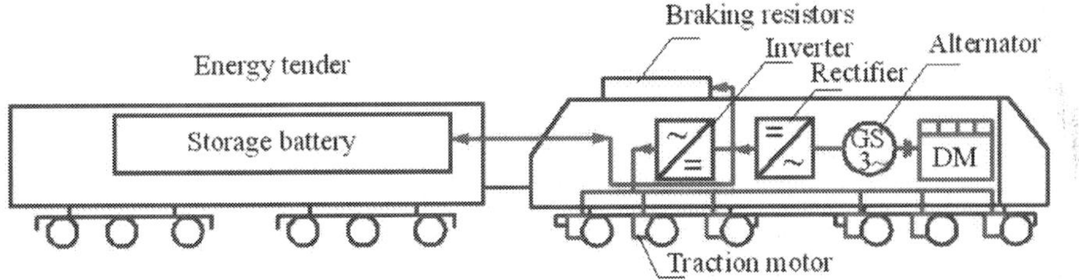

Fig. 16. Circuit diagram of hybrid energy traction system using energy tender vehicle

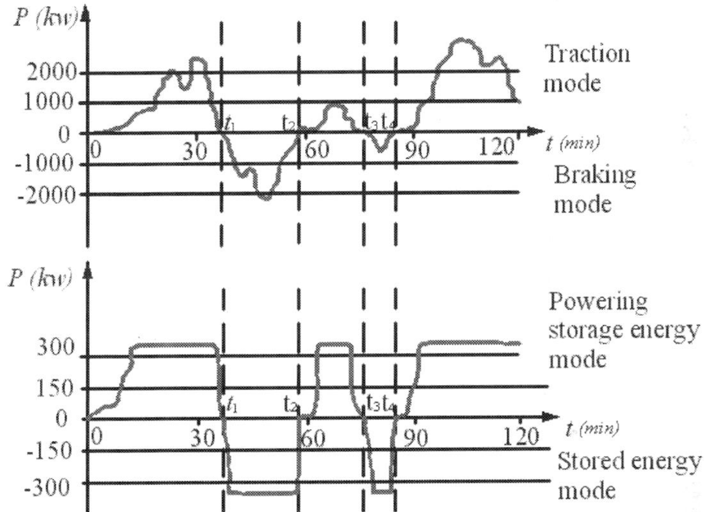

Fig. 17. Timing diagram of hybrid energy traction system using energy tender vehicle

Timing diagram is illustrating the locomotive energy management system, traction and regenerative braking mode. Timing diagram illustrating the hybrid traction system has the following energy storage possibilities: 0- t_1, t_2- t_3-time cycles of using powered storage energy traction and auxiliary equipment mode; t_1- t_2 time cycles of stored energy mode.

9. Energy saving and catenary voltage stabilization systems

Fig 18 presents the diagrams of voltage variation in direct current (DC) contact networks. Their analysis shows that when the load current I_C increases in the complementary (DC) contact network system, the voltage falls significantly (its value diminishes). In comparison to standard voltage of 3000V in the contact network, the values are lesser by 10%, i.e. 300V, whereas they are only -1% lesser in the complementary energy saving voltage stabilization system proposed by the authors. This is achieved by using energy storage batteries parallelly connected to the direct catenary current (b). The batteries are charges from electric trains during the regenerative braking of electric locomotives. Locomotives operating in the traction mode use less electrical energy from traction substation I and substation II because a part of energy is supplied by the energy storage batteries. These batteries do not require a separate voltage source for charging as they are charged by using the kinetic energy of the trains which emerges during the regenerative braking mode of electric trains and electric locomotives. Conventional accumulators, a supercapacitors block or in parallel connected accumulators and a block of supercapacitors may function as energy storage batteries. The complementary system ensures the stabilization of catenary voltage (maintaining it in the set boundaries) when the load current increases (from point A to point B) in the contact network(Precision inductosyn position…, 1996).

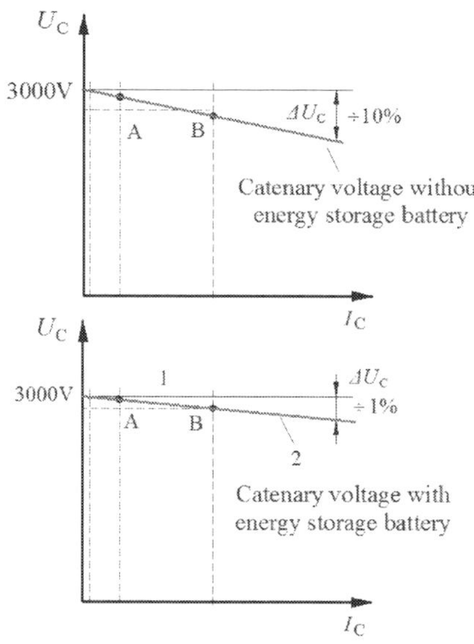

Fig. 18. Parameters of energy saving and variation of catenary voltage in a conventional a) and complementary energy management systems: U_C –catenary voltage; I_C–catenary current

10. Structure and energy management in complementary energy saving and current stabilization systems

Generally, first-generation electrified lines for railways, underground, trams and trolleybuses were exclusively of direct current with voltages of 600, 750, 1500 and 3000V. Although the catenary voltages 600, 750, 1500, 3000V are relatively low and do not meet the present-day requirements since they limit the speed and weight of the trains due to the voltage in the drop line. In order to increase the reliability and stability of DC contact network (present energy system) and save some energy consumed for traction, the authors suggest in parallel connecting energy storage batteries between the contact network and rail. The principled scheme of energy saving and catenary current stabilization structure is given in Fig 19. Energy storage battery in parallel connected to the DC contact network is composed of conventional batteries (CB) and supercapacitors block (SCB).

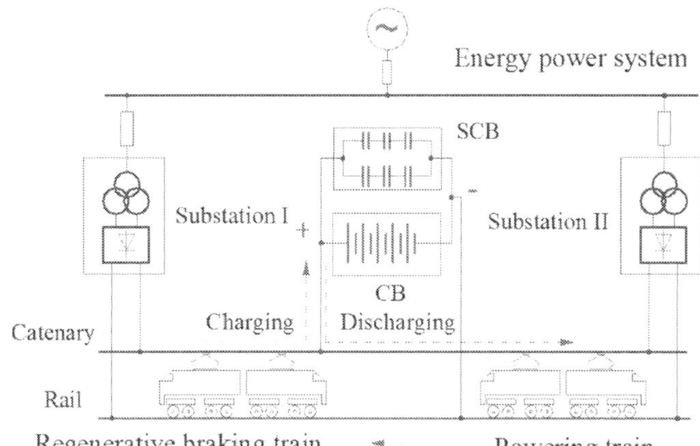

Fig. 19. Complementary principled scheme of energy saving and catenary current stabilization structure: CB- conventional batteries; SCB-super capacitors block

Energy management system is presented in Fig 20. The authors propose using a semiconductor key K, composed of IGBT transistors and diodes, for energy direction control. In the traction-regenerative breaking modes, the energy direction and level of battery charge may be controlled by sending control signals to the electronic key K.

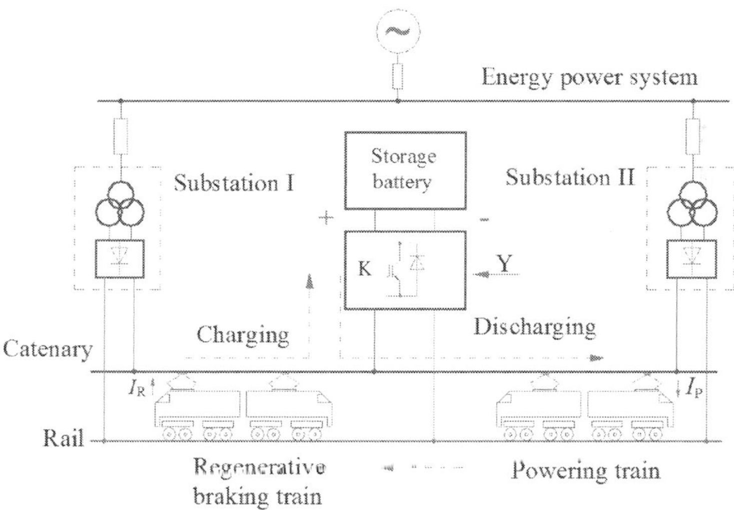

Fig. 20. Scheme of energy management system structure: I_R – regenerative current; I_{P-} traction mode current; K- semiconductor key for energy direction control

The most challenging operating for storage devises on board of traction vehicle are high number of load cycles during the vehicle lifetime, relatively short charge and discharge times as well as high charge and discharge power values. The battery is charged when line voltage goes up so that it limits the line voltage increase. Trains can unlimitedly generate regenerative braking energy when capacitors SCB block and conventional storage batteries CB operate. The regenerative braking energy is consumed by the train itself and by other powering trains. Excessive power is stored in the battery. The charging voltage in the batteries is higher than that of the substation. All charged energy is considered to come from the regenerative braking. The SCB block and conventional store batteries CB enable limiting the voltage increase during the charge. When powered trains are congested at rush hours due to the line voltage tendency to drop, the batteries discharge to reach a voltage balance between the voltages of the SCB-CB block and the substation. The new technical solution is used in conventional batteries with high-performance double layer capacitors (ultra-capacitors). Energy saving system can be used when the vehicles are provided with energy source that allows frequent starting and braking. The system works by charging up these storage devices with electrical energy released when braking. Energy savings and power supply optimization system can reduce the energy consumption of a light rail or metro system by up to 30 percent. Using power supply optimization system for diesel multiple units enables to save more than 35 percent of energy. Alternatively, the stored energy can be used as a performance booster, i.e. to enhance the performance of a vehicle by adding extra power during acceleration.

11. Vehicle catenary – free operation possibilities

In addition to these well-known factors, the municipal authorities are increasingly facing visual pollution caused by power poles and overhead lines obstructing the visibility of landmark buildings and squares. With catenary-free operation, trams can run even through heritage-protected areas, such as parks and gardens, historic market and cathedral squares, where conventional catenary systems are not permitted, thus preserving natural and historic environments. Authors suggest using catenary-free system for trams, light rail vehicles and trolleybuses. In many city centers, the overhead lines and their surrounding infrastructure contribute to visual pollution of historic streets, parks or architectural landmarks. The new system allows catenary-free operation of trams over distances of varying lengths and in all surroundings as well as on underground lines – just like any conventional system with overhead lines. Catenary-free system traction inverter is connected to the storage battery which is charged during vehicle traction motor operation in regenerative braking mode and discharged during traction motor operation in traction mode, where conventional energy lines are discontinued. Energy saver, which stores electrical energy is gained during operation and braking on board of the vehicle by using high-performance double layer capacitor technology. When running on conventional system, trams and light rail vehicles take energy from an overhead electrical line. The authors suggest installing the vehicle (inside or outside) with a storage battery (ultra-capacitors block) which stores the energy gained during regenerative braking operation and is constantly charged up, either when the vehicle is in motion or waiting at a stop, picking up the power from the storage battery. Fig.21 22, 23, show vehicle configuration and catenary-free operation possibilities. The power necessary for catenary-free operation is provided from the battery.

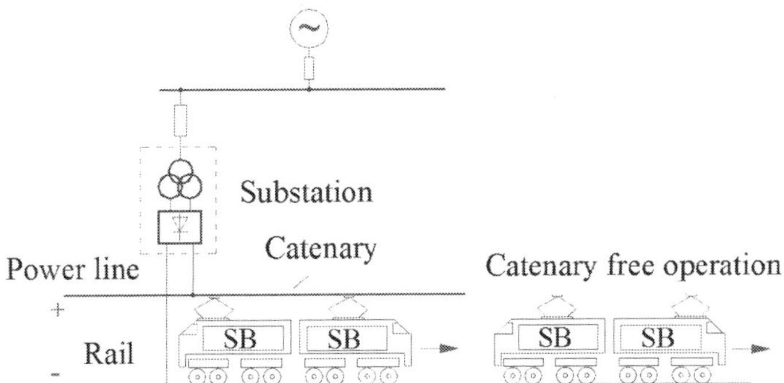

Fig. 21. Circuit diagram of vehicle catenary free operation: SB- storage battery

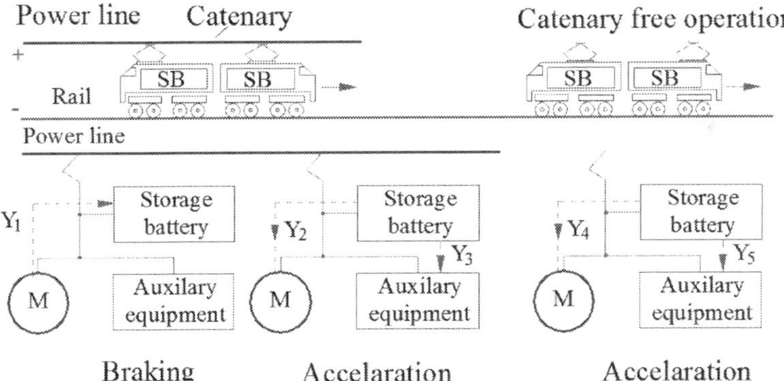

Fig. 22. Catenary-free operation of the vehicle: Y1-Y4-energy management drive signals; M-traction motor

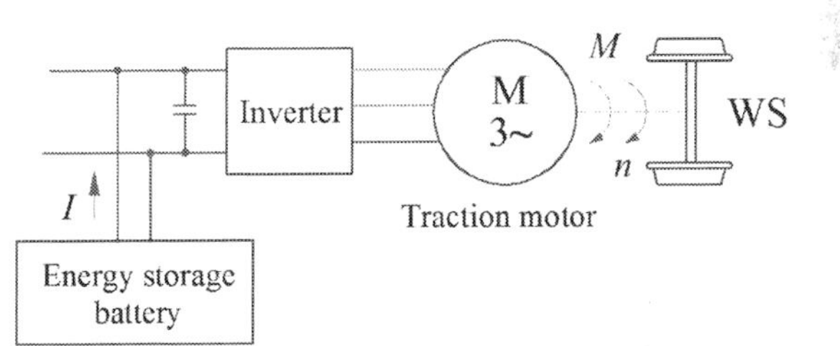

Fig. 23. Circuit diagram of catenary-free operation of the vehicle (traction mode)

The innovative double layer ultra-capacitors store the energy released each time a vehicle brakes and reduce it during acceleration or operation. New technical solution is based on double layer capacitors with along service life and ten times higher performance than conventional batteries. High-performance storage cells are connected in series to create a storage unit. They store the electrical brake energy with relatively low losses(Fuest, K.; Döring, P. Elektrische Maschine und Antriebe, 2000; Stölting, H.-D., Elektronisch betriebene Kleinmaschine, 2002).

12. Hybrid locomotive energy balance

Within the bounds of the present research, the question of qualitative evaluation of regenerative power during hybrid vehicle braking is of fundamental importance.

Vehicle power during braking on horizontal road P_{br} can be expressed by the following equation:

$$P_{br} = k_m \cdot m \cdot a \cdot V \,, \tag{7}$$

Where: k_m - coefficient of rotational masses; m – vehicle mass; a – vehicle acceleration (deceleration); V- vehicle velocity. The power that can be received during regenerative braking is:

$$P_{regen} = k_m \cdot m \cdot a \cdot V \cdot \eta_{regen} \tag{8}$$

Where: k_m - coefficient of rotational masses; m - vehicle mass; a – vehicle acceleration (deceleration); V- vehicle velocity; η_{regen} - efficiency of regenerative braking (can be defined as rate of energy, received during braking up to decrease the kinetic energy of the vehicle). At the same time, regenerative braking power can be considered as electric power which is finally received by the storage element (in this case storage battery):

$$P_{regen} = P_{el} = I_{bat} \cdot U_{bat} \tag{9}$$

Were: P_{el} - electric power received by the battery; I_{bat} - battery current; U_{bat} - battery voltage. The effectiveness of regenerative braking can be estimated using these equations:

$$\eta_{regen} = \frac{P_{regen}}{P_{br}} = \frac{I_{bat} \cdot U_{bat}}{k_m \cdot m \cdot a \cdot V} \,, \tag{10}$$

JSC *Lithuanian Railways* has acquired four new-generation double-deck electric trains type EJ-575. In order to evaluate consumption of energy of the new generation double-deck electric trains type EJ-575, the authors carried out the practical research. The aim of the research is to determine why the electric trains type ER-9M (without energy saving system) and EJ-575 (with energy saving system), running in the same section, consume different amounts of electrical energy. Practical research, and statistical comparisons of results are carried out.

13. Practical researches of train EJ-575 energy management

In order to determine the double-deck electric train EJ-575 energy-management principles and to measure the dynamic electrical and mechanical parameters, the practical experiments were performed. The following channels of parameters measurement are predicted for determining the quantity of energy for traction and electrodynamic braking: the primary traction transformer winding of instantaneous current I_{Tr}, the primary traction transformer winding of instantaneous voltage U_{Tr}, four-quadrant 4Q1, 4Q2 converters, flattening voltage U_d, the first and the second asynchronous traction motor – *speeds of wheelsets* (Braess, H. H.; Seiffert, U., Vieweg Handbuch Kraftfahrzeugtechnik, 2000).

During practical researches dynamic parameters of energy management are measured by using a personal computer, therefore the authors provide 5 *converters* of the abovementioned channels analog signals conversion into discrete in the framework of measurement: *instantaneous current I_{Tr}, instantaneous voltage U_{Tr}*, four-quadrant **4Q1, 4Q2** converters *of flattening voltage U_d*, the first asynchronous traction motor – *speed of wheelset n_1*, the second asynchronous traction motor – *speed of wheelset n_2*. The scheme of the train EJ-575 traction-electrodynamical braking parameters practical research, using a personal computer is given in Fig. 24.

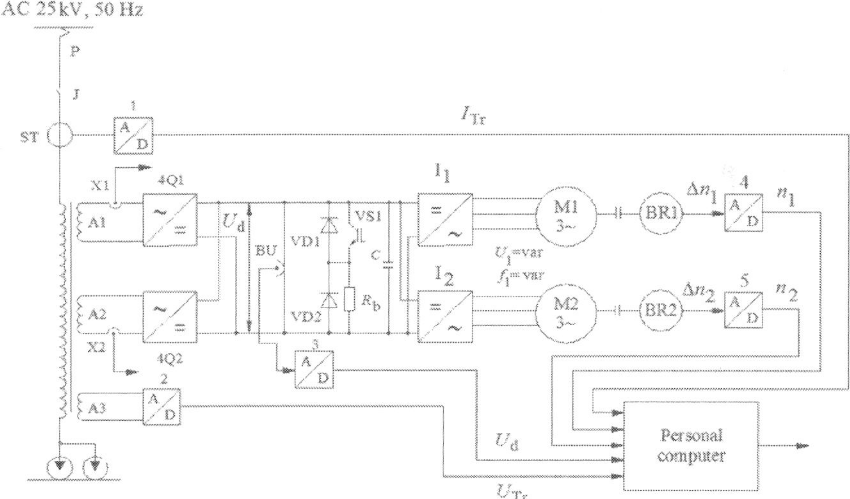

J – high-speed disconnector; P – phantograph; 4Q1, 4Q2 – four-quadrant converters; I_1, I_2 – inverters; M1, M2 – AC asynchronous traction motor; C – energy accumulation condenser; X1, X2 – secondary traction transformer windings current sensors; ST – primary winding traction transformer current sensor; A1, A2 – secondary traction transformer windings; A3 – secondary traction transformer winding for measurement of contact network voltage (25 kV); BU – flattening voltage sensor; R_b – dynamic braking resistor; VS1 – IGBT-transistor braking current (braking force) value regulator; BR1, BR2 – speed sensors of traction motors; VD1, VD2 –diodes; C – capacitor; 1, 2, 3, 4, 5 – analogical-digital converters; Δn_1, Δn_2 – asynchronous traction motors speed variation; WS1, WS2 – wheelsets

Fig. 24. The scheme of the train EJ-575 traction-electrodynamical braking parameters practical research, using a personal computer:

EJ-575 energy-management parameters are determined on the 6th of October 2010 for the train No EJ-818 of Vilnius–Kaunas district in traction-electrodynamic braking modes. The measured dynamic parameters are displayed on monitor of personal computer. For the convenience of the research results analysis the controlled parameters are provided in one system of coordinates. I_{Tr} variation of current is provided in real values, the parameters of instantaneous contact network voltage U_{Tr}, displayed on monitor of personal computer, must be multiplied by 100, flattening voltage U_d values must be multiplied by 10; single-carriage asynchronous traction motor of the train EJ-575 (of 1, 2 wheelsets) variation of speed is charted by marking n_1, n_2, and shows the instantaneous values of train speed, km/h. The values of dynamic parameters, measured with personal computer during the practical research are given in Fig. 25.

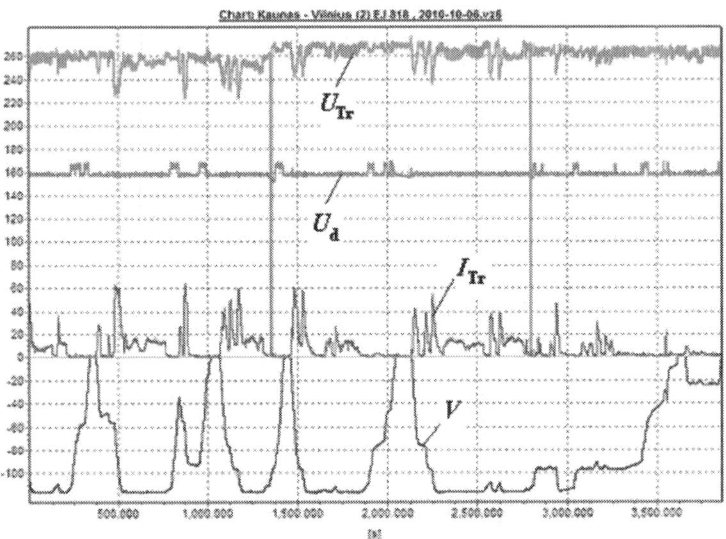

I_{Tr} – diagram of primary traction transformer winding current variation (A); U_{Tr} – diagram of contact network voltage variation(V); flattening voltage U_d variation diagram (V); V – diagram of speed variation of the train (km/h); t – time (s)

Fig. 25. Parameters values of the double-deck electric train EJ-575 energy-management in traction-electrodynamic braking modes:

14. Results of practical research on energy control in EJ-575

Variation range of instantaneous active power P used in traction cycles 1T and 2T is up to 2×1250 kW (two separate windings of traction transformers). Variation range of instantaneous active power P in electrodynamic braking cycles 1S–2S is 25–60 kW. During electrodynamic braking, the amount of energy is used only for power supply to ancillary devices. The electrodynamic braking of the train is carried out using kinetic energy.

In traction and electrodynamic braking cycles, the amount of consumed contact network energy $P(t)$ converted into useful work is described by the respective areas delineated by curves. The amount of energy is determined by integrating the respective cycles, following the given formulas (Bureika 2008; Dailydka, Lingaitis ... 2008):

$$W = \int_{0}^{t} P dt ;\qquad\qquad(11)$$

$$W = \int\limits_0^{t_1} Pdt + \int\limits_0^{t_2} Pdt + \int\limits_0^{t_3} Pdt + ... + \int\limits_0^{t_i} Pdt . \tag{12}$$

The diagrams showing variation of contact network energy $P(t)$ and electrodynamic braking $P(t)$ energy consumed in electric double-deck train EJ-575 are given in Fig. 26.

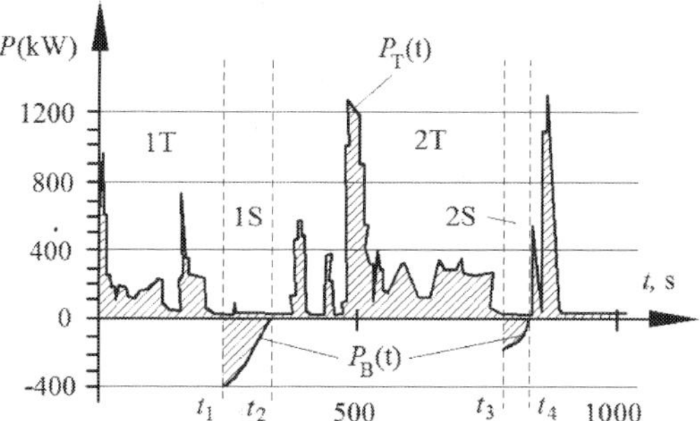

$0-t_1$ – traction cycle 1T; t_1-t_2 – electrodynamic braking cycle 1S; t_2-t_3 – traction cycle 2T; t_3-t_4 – electrodynamic braking cycle

Fig. 26. Diagrams showing variation of contact network energy $P(t)$ and electrodynamic braking $P(t)$ energy consumed in electric double-deck train EJ-575:

Variation range of contact network energy $P_T(t)$ consumed in traction cycles 1T and 2T is represented in traction cycles $0-t_1$ and t_2-t_3; Variation range of electrodynamic braking $P_B(t)$ energy is given in cycles t_1-t_2 and t_3-t_4. The amount of electrodynamic braking $P_B(t)$ energy converted into useful work is described by the respective areas delineated by curves in cycles t_1-t_2 and t_3-t_4.

15. Results of research

Comparison of research results on electrical energy consumption in electric trains ER-9M and EJ-575 in 2010 is presented in the diagrams below (Fig. 27, Fig. 28). They are given following the statistical data of JSC *Lithuanian Railways*.

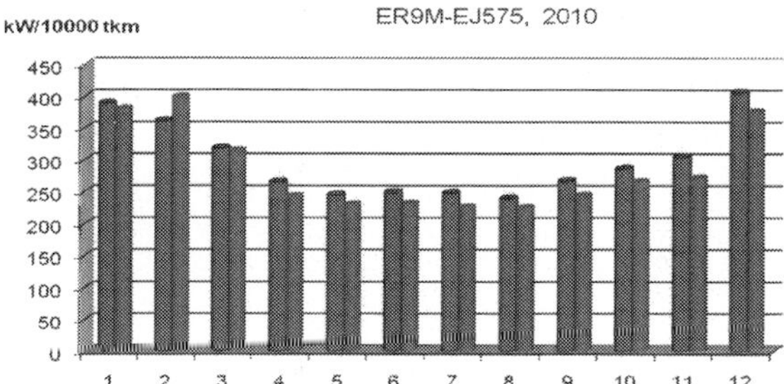

Fig. 27. Electrical energy input (kW/10000 tkm) of electric trains ER-9M with DC traction motors and EJ-575 with AC traction motors in 2010

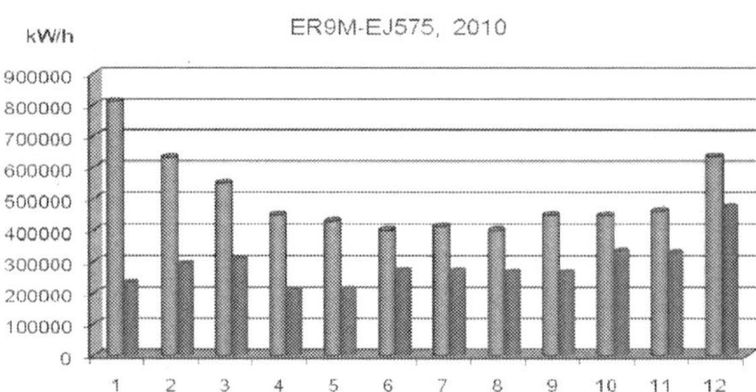

Fig. 28. Electrical energy input (kW/km) of electric trains ER-9M with DC traction motors and EJ-575 with AC traction motors in 2010

16. Conclusions

1. Electrodynamic braking is the main braking technique used for modern electrically-driven locomotives.
2. The use of supercondensers in the locomotives with electric drive expands the regenerative braking range to full stopping. This creates the conditions for full use of kinetic energy of the train.
3. Using supercondensers enables replacing the diesel motors of powerful locomotives and ships.
4. Supercapacitors were chosen to act as energy buffer.
5. The use of regenerative braking of electric locomotives for high-speed trains under the conditions of heavy railway traffic allows 25–40 % of electric power to be returned to the power system. The required regenerative braking forces can be obtained in a wide range, with a possibility to return energy to energy supply in a high-speed range and to store energy in a low-speed range.
6. All diesel electric powered locomotives should use hybrid traction technology.
7. Hybrid traction technology locomotives can use regenerative braking of high-speed and a low-speed range.
8. The energy used by hybrid traction technology locomotives is reduced by 25–30 %.
9. The offered regulation algorithm allows obtaining various types of flat characteristics enabling asynchronous traction motor to be extensively used in traction, recuperation and dynamic braking modes of operation.
10. A circuit scheme of using hybrid traction technology with energy storage tender and catenary-free operation was proposed.
11. It is possible to use the regenerative braking power in diesel electric locomotives for starting engine, acceleration, and operation mode.
12. Energy savings and power supply optimization possibilities were proposed.
13. The electrodynamic braking system installed in EJ-575 enables to stop the train without friction braking.
14. The electrodynamic braking system installed in EJ-575 enables a complete use of kinetic energy of the train in the braking cycles without contact energy network.
15. The use of kinetic energy of the train saves 25–30 % of electrical energy used for traction.

16. Flattening voltage U_d in ER-9M is step-controlled and for this reason, the currents of DC traction motors and energy losses during starting are great.
17. Due to the step-controlled flattening voltage U_d of traction motors in ER-9M, braking alternates in a step manner, i.e. unevenly; thus the passengers   experience discomfort and automatic coupling is affected by dynamic forces.
18. Kinetic energy of the train is not used for braking in ER-9M. The train is braked using friction, which increases operational expenses.
19. The speed of asynchronous traction motors of EJ-575 is evenly controlled by alternating power supply voltage and values of frequency.

17. References

P. Barrade, *Series connexion of Supercapacitors : comparative study of solutions for the active equalization of the voltage*, École de Technologie Supérieure (ETS), Montréal, Canada, 2001

J. D. Boyes and H. H. Clark, *Technologies for energy storage flywheels and super conducting magnetic energy storage*, IEEE, 2000.

Braess, H. H.; Seiffert, U. Vieweg Handbuch Kraftfahrzeugtechnik. Friedrich Vieweg & Sohn Verlagsgesellschaft GmbH, Wiesbaden. 2000.

Fuest, K.; Döring, P. Elektrische Maschine und Antriebe. Lehr- und Arbeitsbuch. Vieweg. Wiesbaden. 2000.

R. G. V. Hermann, *High performance double-layer capacitor for power electronic applications*, in Second Boostcap meeting, Montena Components SA, Fribourg, Switzerland, 2001.

Liudvinavičius L. Lingaitis L. P. 2010. *New locomotive energy management systems.* / Maintenance and reliability = Eksploatacja i niezawodność / Polish Academy of Sciences Branch in Lublin. Warszawa. ISSN 1507-2711. No 1, 2010, p. 35-41.

Liudvinavičius, Lionginas; Lingaitis, Leonas Povilas; Dailydka, Stasys; Jastremskas, Virgilijus. *The aspect of vector control using the asynchronous traction motor in locomotives.*, Transport. Vilnius: Technika. ISSN 1648-4142. Vol. 24, No 4, 2009, p. 318-324.

Liudvinavičius L. Lingaitis L.P: *Electrodynamic braking in high-speed rail transport.*, Transport, Vol XXII, No 3, 2007, p.p.178-186.

Lingaitis L. P., Liudvinavičius L.: *Electric drives of traction rolling stocks with AC motors.* Transport, Vol XXI, No 3, 2006, p.p. 223–229.

Precision inductosyn position transducers for industrial automation, aerospace and military application. Farrand Controls information Issue. USA. 1996. 15 p.

A. Rufer and P. Philippe, *A supercapacitor-based energy storage system for elevators with soft commutated interface*, IEEE Transactions on industry applications, vol. 38, no. 5, pp. 1151-1159, 2002.

Sen P. C. Principles of Electric Machines and Power Electronics. New York-Chichester-Brisbane-Toronto-Singapore-Weinheim. John Wiley&Sons. 1996.

Strekopytov, V. V., Grishchenko , A. B.; Kruchek V. A. Electric drives of the locomotives. Moscow: Marshrut.2003. 305 p.

Stölting, H.-D. Elektronisch betriebene Kleinmaschine. Vorlesungsmanuskript. Universität Hannover. 2002.

Takashi Kaneko.,et al.,. *Easy Maintenance and Environmentally-friendly Train Traction System Hitachi Review,Vol.53, No 1,2004, p.p 17-19p.*

Yamaguchi, J. Blue skies at Makuhari. *Automotive Engineering International*, 2006, No 1, p.55-62.

11

An Adaptive Energy Management System Using Heterogeneous Sensor/Actuator Networks

Hiroshi Mineno[1], Keiichi Abe[2] and Tadanori Mizuno[2]
[1]Faculty of Informatics, Shizuoka University
[2]Graduate School of Science and Technology, Shizuoka University
Japan

1. Introduction

Global energy consumption has been increasing over the past half century, mainly due to increasing populations and economic development around the world. Although the development of low-consumption appliances, highly efficient heat pump systems (such as Ecocute), and smart meters that identify consumption in more detail than conventional meters contribute in no small part to reducing the emission of greenhouse gases in homes and office buildings, it is important to understand that the rapid growth in ubiquitous comfort services is resulting in higher power consumption. In the future, there may be more convenient appliances coming on the market to create a ubiquitous smart infrastructure.

Even though individual appliances might be ultra-low power, the amount of energy consumption in buildings will increase as the number of these appliances increases. There has been research in the field of home/building energy management systems (HEMS/BEMS) (Cao et al., 2006; Inoue et al., 2003; Kushiro et al., 2003; Zhao et al., 2010), but integrating new devices on the market is not easy. Moreover, it is difficult to develop appropriate systems for different lifestyles. For example, a building's carbon footprint is the product of complex interplay between the buildings' structural and infrastructure characteristics, operational patterns and business processes, weather and climate dynamics, energy sources, and workforce commute patterns. Because these disparate factors can change daily, any recommendation based on a snapshot will rapidly become invalid.

Therefore, we believe the key feature of next-generation HEMS/BEMS is adaptability. If the systems are able to use information from multi-vendor sensors/actuators in heterogeneous networks, we can develop more adaptable energy management systems for visualizing and controlling the living climate appropriately. We feel this would enable us to create ubiquitous services for a more convenient, eco-friendly lifestyle.

We have developed an adaptive energy management system (A-EMS) for controlling energy consumption by converging heterogeneous networks (Mineno et al., 2010) such as power line communications (PLC), Wi-Fi networks, ZigBee, and future sensor networks. We created a prototype system that enables users to freely configure a cooperative network of sensors and home appliances from a mobile device. Although there are many similar technologies for integrating multi-vendor devices (IEEE 1451; Sensor Model Language; Device Kit; Chen, et al., 2009), we used P2P Universal Computing Consortium (PUCC) technology, which enables sensors and other devices to connect to each other. PUCC has the advantage of detecting services and devices using a P2P network. We developed our middleware with Java, which can work as a bundle on an Open Services Gateway initiative (OSGi) framework (OSGi Alliance). Experimental results demonstrated that the proposed system can easily detect wasted electrical energy that has never been noticed before.

The rest of this chapter is organized as follows. Section 2 outlines heterogeneous network convergence as related work. In Section 3, we describe the features of our proposed A-EMS in detail. Section 4 discusses the implementation of our prototype and section 5 shows the experimental results. Section 6 concludes the paper with a brief summary and mentions future work.

2. Related work

There are currently many electrical devices connected to networks. However, many types of networks coexist, and they all use different communication protocols. Therefore, a technology to integrate different networks is needed to provide services and control various devices connected to different networks. In the same vein, sensor/actuator devices also have a networking function and can be used to construct a smart system (Tzeng et al., 2008; Han et al., 2010; Park et al., 2007; Son et al., 2010; Suh et al., 2008). However, most services for and studies on sensor networks are limited to a particular sensor network. If it were possible for sensors/actuators to communicate with several types of networks, various devices could use heterogeneous sensor data.

Because the Internet Protocol (IP) horizontally integrates the control mechanism between heterogeneous networks and their extensions, heterogeneous network convergence would provide a tremendous opportunity for future advanced infrastructure management. The trend with sensor/actuator networks is the same. Recent developments with small sensor nodes that can be connected to a network have spurred studies on integrating non-IP-based sensor networks, which function by connecting several nodes, in which IP-based sensor gateways are being developed as common interfaces for connecting devices.

To create IP-based sensor gateways, we can use a number of proposed standards to describe devices and enable their integration (Chen et al., 2008). For example, IEEE 1451 describes a set of open, common, network-independent communication interfaces for connecting devices to microprocessors, instrumentation systems, and control/field networks. SensorML focuses on creating measurement models using sensors and instructions for deriving more accurate information from observations. Device Kit, following SODA (Chen, et al., 2009), is an OSGi-enabled technology that can interface with hardware devices using Java. It enables the development of applications for devices when hardware-specific information is unknown. Although these standards model devices all have various advantages, we used PUCC specifications because they enable sensors and all other devices to connect to each other. PUCC has the advantage of detecting services and devices through a P2P network such as an UPnP network.

3. Adaptive energy management system (A-EMS)

3.1 Installation level

Although there has been previous research on HEMS/BEMS, it has been limited to "detect something and notify or control predefined action." These systems are not versatile.

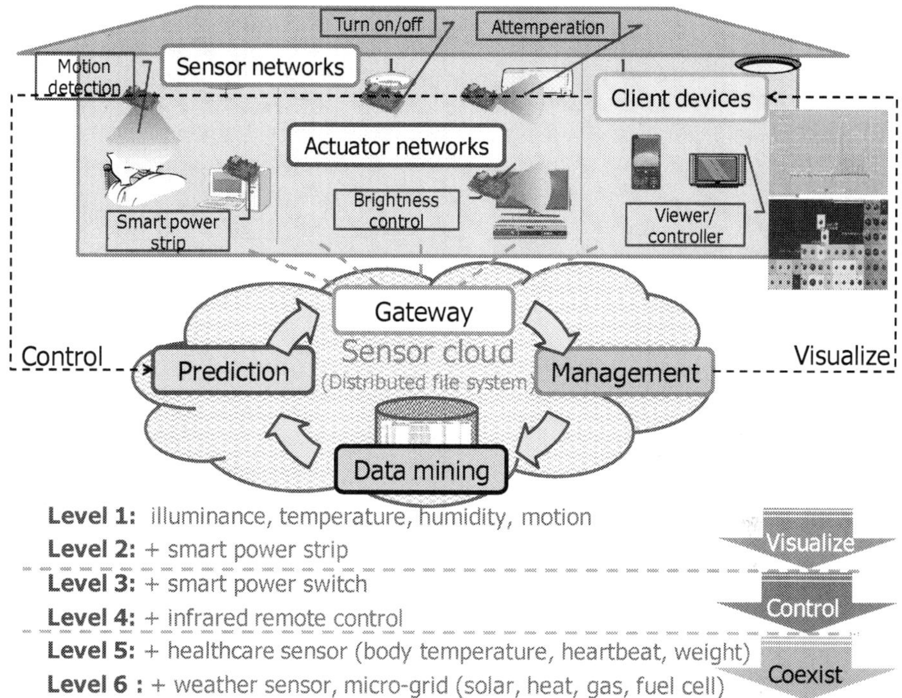

Fig. 1. Overview of proposed adaptive energy management system (A-EMS) and its implementation levels.

Moreover, it is not easy to integrate a vendor's new device after deployment, and it is difficult to develop appropriate systems for users with different lifestyles. Fig. 1 shows an overview of our proposed A-EMS and its implementation levels. We designed the A-EMS on the basis of a sensor cloud as a feedback system. In the proposed system, a PUCC P2P network is established among IP-enabled nodes. If the sensor gateway of another domain network acts as the PUCC P2P node, the PUCC P2P network incorporates the heterogeneous network and allows P2P nodes and sensor gateways to access the connected network adaptively.

The system can be installed in a phased manner depending on the required energy management level and available sensor/actuator devices. Level 1 visualizes the living climate. We can visualize the transition of unknown information, such as living climate data and someone's presence or absence, by installing different types of sensor nodes that sense illuminance, temperature, humidity, and motion. Level 2 visualizes the energy consumption. We can determine the effect of energy-saving actions by revealing the correlation between life pattern and energy consumption via additionally installed smart power strips instead of existing power strips. In this work, we assume the smart power strip is an electrical meter that can remotely report the power consumption in addition to power factor, electrical current, and pressure of connected devices. Level 3 controls the legacy appliances. If the smart power strip can turn on/off the connected devices remotely, we can execute basic living climate control for energy saving, such as turning off the standby power of appliances. Level 4 controls the appliances that have an infrared remote (IR) control function. There are many devices in homes that have this function, so we can include multi-vendor devices as controlled objects if such an actuator node is installed in our system. Level 5 achieves coexisting. If healthcare sensors, such as body temperature, heartbeat, or weight, are installed in the system, it can execute personal living climate control based on different comfort levels and conditions. Then, at level 6, the living climate control is consciously aware of the balance between natural and artificial control because it takes advantage of information on heterogeneous networks as well as information from the weather sensor and micro-grid (solar, heat, gas, and fuel cell).

3.2 PUCC overlay sensors/actuator network

The PUCC is a standard-setting organization that develops technologies to connect and operate many types of devices on P2P overlay networks. By forming an overlay network using PUCC protocols, we can seamlessly connect devices in a heterogeneous network environment. PUCC protocols are defined on an application layer as an upper-level protocol of existing communication protocols like TCP/IP and ZigBee.

The PUCC platform provides these protocols as general-purpose middleware independent of any specific application. Various P2P protocols necessary for P2P communication can be implemented in this platform. A middleware application programming interface (API) enables access to P2P protocols. By installing PUCC middleware on every device or gateway that communicates with other devices, various functions are provided in addition to P2P communication (Kato at el., 2009). We developed a library for communication using PUCC protocols that provides the API which device can participate in the PUCC network and uses methods to enable cooperative behavior with other devices. Methods that are central to our system include the discovery, subscribe, notify, and invoke methods.

The discovery method enables us to obtain information about a device in the network. This information is described in a PUCC format. The subscribe method enables us to know of changes in devices states. The device monitors its own status, and if it changes, it announces the change by using the notify method. The invoke method enables us to execute services involving a device: e.g., turning on an appliance.

The PUCC P2P and gateway nodes have metadata to describe information on published services (Fig. 2). If the gateway node has several devices/nodes connected to another domain network, the gateway might describe several services as either one or several sets of metadata. A metadata sample is shown in Fig. 3.

3.3 Mutually complementary communication

We proposed the use of a mutually complementary communication protocol (MCCP) for indoor sensor/actuator networks based on multi-interface communications (Sawada et al., 2011). The MCCP uses several metrics of link quality indicator (LQI) in each interface to cognize the state of the path and tracks the variation of the network topology for transmitting data. Each node periodically informs the LQI values of each interface, each node is able to select the better interface for transmitting data to the neighbor nodes. The protocol basically follows the link state routing protocol which is well used in the computer communication network.

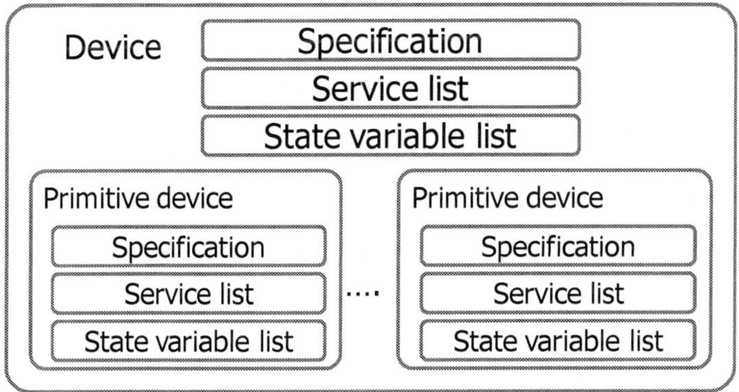

Fig. 2. Overview of PUCC metadata.

```
<Device type="http://pucc/sensor" id="sensor01" name="sensor in labo">
<Specification>
 <Manufacuturer> xxx company </Manufacuturer>
      :  (snip)
</Specification>
<StateVariableList>
 <StateVariable name="temperature" datatype="double" sendEvents="yes">
 <AllowedValueRange>
  <Max> 100 .0</Max>
  <Min> -50 .0</Min>
 </AllowedValueRange>
      :  (snip)
</StateVariableList>
<ServiceList>
 <Service name="getTemperatureValue" type="http://getTemperature" />
 <InputParameterList />
 <OutputParameterList>
  <Parameter name="temperature" type="double" />
 </OutputParameterList>
 </Service>
      :  (snip)
</ServiceList>
</Device>
```

Fig. 3. Sample of PUCC metadata.

The LQI value varies according to the communication environment, that is, the high LQI value shows the environment is good for communication, and the low LQI value shows the environment is bad for communication. Therefore, simply comparing the LQI of all interfaces and select one of them which shows the highest LQI value is one method of switching the interface. The MCCP converts LQI into packet reception rate (PRR) for comparing the interface in the same scale.

In routing, the MCCP follows not only the shortest path routing method but also a policy to select a path which has totally the most high quality links. To increase the choices of the path selection, the protocol is used to consider redundancy paths to a destination as much as possible. Moreover, the MCCP adapts to the variation of topology by updating the routing table dynamically. The network consists of a top router, routers, and end devices. The top one not only routes data to other routers but also manages the whole network. The role of the router is only to route data to other routers. End devices, which are sensor or actuator nodes, send data directly to a router. One router and some end devices are connected in a star topology and make up a subnet. Each subnet is connected with each router in a tree topology.

Each router, including the top one, has its own neighbour table, which is constructed by exchanging hello packets with each one-hop neighbour routers. The hello packet, the structure of the neighbour table, and their usage basically follows the TBRPF neighbour discovery (TND) protocol (RFC 3684). The switching method, which means how to select the interface for transmitting, uses the field of 'Metric' of the neighbour table. The neighbour table is also used for finding out the disconnected link with the neighbour router. If a router finds a disconnected link, it sets the link state information to its neighbour table, and advertises the information to its routing path.

The MCCP uses a table-driven routing method. Each router in the network has its own routing table. The routing table is constructed when the router joins the network, then it updates its own routing table dynamically by exchanging topology update packets periodically with its neighbour routers. When a router joins the network, it constructs a routing table as follows. First, when the router finishes its initialization, it broadcasts a 'Top Router Join' packet to send it to the top router by multi-hop. When the top router receives the 'Top Router Join' packet, it broadcasts a 'Top Router Offer' packet to send it to the router by multi-hop. At this time, the router identifies the shortest path to the top router by counting hop counts of the 'Top Router Offer' packet. The router then sends a 'Top Router Confirm' packet to the top router to advertise that the router could join the network. After joining the network, routers exchange topology update packets with neighbour routers to keep the latest link state consistent.

A network of the MCCP is able to function as an overlay network that consists of multiple networks of different interface. With the overlay approach, user applications do not have to factor in which interface to use when transmitting data. Moreover, as the number of interfaces, redundant paths also increase.

4. Prototype development

4.1 Prototype hardware platform

We gradually developed various sensor/actuator nodes and several enhancement modules to use in our proposed A-EMS. Fig. 4 shows some of the developed prototype hardware.

The ZigBee sensor node (a), the first we developed, is equipped with three sensors for temperature, illuminance, and motion. It can be operated with 4 AAA batteries or an AC adapter. We used Renesas Technology Co.'s MCU as the ZigBee modem module. The user application and device driver programs are also implemented on this MCU.

The power consumption measurement module (b) is one of the enhancement modules we developed that can be connected through the UART serial interface with a ZigBee sensor node. We used a PIC16F877 as the MCU of this enhancement module. This module can measure instantaneous voltage and current values at the same time with a transformer (VT) and a current transmission (CT), respectively, as well as active power, reactive power, apparent power, current consumption, and moment of force. The average power consumption is calculated for the product of the instantaneous value of the measured voltage for four cycles and the instantaneous value of the measured current. The measured precision is within ±2% by using various loads from a low to high power factor that are then compared with Yokogawa Electric Works, Ltd.'s WT230, which is a highly accurate power meter.

The router node (c) consists of a main application board with a ZigBee module and a power line communication (PLC) module. The application layer and the device driver program manage these modules using a real-time OS, Renesas Technology Co.'s M3T-MR30/4, to conform to the μITRON 4.0 specifications. Our connected medium-speed PLC modem enables PLC at about 400 kbps with a frequency band of 2 to 9 MHz.

Recently, we developed power-saving, small-sized IEEE 802.15.4 sensor nodes (d) (e) (f). They work in combination with replaceable boards such as network, living climate sensor, IR remote control, and smart power strip boards. We used a Renesas Technology Co.'s R8C as an MCU. These nodes can be operated with either CR2 batteries or an AC adapter. They can also be connected to additional sensors, such as soil moisture and CO2, via I2C bus.

The sensor node (d) is equipped with four types of sensors for temperature, humidity, illuminance, and motion. The standby power is about 37 μA and the operating power is about 26.8 mA. If we assume intermittent operation with 10-minute intervals, 80% battery efficiency of the 750 mA CR2 battery, and 2.7 V stable operating voltage, the sensor node runs approximately 602 days.

The IR control node (e) has one IR receiving part and four IR emitting parts. This node, which can record 20 IR signal patterns via the IR receiving part, is used to operate appliances that operate via IR control, including TVs, HDD recorders, and air conditioners.

The smart power strip node (f) has the same function as the power consumption measurement module (b). This node can remotely turn on/off connected appliances by attaching a solid state relay (SSR) module. We used a Panasonic Electric Works Co.'s AQA611VL as the SSR. The load current was 40 A.

Fig. 5 shows the block diagram of our developed network board. Each type of sensor node consists of the combination of replaceable boards and it has self-networking and communication abilities. The network infrastructure is deployed through router nodes that can create a multi-hop network by wireless or power line communication. The system is useful for improving network facilities in old buildings or houses that do not have enough network cables.

4.2 Prototype implementation

We implemented a prototype system based on our design shown in Fig. 6. The left side of the figure shows the overlay sensor/actuator network. The sink and coordinator nodes are connected to the sensor gateway, which integrates sensing data from different sensor networks and either stores it or passes it on to client devices or other devices equipped with a PUCC application. To integrate multiple sensor devices, which can be of different types, all sensing data that is not required for real-time features is stored in an integrated DB (a PostgreSQL in this prototype). Information about the sensor device is stored in the metadata and a table for the sensing data is created in the integrated database (DB).

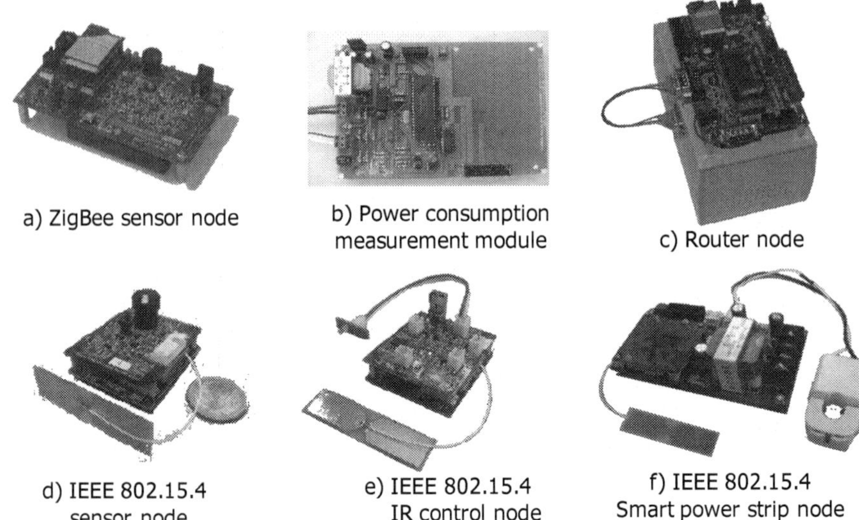

a) ZigBee sensor node

b) Power consumption
measurement module

c) Router node

d) IEEE 802.15.4
sensor node

e) IEEE 802.15.4
IR control node

f) IEEE 802.15.4
Smart power strip node

Fig. 4. Developed hardware platform.

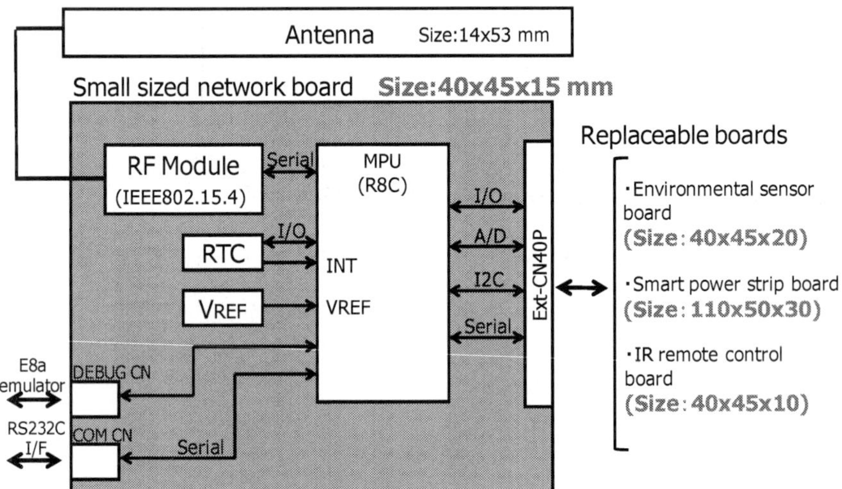

Fig. 5. Developed hardware platform.

The actuator controller module monitors the sensing data selected by the client device and checks if the event occurrence condition is satisfied. If the sensing data satisfies the occurrence condition, the controller module informs the service management module. In addition, if a sensor device that has been monitored is pulled out of the network, this module informs the service management module that event detection has become impossible.

The actuator controller module enables composite event detection using sensing data that belongs to a different sensor gateway. This module executes parsing, creates an event tree for a composite event, and facilitates the monitoring of the selected sensor device. If the sensing data satisfies the event occurrence condition, this module checks whether the value satisfies the service execution condition based on the event tree. If it does, the service management module is informed. If not, this module tells other sensor gateways that have not yet satisfied the event condition that its sensor gateway satisfied the event occurrence condition. On the other hand, if the sensor data falls outside the range of the event occurrence condition, this module informs every sensor gateway related to the composite event.

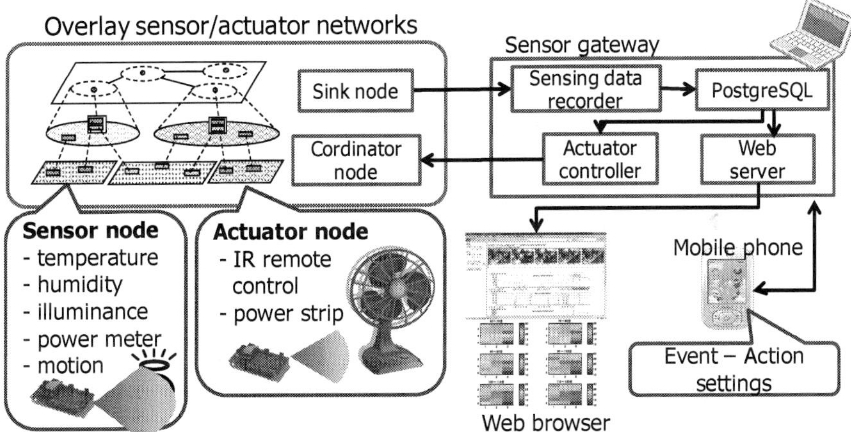

Fig. 6. Prototype implementation.

The sensor gateway processes messages generated when a gateway communicates with other PUCC nodes, such as a client device, through the PUCC platform. On the basis of the sent message, the appropriate process is executed to refer to a DB table or request a definition of services and events from the service management module.

Although the PUCC P2P nodes function as the sensor gateway and the mobile phone in this prototype system, we can freely configure the cooperative behavior of sensors and actuators from any mobile device. It is easy to integrate other new devices after deployment and to develop appropriate systems for indivisuals with different lifestyles through the PUCC platform.

4.3 Controller user interface

As shown in Fig. 7, we implemented the user interface on a client's device. In this case, we controlled the camera based on the status of the sensor nodes. Each element in the system works as described below. We used a Nexus One smart phone equipped with the Android OS and installed our developed PUCC middleware. With this middleware, the user can configure an event detected by a sensor and the service provided by a device.

The client device finds the devices by using a discovery method and generates a GUI from the collected information. The client can then use the GUI to define the event condition (such as temperature > 30 degrees) that sensors can detect. After defining the event condition, the user can also select a service for this condition. When the user finishes the configuration, the middleware sends the event condition to the sensor gateway. When sensors satisfy it, the middleware receives the notify message and then sends an invoke message to the home appliance gateway to execute the service.

The sensor gateway has metadata, information about the device itself, and sensors collecting data, so the client device is able to know what types of sensors are connected to the gateway. When the gateway receives a subscribe message from other devices, it analyzes the event condition and determines if the sensor data matches the condition. If it does, the gateway notifies the user or other gateways with a notify method.

The home appliance gateway has metadata, information about the device itself, and a connected appliance, so the client device is able to know what kind of service is at the connected appliance. When this gateway receives an invoke message from other devices, it analyzes the messages and determines which service to execute.

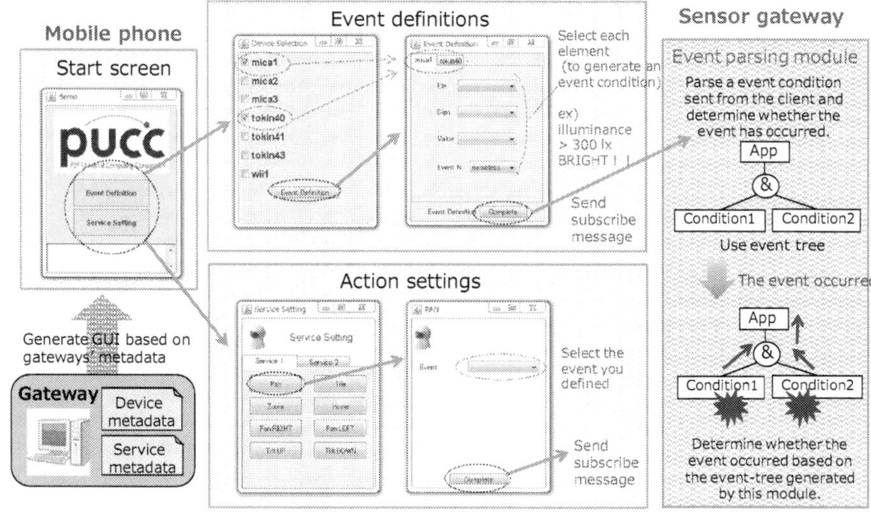

Fig. 7. Controller user interface on mobile phone.

4.4 Visualization user interfaces

We developed several Web interfaces to convert the information gathered by sensors into user-friendly formats, as shown in Figs. 8 and 9. When a user accesses the interface to view selected records (Fig. 8), he or she inputs the ID for the selected sensor, a start and finish period, and the number of records requested (distributed evenly in the time period) with a check box that can handle two or more selsections. Options include light, motion sensor, temperature, humidity, battery power, power consumption, measured voltage, and power factor. This sample shows the trend charts of light, motion, temperature, and power consumption with an amChart [9] that can display the graph with JavaScript and Flash.

The volume of information from the sensors is huge, so when all the information is downloaded for each case, the infrastructure between the Web server and the PostgreSQL server is overloaded. To solve this problem, we implemented a function inside the PostgreSQL server that requests only the required amount of data. The ID, term, and amount of desired data are then transmitted to the PostgreSQL server from the Web server PHP. The selection function gathers the requested information from the tables on the PostgreSQL database in accordance with the timestamp and then sends it back to the PHP. The PHP makes an XML file from the PostgreSQL data and reads it with a drawing program.

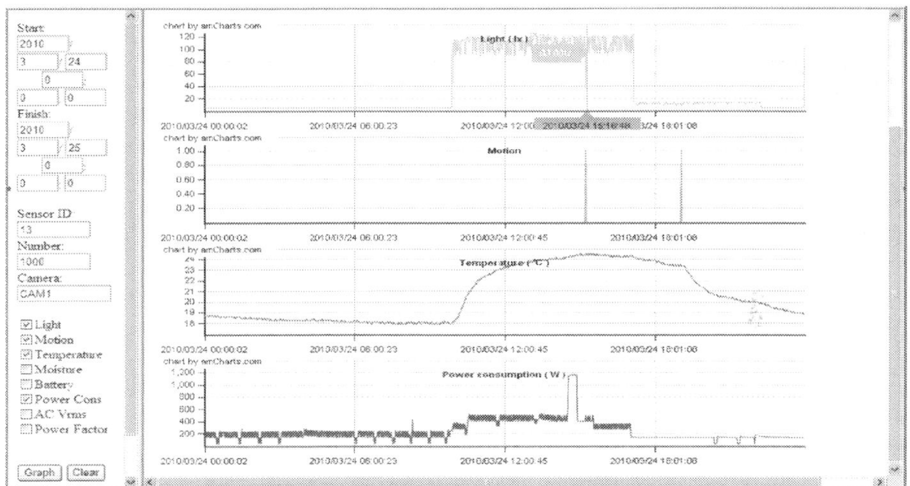

Fig. 8. Visualization user interface for each sensor node.

Fig. 9 shows another visualization user interface for grasping an entire area. The color of each rectangle indicates the measured temperature of each sensor node. Blue is colder and red is hotter. The size of the center circle within each rectangle shows the activity level based on the detection conducted every 15 minutes. The blue color of each center circle depicts the detection of the motion sensor. The small circle on the upper left of each rectangle shows the someone's presence or absence at each work space predicted by simple data mining. This interface reports increasing activity as the circle gets bigger. Users can observe the living climate at any time through the Web browser.

5. Experimental results

5.1 Experimental environment

We deployed our developed nodes in our laboratory as shown in Fig. 10. We installed 54 ZigBee sensor nodes (Fig. 4(a)) on the ceiling, ten power consumption measurement nodes (Fig. 4(b)), six ZigBee router nodes, and a sensor gateway connected to the ZigBee coordinator and the sink nodes. The appliances connected to the power consumption measurement nodes included printers, a refrigerator, a microwave, an electric pot, a plasma display, circuit breakers, and power strips. We installed two types of power consumption measurement nodes: one for plugging appliances (Fig. 11(a)) and the other for attaching to a

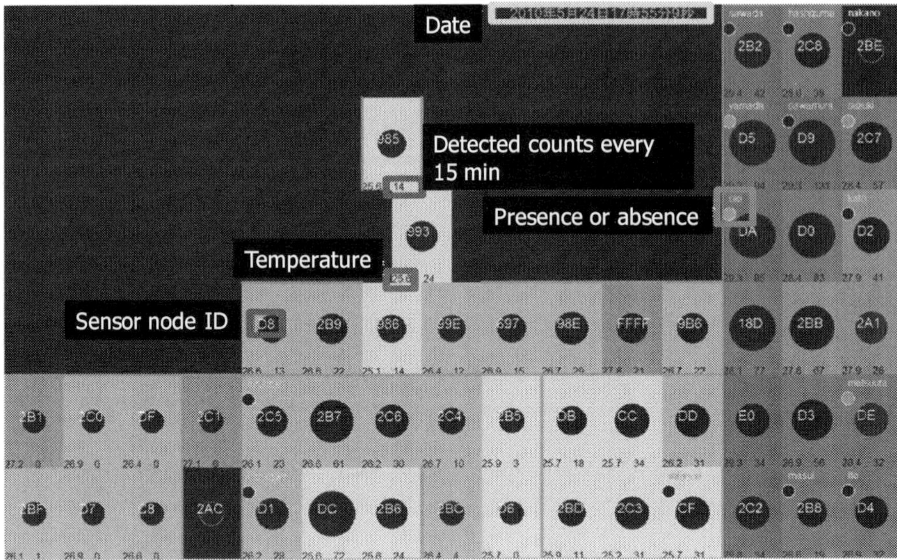

Fig. 9. Visualization user interface for grasping entire area.

circuit breaker (Fig. 11(b)). This experimental environment corresponds to the second installation level of our A-EMS.

To evaluate the accuracy of our developed power consumption measurement node, we also installed another electric power meter (SyoeneNavi, CK-5 and WHM3-SP01, made by Chugoku Electrical Instruments) in the circuit breaker in room J1407.

5.2 Validation of our power consumption measurement module

We measured the accuracy of our power consumption measurement node with SyoeneNavi. Fig. 12 shows the measured power consumption of a circuit breaker in room J1407 from midnight to 11 p.m. on March 23, 2010. Fig. 12(a) shows the power consumption result without any tuning. As for the transition time of power consumption, almost the same results are obtained, though an offset average of a 17 Wh increase is caused overall compared with SyoeneNavi. Fig. 12(b) shows subtracks of offset processing. There was a difference from -33.6 Wh to 24 Wh at the location in which a power consumption of 400 Wh was exceeded through measurements of SyoeneNavi. The result for power consumption of less than 400 Wh with our developed node and SyoeneNavi were the same. It was 5,953 Wh in our developed node and 5,876 Wh in the SyoeneNavi measurement when the entire amount of power consumption on this day was calculated. To make the measurement equal to SyoeneNavi, we should add the offset processing when the error margin rate based on SyoeneNavi is within 1%.

5.3 Evaluation of detailed breakdown function of electric energy

We evaluated the function that breaks down in detail the amount of power consumption. Fig. 13 shows a detailed breakdown of hourly electric energy for each time zone. The bar

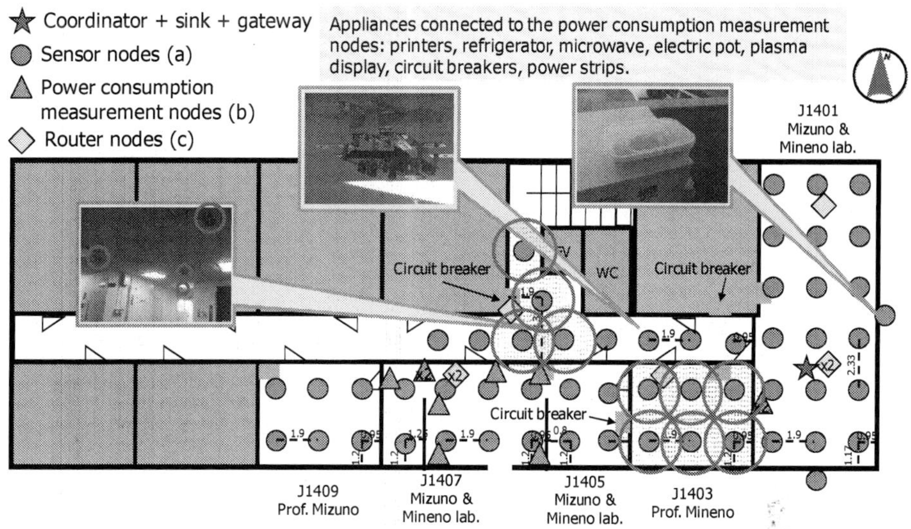

Fig. 10. Experimental environment in our laboratory.

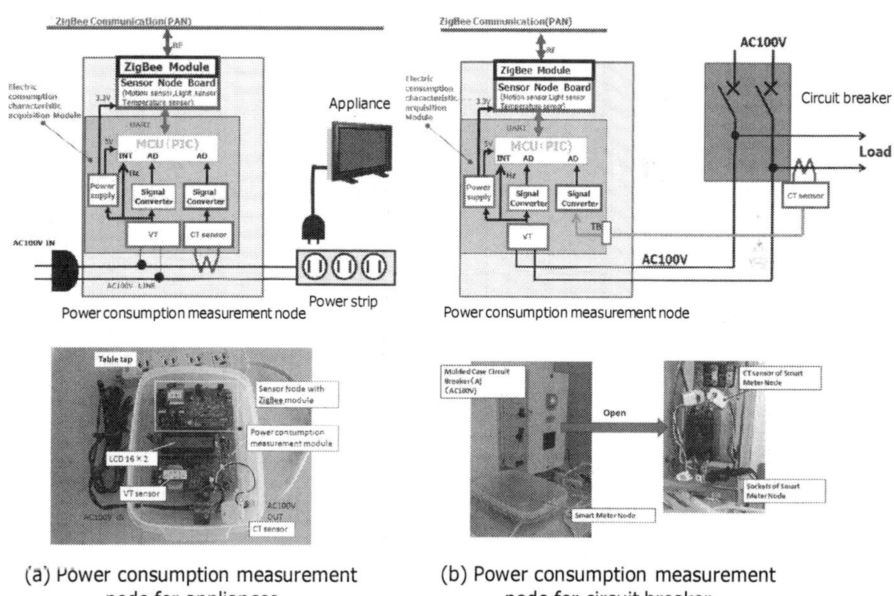

(a) Power consumption measurement node for appliances

(b) Power consumption measurement node for circuit breaker

Fig. 11. Power consumption measurement nodes.

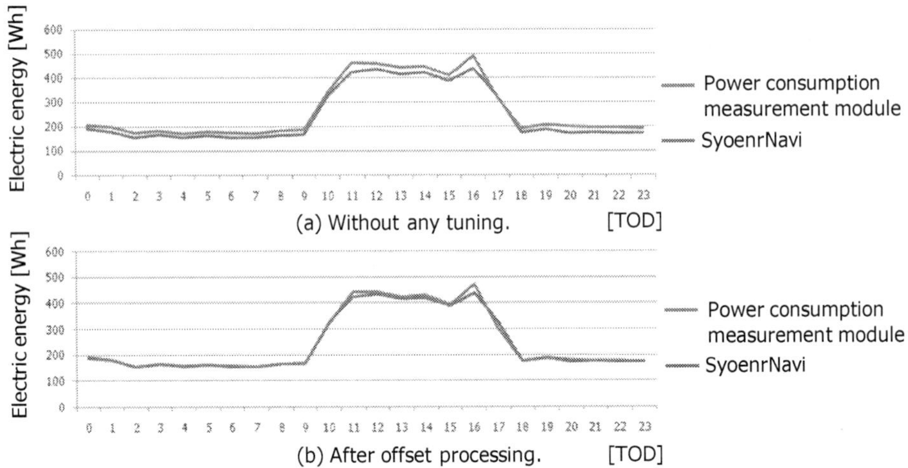

(a) Without any tuning.   [TOD]

(b) After offset processing.   [TOD]

Fig. 12. Validation of our developed power consumption measurement module.

graph indicates individual energy consumption of the connected appliances (refrigerator, electric pot, power strips, etc.) to the power consumption measurement nodes. The maximum value is the measured electric energy at the circuit breaker in room J1407.

We can see an individual's behavior such as turning on a PC, LCD, or printer, from 9 a.m. to 6 p.m. The power consumption of the equipment connected to the power consumption measurement nodes is indicated as part of the power strips. The other parts in Fig. 13 show the power consumption of equipment connected to the buried outlets that were not directly connected to the power consumption measurement node, such as a miniature heater, the access point of a wireless LAN, two cordless handsets, a plasma display, and so on. The data from 6 o'clock to 11 p.m. suggests that the individual's behavior is not seen because the electric energy is lower than 190 Wh. The consumer electronic products with the highest consumption are a refrigerator and an electric pot. The average electric energy of the refrigerator was about 100 Wh and the electric pot was about 62 Wh.

The result of visualizing the individual energy consumption led us to conclude that reducing the power consumption of an electric pot is an energy-saving action, as has been widely alleged. When nobody is using it, the power supply of an electric pot should be unplugged and kept warm for a shorter time. Fig. 13(b) shows the effect of energy-saving actions, such as unplugging an electric pot from 5 p.m. to 9 p.m. The amount of power saving per month was about 29.5 KWh. This enables us to save 649 yen a month based on the charge unit price of 22 yen, the average unit price of the nationwide electric power company. This amounts to electric bill savings of about 8,000 yen per year.

The experimental results showed that a detailed breakdown of the amount of electric energy for each time zone can reveal the waste of power consumption in our daily life. It also gives us the chance to think of better energy consumption habits.

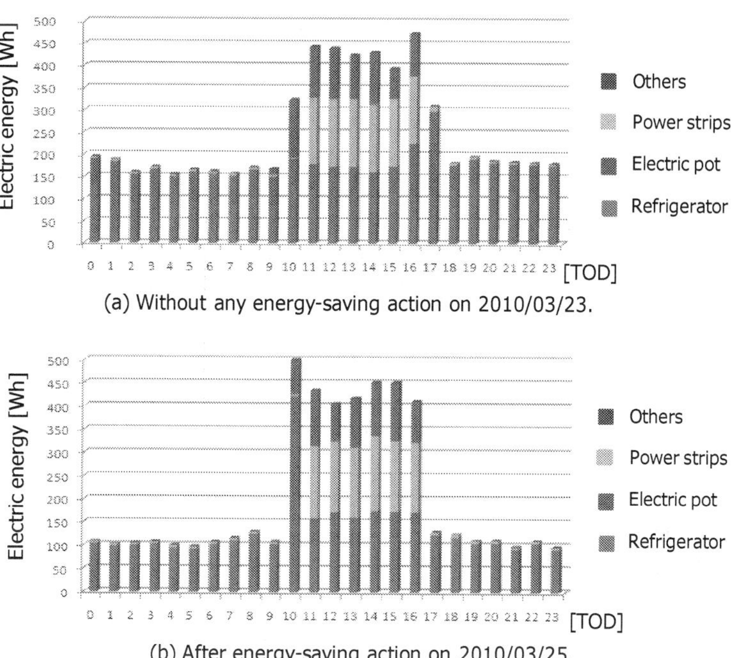

(a) Without any energy-saving action on 2010/03/23.

(b) After energy-saving action on 2010/03/25.

Fig. 13. Detailed breakdown of electric energy data at circuit breaker.

6. Conclusion

We proposed an adaptive architecture for A-EMS that can control energy consumption by maintaining a certain installation level through converged heterogeneous networks. We developed various types of evaluation nodes and several enhancement modules to deploy the proposed A-EMS. We also developed a prototype system that enables users to configure a cooperative network of sensors and actuators from a mobile device. If several multi-vendor sensors/actuators in heterogeneous networks are able to communicate with each other, information that sensors have could be used extensively by many ubiquitous services for a more eco-friendly lifestyle.

Experimental results for our prototype system showed that it is possible to adaptively improve energy efficiency using the proposed A-EMS. For example in our laboratory, the amount of power saving per month was about 29.5 KWh. This enables us to save 649 yen a month based on the charge unit price of 22 yen, the average unit price of the nationwide electric power company. This amounts to electric bill savings of about 8,000 yen per year.

In our future work, we plan to deploy a controlling function based on data mining to activate installation level 3 or more. We will quantitatively evaluate the reduction in energy consumption, and we aim to make out A-EMS a more versatile system by deploying different types of sensors and actuators, developing device searching that uses the location information of the device, and increasing the number of services a user can request.

7. References

Cao, L.; Tian, J. & Zhang, D. (2006). Networked Remote Meter-Reading System Based on Wireless Communication Technology, *IEEE International Conference on Information Acquisition*, pp.172-176, 2006

Inoue, M.; Higuma, T.; Ito, Y. & Kubota, H. (2003). Network Architecture for Home Energy Management System, *IEEE Transactions on Consumer Electronics*, Vol.49, pp.606-613, 2003

Kushiro, N.; Suzuki, S.; Nakata, M.; Takahara, H. & Inoue, M. (2003). Integrated Residential Gateway Controller for Home Energy Management System, *IEEE Transactions on Consumer Electronics*, Vol.49, No.3, pp.629-636, 2003

Zhao, P.; Suryanarayanan, S. & Simoes, M. (2010). An Energy Management System for Building Structures Using a Multi-Agent Decision-Making Control Methodology, *Proceedings of IEEE Industry Applications Society Annual Meeting (IAS)*, pp.1-8, 2010

Mineno, H.; Kato, Y.; Obata, K.; Kuriyama, H.; Abe, K.; Ishikawa, N. & Mizuno, T. (2010). Adaptive Home/Building Energy Management System Using Heterogeneous Sensor/Actuator Networks, *Proceedings of IEEE International Conference on Consumer Communications & Networking Conference (CCNC)*, 2010

IEEE 1451 Standard, http://ieee1451.nist.gov/

Sensor Model Language, www.opengeospatial.org/standards/sensorml/

Device Kit Website, www.eclipse.org/ohf/components/soda/index.php

Chen, C. & Helal, A. (2009). Device Integration in SODA using the Device Description Language, *IEEE International Symposium on Applications and the Internet (SAINT)*, 2009

Peer to Peer Universal Computing Consortium, http://www.pucc.jp/

OSGi Alliance, http://www.osgi.org/

Tzeng, C.; Wey, T. & Ma, S. (2008). Building a Flexible Energy Management System with LonWorks Control Network, *Proceedings of IEEE International Conference on Intelligent Systems Design and Applications (ISDA)*, pp.587-593, 2008

Han, D. & Lim, J. (2010). Smart home energy management sytem using IEEE 802.15.4 and ZigBee, *IEEE Transactios on Consumer Electronics*, Vol.56, No.3, pp.1403-1410, 2010

Park, H; Burke, J. & Srivastava, M. (2007). Design and Implementation of a Wireless Sensor Network for Intelligent Light Control, *Proceedings of IEEE International Conference on Information and Processing in Sensor Networks (IPSN)*, pp.370-379, 2007

Son, Y.; Pulkkinen, T.; Moon, K. & Kim, C. (2010). Home Energy Management System based on Power Line Communication, *IEEE Transactions on Consumer Electronics*, Vol.56, No.3, pp.1380-1386, 2010

Suh, C. & Ko, Y. (2008). Design and Implementation of Intelligent Home Control Systems based on Active Sensor Networks, *IEEE Transactinos on Consumer Electronics*, Vol.54, No.3, pp.1177-1184, 2008

Chen, C. & Helal, A. (2008). Sifting through the Jungle of Sensor Standards, *IEEE Pervasive Computing Magazine*, Vol.7, No.4, 2008

Kato, Y.; Ito, T.; Kamiya, H.; Ogura, M.; Mineno, H.; Ishikawa, N. & Mizuno, T. (2009). Home Appliance Control Using Heterogeneous Sensor Networks, *Proceedings of IEEE Consumer Communications & Networking Conference (CCNC)*, 2009

Sawada, H.; Kuriyama, H.; Nakano, Y.; Mineno, H. & Mizuno, T. (2011). Mutually Complementary Communication Protocol for Indoor Sensor/Actuator Networks, *Proceedings of IEEE/IPSJ Internationa Symposium on Applications and the Internet (SAINT)*, 2011

Ogier, R.; Templin, F. & Lewis, M. (2004). Topology Dissemination Based on Reverse-Path Forwarding (TBRPF), *RFC 3684*, 2004

12

Smart Grid and Dynamic Power Management

Dave Hardin
EnerNOC, Inc.
USA

1. Introduction

Historically, energy has been relatively inexpensive. Efforts to manage the efficient use of electrical energy have been of secondary importance and often limited to initial architectural and design considerations. Inexpensive and widely-available energy has led to unprecedented economic growth but the costs and risks are increasing: the costs of fossil fuels, the costs to the environment, and the risks to foreign supplies.

With the passage of the Energy Independence and Security Act of 2007, the United States embarked on a path to modernize the electrical grid as described in Title XIII – Smart Grid. (US Title XIII, 2007) This modernization is transforming how energy is generated, transmitted, distributed and consumed in residential, commercial and industrial facilities but it is not changing the basic electrical constraints of the system.

Electrical supply and demand must remain in balance at all times. This balance has traditionally been attained through dispatching generation and day-ahead scheduling along with sufficient capacity reserves. Temporal load change typically follows a macro pattern based on diurnal or daily variation. Power usage increases during the day and decreases at night. It is this cycle, or load curve, which drives modern grid operations. Sufficient reserve capacity is required to meet any demand peaks. Generation failures and circuit trips also require that reserves be brought on-line. When, for any reason, supply does not equal demand, the grid can collapse resulting in a blackout.

2. What is Smart Grid

The U.S. Energy Independence and Security Act of 2007, TitleXIII and the NIST (National Institute of Standards and Technology) Smart Grid Framework (SG Roadmap, 2010) describe the goals and objectives of Smart Grid. EISA Title XIII defines the following characteristics of Smart Grid:

1. "Increased use of digital information and controls technology to improve reliability, security, and efficiency of the electric grid.
2. Dynamic optimization of grid operations and resources, with full cyber-security.
3. Deployment and integration of distributed resources and generation, including renewable resources.
4. Development and incorporation of demand response, demand-side resources, and energy-efficiency resources.

5. Deployment of `smart' technologies (real-time, automated, interactive technologies that optimize the physical operation of appliances and consumer devices) for metering, communications concerning grid operations and status, and distribution automation.
6. Integration of `smart' appliances and consumer devices.
7. Deployment and integration of advanced electricity storage and peak-shaving technologies, including plug-in electric and hybrid electric vehicles, and thermal-storage air conditioning.
8. Provision to consumers of timely information and control options.
9. Development of standards for communication and interoperability of appliances and equipment connected to the electric grid, including the infrastructure serving the grid.
10. Identification and lowering of unreasonable or unnecessary barriers to adoption of smart grid technologies, practices, and services."

FERC (Federal Electricity Regulatory Commission) outlined the top eight (8) U.S. National Smart Grid priorities as:

"Wide-area situational awareness: Monitoring and display of power-system components and performance across interconnections and over large geographic areas in near real time.

Demand response and consumer energy efficiency: Mechanisms and incentives for utilities, business, industrial, and residential customers to cut energy use during times of peak demand or when power reliability is at risk.

Energy storage: Means of storing energy, directly or indirectly.

Electric transportation: Refers, primarily, to enabling large-scale integration of plug-in electric vehicles (PEVs).

Cyber security: Encompasses measures to ensure the confidentiality, integrity and availability of the electronic information communication systems and the control systems necessary for the management, operation, and protection of the Smart Grid's energy, information technology, and telecommunications infrastructures.

Network communications: The Smart Grid domains and subdomains will use a variety of public and private communication networks, both wired and wireless.

Advanced metering infrastructure (AMI): Currently, utilities are focusing on developing AMI to implement residential demand response and to serve as the chief mechanism for implementing dynamic pricing.

Distribution grid management: Focuses on maximizing performance of feeders, transformers, and other components of networked distribution systems and integrating with transmission systems and customer operations."

The U.S. NIST and the Smart Grid Interoperability Panel (SGIP) created the Smart Grid Conceptual Model (SGIP CM, 2010) which describes the seven (7) primary domains that comprise Smart Grid: Bulk Generation, Transmission, Distribution, Customer, Markets, Operations and Service Provider. (See Figure 1)

"The Smart Grid Conceptual Model is a set of views (diagrams) and descriptions that are the basis for discussing the characteristics, uses, behavior, interfaces, requirements and standards of the Smart Grid." (SGIP CM, 2010)

The two domains with the greatest direct impact on the electrical supply chain are the customer (See Figure 2) and bulk generation (See Figure 3) domains as they form the core drivers for change in the electrical system.

The other domains will, in general, need to adapt to the changes in these two domains but all domains are interconnected and therefore affect each other. Changes occurring in the wholesale and retail markets will directly impact other domains. New services and service

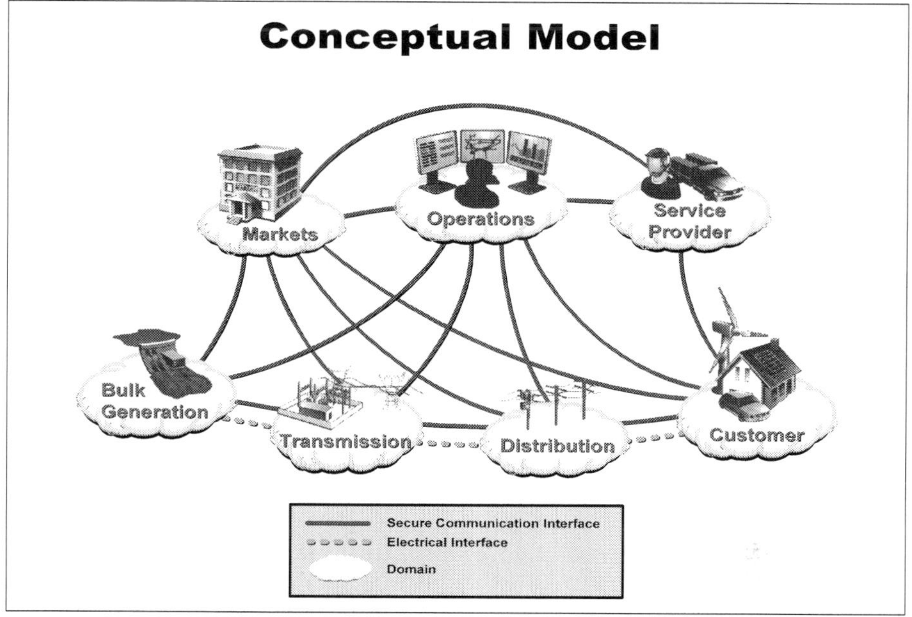

Fig. 1. Smart Grid Conceptual Model

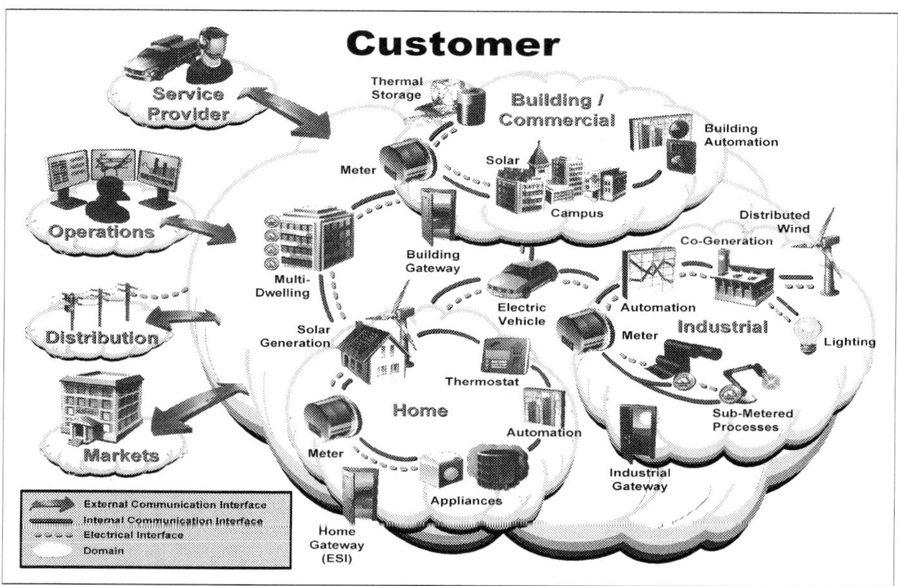

Fig. 2. Customer

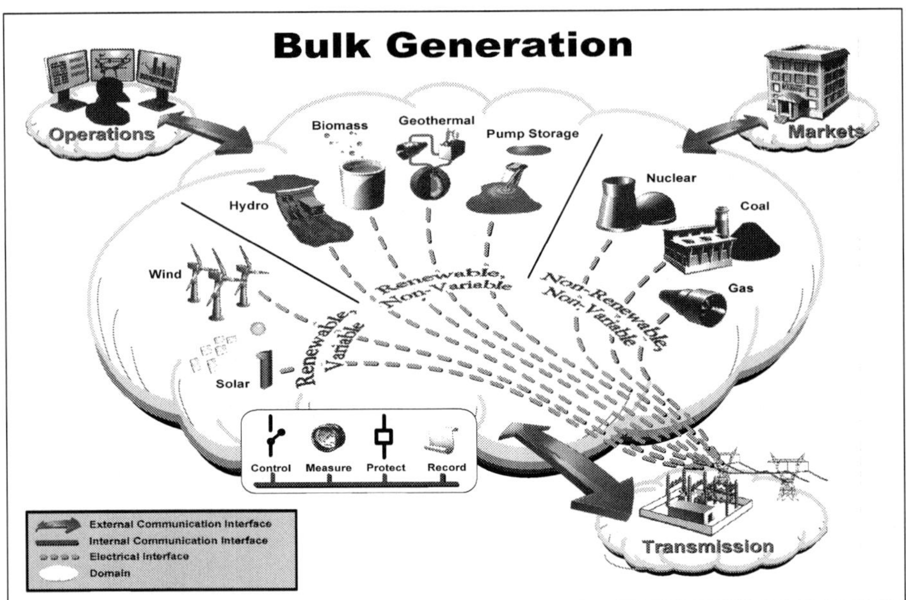

Fig. 3. Bulk Generation

providers will enable new capabilities which will be consumed by other domains. The operations domain integrates and balances network resources with the objective of achieving safe, secure and reliable real-time operations of the power system.

The bulk generation domain is categorized into: 1) non-renewable, non-variable, 2) renewable, non-variable and 3) renewable, variable generation. The first two categories represent traditional generation that can be dispatched when needed. The third category represents a new challenge for the grid.

Within the bulk generation domain, large quantities of renewable generation need to be integrated into the grid. The ideal generation would be in the form of renewable, non-variable. This would permit the generation source to be dispatched by the regional balancing authority. Renewable, variable generation such as wind and solar require fast-responding reserve generation such as spinning reserves or natural gas turbines to take over when the wind stops blowing or the sun becomes blocked by clouds. This requirement adds significant costs and impedes the growth of variable renewables, even if the occurances are rare. Renewable generation on the grid currently amounts to 4% of the overall generation. The goal of increasing this to 30% will result in a grid that has significantly more variability than the current grid. Could a more cost effective and reliable approach include bringing customer energy curtailment resources into the feedback loop through the use of dispatchable high-performance demand response?

3. Smart Grid feedback loops

Bringing the customer further and further into the energy loop is an important facet of Smart Grid development that requires more analysis.

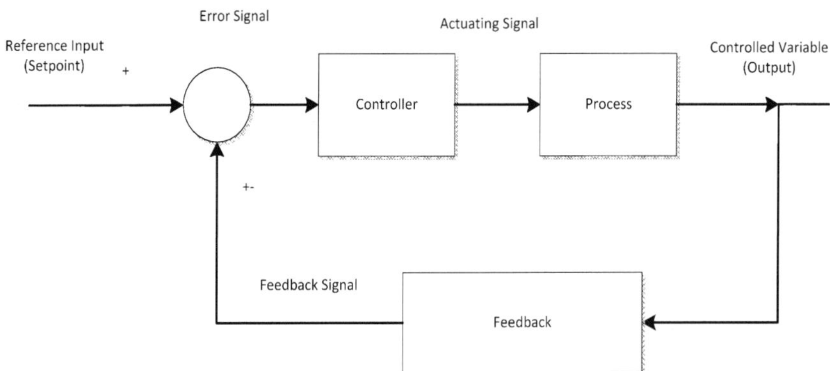

Fig. 4. Balancing Feedback Loop

Smart Grid is a system of systems tied together with large, wide-area feedback loops. These feedback loops constitute the basic behavioral operating unit of a system of systems. (Meadows, 2008) They can bring either stability or instability to the system. They can create growth or shrinkage of the system.

Feedback loops return an amplified portion of the output signal back around to the input where it either adds or subtracts from the input signal. This simple basic structure forms the foundation for automatic control theory which is widely applied within a number of domains including manufacturing automation, aircraft control and automotive systems.

If the fedback signal tends to subtract from, or offset, the input signal and decrease the output, it is a negative or balancing feedback loop. If the fedback signal adds to the input signal, it is a positive or reinforcing loop. The system and feedback loop have transfer functions, usually expressed in terms of Laplace transforms, which relate the output signal to the input signal. The behavior of the loop when any given input signal is applied can then be determined. The transfer function has solutions called poles and zeros under which it either drives the loop toward oscillation or becomes zero. Both conditions have negative impact if the loop is a balancing loop.

An example of a simple on/off balancing loop is the home thermostat. The desired balance point is the temperature setpoint. The feedback signal is the room temperature. When the room temperature reaches the setpoint temperature the heater is turned off until the temperature decreases below the setpoint. This digital loop inherently oscillates and relies upon the high capacity and slow response of the room and heating system to achieve acceptable results.

Reinforcing feedback loops amplify the output by building upon themselves resulting in exponential growth or collapse. The rate of growth is determined by the amount of feedback or gain.

An example of a simple reinforcing loop is compound interest where the interest earned on a financial account is fedback into the account resulting in the exponential growth of the account value over time.

A fundamental property of feedback loops is that they have a propensity to oscillate. This oscillation is caused by loop time-delays, or deadtime, that lead to the phase-shifting of feedback signals. If the resulting phase-shift is equal to 180 degrees, then a negative feedback signal turns into a positive feedback signal. This causes balancing loops to become reinforcing loops and if the strength of the feedback is sufficient (i.e. product of loop gains >= 1.0) , they become unstable and oscillate. Sufficient upfront system design is required so that this condition does not arise. (Shinskey, 1979)

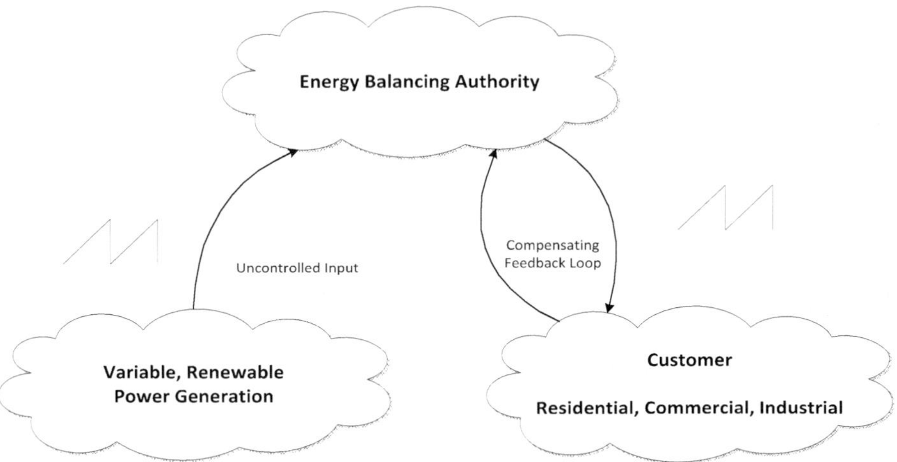

Fig. 5. Dynamic Power Feedback Loop

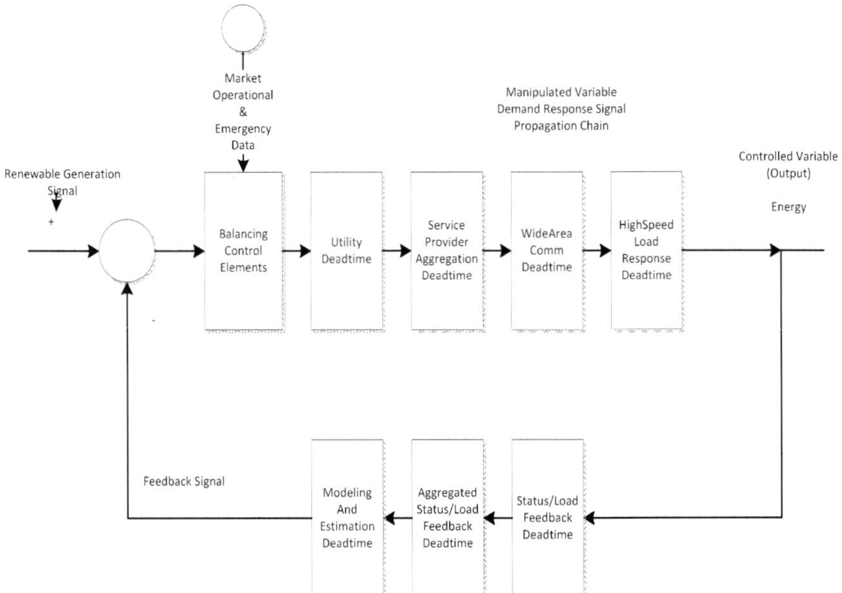

Fig. 6. Simplified Renewable Generation Demand Response Loop Diagram

Complex systems of systems are affected by large numbers of interacting feedback loops. Some of these loops have little effect on overall system behaviour while other loops can dominate system behaviour. In this context, an important and dominant smart grid feedback loop is the one that connects variable, renewable generation, such as wind and solar energy, with the power consumption of the customer. This loop is shown in Figure 5.

The balancing authority has the responsibility to maintain the electrical grid in balance at all times with the power supply equal to the power demand. As power from wind and solar generation is fed onto the grid, the balancing authority only has the ability to dispatch a decrease in renewable generation by disconnecting it from the grid but not the ability to increase its power output. Compounding this issue is the power variability due to wind and solar fluctuations. This is in contrast to traditional grid operations which typically vary over a 24-hour period with reserve generation capacity being brought online or taken offline based on demand load. PJM studies have shown significant impact on the bulk electrical system due to wind variability with a corresponding impact on the LMP (locational marginal pricing) wholesale energy price. (See Figure 7)

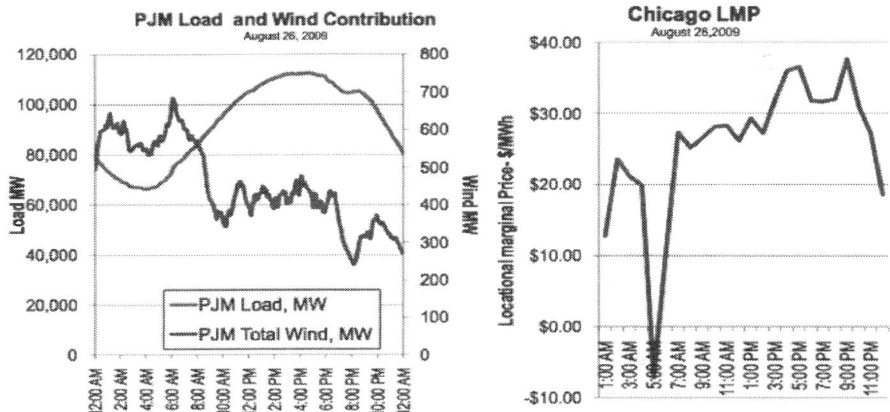

Source: PJM

Fig. 7. Wind Variability

The balancing authority can compensate for this variability by dispatching fast-responding generation, adding sufficient power storage capacity and fast ramp-down of customer load. All of these options however have an associated cost and response time. (Hirst & Kirby, 2003) Fast-responding generation in the form of spinning reserves or natural gas turbines are effective but very costly. Bulk storage represents a very good solution in theory but economical grid-scale storage systems are still being developed. Reducing customer load through energy demand response represents a solution that has already been proven successful in its ability to provide the dispatchable curtailment of large quantities of power but its use as a high-speed compensator for renewable variability represents an area of growth and opportunity. (Kalisch, 2010)

The feedback system consisting of renewable generation, a balancing authority, utilities, service providers and customers can be described by a simplified renewable generation demand response feedback loop diagram, Figure 6. This loop does not include fast-response generation or power storage elements.

The demand response loop is being driven by the uncontrolled renewable generation signal. If the renewable generation decreases, then the signal to the loop calls for a decrease in customer demand. The signal then propagates through several control and time-delay elements before aggregated customer power ramp down occurs. The feedback signal provides near real-time information including state and status data along with the actual power curtailment. Based on this information, the loop balances the curtailment with the generation. In order to remain stable and not oscillate, the loop needs to respond faster than the renewable generation driving signal.

This is very similar to industrial supervisory setpoint control. Supervisory control often utilizes a cascade loop consisting of a primary outer-loop which sends a setpoint signal to a secondary inner loop. The inner loop has faster response dynamics than the outer loop allowing it to track changes in the supervisory setpoint without becoming unstable and oscillating. Many considerations, such as loop windup, need to be taken into account due to the interactions between these two loops. (Skinskey, 1979)

The importance of inner loop response time means that the time-delays and latencies within a demand response loop need to be minimized as much as possible so that the overall loop response can be minimized. This includes both communication latencies and process delays. Applying this concept to demand response for renewable energy, the resulting loop dynamics determine how fast and effective the demand response loop will be in compensating for variations in renewable generation. The faster the loop response, the more effective demand response will be in mitigating real-time variance in renewable generation. One of the primary elements that contributes time-delays is customer load response.

4. Customer load response

Customer demand response can be characterized by the magnitude and speed of load response. This applies to both dynamic pricing and demand response event signalling. Four categories have been identified for classifying demand response performance. Each category, described below, will have different feedback loop dynamics and will affect the customer in different ways. Systems with large energy storage capacity are ideal for demand response applications in all categories listed.

Category 1: soft demand response

The response time required in soft demand response is often flexible and can vary from hours to days. Soft demand response events are targeted at the daily power consumption macro cycle which is driven by higher usage during the day followed by lower usage during the night. Energy curtailment can typically be planned and scheduled in advance.

Load response strategies include both load shedding as well as load shifting. Load shedding involves curtailing equipment that is not mission critical and load shifting is the rescheduling of energy-intensive operations to a different time period. This includes production lines and processing equipment.

Equipment typically curtailed includes:
1.   External and internal lighting including parking lot lighting
2.   External water fixtures
3.   Air handlers

4. Anti-sweat heaters
5. Chiller controls and chilled water systems
6. Defrost elements
7. Elevators and escalators
8. HVAC (Heating, Ventilation and Air Conditioning) Systems
9. Irrigation pumps
10. Motors
11. Outside signage
12. Pool pumps and heaters
13. Refrigerator systems
14. Water heating systems

The load response times of these systems vary from seconds to hours. Longer response times can be accommodated through pre-ramp down control strategies while equipment with faster response times can be actuated directly.

Category 2: firm demand response

The response time required in firm demand response varies between five (5) minutes and ten (10) minutes. This aligns with ten-minute wholesale ancillary markets.

Firm demand response provides the grid balancing authority with the ability to balance a reduction in generation capacity with a compensating reduction in load. This category is appropriate for balancing variable renewable generation that has sufficient inertia, capacity or prediction.

Examples of equipment typically capable of firm demand curtailment include:
1. External and internal lighting including parking lot lighting
2. External water fixtures
3. Air handlers
4. Elevators and escalators
5. Irrigation pumps
6. Motors
7. Outside signage
8. Pool pumps

Category 3: near realtime demand response

Near realtime demand response requires response times of one (1) minute to five (5) minutes. These are appropriate for fast responding ancillary energy markets driven by significant quantities of variable renewable generation.

Only equipment capable of high speed ramp down can participate in near realtime demand response. Typical examples include:
1. External and internal lighting including parking lot lighting
2. External water fixtures
3. Air handlers
4. Irrigation pumps
5. Motors
6. Outside signage
7. Pool pumps

Category 4: realtime demand response

Realtime demand response require response times from one (1) second to one (1) minute. These applications include power frequency and load regulation as well as emergency response to grid faults. Realtime response requires very high speed equipment shutdown capability as provided by motor-driven equipment or lighting.

In general, the ease with which a customer can react will decrease moving from category 1 to category 4. In order to achieve five (5) minute down to one (1) minute response, the decision making processes involved in load shedding, shifting or shaping must be automated and streamlined in order to provide a high degree of determinism and reliability. Demand response signals will contain both discrete and continuous information. Discrete information will often be in the form of dispatch triggers that initiate action. Continuous information will be in the form of value metrics such as dynamic pricing which will be used as input into decision-making algorithms.

5. Commercial and industrial dynamic power management strategies

The electrical energy consumed and produced within commercial and industrial (C&I) facilities represents a major percentage of the overall electrical energy consumed in the United States. The Department of Energy (DOE) estimates (US EIA, 2011) that 50% of the electrical energy produced in the United States is consumed within the commercial and industrial sectors. Residential homes consume an additional 22%. Commercial and industrial facilities have large power footprints distributed over a relatively small number of sites resulting in power densities that provide economies of scale and increase the potential impact these facilities can have on the bulk electric system.

This potential impact is offset by the primary business objective of commercial and industrial facilities to provide products and services for their customers. Electrical power is one of many resources necessary to produce these products and services. The level of interaction of any specific C&I customer with demand response signals can be directly related to the economic impact that electrical energy has on its operations coupled with the operational flexibility of rescheduling production. The more energy required producing products and services, the more effectively dynamic power management techniques can be applied.

Large commercial and industrial facilities consist of complex processes through which raw materials and other resources are combined and transformed into useful products. The ISA-SP95 standard consists of a four (4) layer model which describes how and where decisions are made concerning manufacturing processes. (ISA 95) (See Figure 8)

The four layers include:

- Level 4 – Business
- Level 3 – Operations
- Level 1 and 2 – Control

Dynamic power management decisions can occur within each of these layers. Decisions at Level 4 represent business decisions where the response to grid signals can be planned and optimized in context with the business as a whole. Decisions at Level 3 represent operational decisions where the response to grid signals is determined by supervisory systems in context with manufacturing operations. Decisions at Level 1 and 2 represent control decisions where the response is determined by control system logic running in programmable logic controllers and other automation devices.

Each level is characterized by both the amount of load reduction available coupled with the ramp rate of that load reduction. Decisions made at higher levels can typically provide more load reduction but require longer time intervals while decisions made at lower layers can provide faster response but provide less load reduction. The overall response of a facility will be determined by the contributions of all levels.

ISA 95 Levels – Distinct Sets of Activities

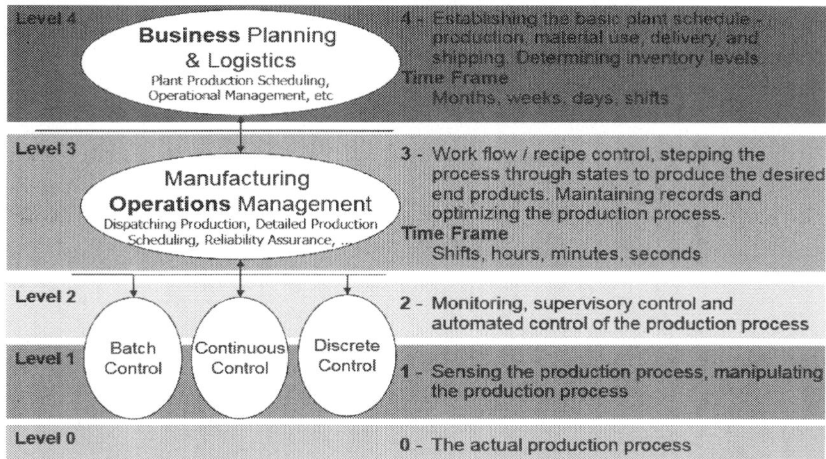

Fig. 8. ISA 95 Levels[1]

Demand response signals enable C&I customers to locally manage and optimize their energy production and usage, dynamically in real-time, as an integral participant in the electrical supply chain. These interactions permit customers to adapt to changing conditions in the electric system but they also require the use of advanced automation and applications in order to fully achieve the potential benefits.

An example of a typical interaction involves a manufacturer that bids demand response load reduction into a 5-min reserves ancillary market of the local balancing authority through a local service provider. These contingency reserves provide fast ramping of demand resources in the event of a generator or line trip. The manufacturer interfaces grid dispatch signals from the service provider directly to the industrial automation system in order to execute fast-ramp down of several large loads that can be interrupted without affecting the production line. The service provider receives the dispatch event and cascades the event to all participating industrial sites. In some cases, there will be fewer participants localized within a constrained region but in other cases, there will be large numbers of participants spread over a large region. Each site must receive the signal in a timely fashion to maximize its ability to reduce load in the short time window provided. The on-site dynamic power management system monitors the event and feeds back real-time event performance to the service provider. The service provider in turn summarizes and feeds back to the balancing authority concerning overall reserve capacity provided.

This is one of many scenarios and markets that will require C&I customers to respond rapidly and efficiently to demand response signals originating from the grid.

---

[1]Used with permission, Dennis Brandl, 2011

6. Smart grid technology trends

Smart Grid enables two technologies that have a direct impact on the dynamic management of energy. These are; 1) microgrids and distributed energy generation and 2) transactive energy. Most C&I facilities are consumers of electrical energy but only a subset generate power on-site. Distributed generation permits more facilities to generate on-site energy and become self-contained microgrids (Galvin & Yeager 2008) connected to the electrical system. These microgrids will benefit both the electrical distribution system as well as the facility while helping to optimize the system-wide generation and consumption of energy.

Microgrids are self-contained, grid-connected energy systems that generate and consume on-site power. These systems can either import power from, or export to, the grid as well as having the capability to disconnect (or island) from the grid. The decision making process required to determine the best mode of operation requires taking into consideration both local operations as well as grid operations.

When external power cost is relatively high, a strategy based on exporting excess power generation and minimizing imported power would be the best course of action. If the cost of external power goes below the cost of self-generated power, then maximizing the power imported from the grid while decreasing on-site generation would be a suitable strategy. If an emergency or fault occurs on the external grid, the microgrid load can be curtailed or disconnected from the grid and reconnected when conditions permit.

The infrastructure needed to manage power supply and demand in context with the power grid enables the economically-viable expansion of on-site microgrid generation to include renewables and storage. These distributed energy resources (DER) are then presented as assets to the grid while being maintained and supported within the microgrid. Renewable generation includes not only solar and wind farms but also power harvested from process by-products or process energy stored as heat or pressure.

Today's centralized control of the power grid will evolve toward distributed control with more localized, autonomous decision making. These decision-making "software agents" will interact with other agents to optimize the energy utilization of connected devices and systems. These interactions, known as transactive energy, will be in the form of transactions with other systems which will be based on local economics and context.

Wholesale markets provide customers and service providers with the ability to bid large resources (typically greater than 1 MW) while retail markets will enable smaller energy transactions to occur as they become economically viable. These can be considered "micro transactions" and will occur between energy providers and consumers.

The microgrid is one form of autonomous system but as transactions involving the buying and selling of retail power evolve toward smaller and smaller entities, decision making will become more and more granular. Energy transactions could occur between components within microgrids, between microgrids, between microgrids and even smaller self-contained energy systems such as "nanogrid" homes and buildings.

Transactive energy does not change the requirement that the power grid must operate in a stable state of equilibrium with supply equal to demand. Autonomous market-driven behaviour creates system oscillations and instabilities through positive reinforcing feedback cycles. This behaviour can be very detrimental for grid-scale operations and must be managed proactively to avoid negative side effects.

As with variable renewable generation, an increase in the use of value-based economic or market-derived signals, such as dynamic pricing, to modulate energy consumption will increase the dynamics of the power grid. These value-based signals need to be injected into the customer feedback loop so that acceptable stability is maintained. Techniques must be implemented that limit the operating range within which market activity is permitted.

These techniques need to not only limit the acceptable operating range but must also limit by rate-of-change and duration.

7. Conclusion

Dynamic power management is a key enabler for the integration of large quantities of renewable power generation onto the electrical grid. These renewable energy resources will significantly increase the variability of electrical power and impact the dynamics and stability of the power grid. Maintaining a reliable and stable grid will require that these dynamics be balanced in real-time.

Smart Grid enables customers to dynamically manage power usage based on electrical grid operating conditions and economics. Through systems integration, grid stability and reliability are enhanced while the customer benefits from lower costs and more reliable electrical power.

An important method for providing grid balancing is through the use of compensating negative feedback loops which leverage customer demand to offset variation in supply. These feedback loops will have an inherent tendency to oscillate if not designed and operated within acceptable boundary constraints relating to closed loop gain and phase shift caused by time delays and latencies.

These closed loop constraints subsequently bind the time requirements for customer load response. This increases the importance of determistic response time when integrating customer demand response and dynamic power management strategies with real-time power grid operations.

Customer demand response is not limited to load reduction. Comprehensive dynamic power management strategies integrate on-site convertible process energy storage, distributed renewable generation and CHP (combined heat and power) co-generation into a portfolio of distributed energy resources (DER) with a range of response and load capability. Resources that provide fast-enough response can participate as active elements in the closed renewable generation demand response feedback loop.

8. Acknowledgement

The author would like to acknowledge the extensive and excellent work being carried out by the U.S. Department of Energy, the U.S. National Institute of Standards and Technology and the many individuals and organizations actively involved in the Smart Grid Interoperability Panel.

9. References

Galvin, Robert & Yeager, Kurt. (2008). *Perfect Power: How the Microgrid Revolution Will Unleash Cleaner, Greener, More Abundant Energy*. McGraw Hill, ISBN: 978 0071548823.

Meadows, Danella H. (2008). *Thinking in Systems: A Primer*. Chelsea Green Publishing, ISBN: 1603580557.

Fox-Penner, Peter. (2010). *Smart Power: Climate Change, the Smart Grid, and the Future of Electric Utilities.* Island Press, ISBN: 978-1-59726-706-9.

Shinskey, F Greg. (1979). *Process Control Systems.* McGraw-Hill, ISBN: 0-07-056891-X.

Hurst, Eric & Kirby, Brendan. (2003). *Opportunities for Demand Participation in New England Contingency-Reserve Markets,* New England Demand Response Initiative.

ISA 95, *Manufacturing Enterprise Systems Standards,* The International Society of Automation, 67 Alexander Drive, PO Box 12277, Research Triangle Park, NC, 27709.

Kalisch, Richard. (2010). *Following Load in Real-Time and Ramp Requirements.* Midwest ISO.

SG Roadmap. (2010). *NIST Framework and Roadmap for Smart Grid Interoperability Release 1.0,* NIST Special Publication 1108.

US EIA (Energy Information Administration) (2011). *Electric Power Annual.*

SGIP CM. (2010). NIST SGIP *Smart Grid Conceptual Model Version 1.0.*

US Title XIII. (2007). *Energy Independence and Security Act of 2007,* United States of America.

Yin, Rongxin, Peng Xu, Mary Ann Piette, and Sila Kiliccote. "*Study on Auto-DR and Pre-cooling of Commercial Buildings with Thermal Mass in California.*" Energy and Buildings 42, no. 7 (2010): 967-975. LBNL-3541E.

Kiliccote, Sila, Pamela Sporborg, Imran Sheikh, Erich Huffaker, and Mary Ann Piette. *Integrating Renewable Resources in California and the Role of Automated Demand Response.,* 2010. LBNL-4189E.

Kiliccote, Sila, Mary Ann Piette, Johanna Mathieu, and Kristen Parrish. "*Findings from Seven Years of Field Performance Data for Automated Demand Response in Commercial Buildings.*" In 2010 ACEEE Summer Study on Energy Efficiency in Buildings. Pacific Grove, CA, 2010. LBNL-3643E.

Rubinstein, Francis, Li Xiaolei, and David Watson. *Using Dimmable Lighting for Regulation Capacity and Non-Spinning Reserves in the Ancillary Services Market. A Feasibility Study.,* 2010. LBNL-4190E.

Demand Management and Wireless Sensor Networks in the Smart Grid

Melike Erol-Kantarci and Hussein T. Mouftah
School of Information Technology and Engineering, University of Ottawa
Ontario, Canada

1. Introduction

The operation principles and the components of the electrical power grid are recently undergoing a major renovation. This renovation has been triggered by several factors. First, the grid recently showed signs of resilience problems. For instance, at the beginning of 2000s, California and Eastern interconnection of the U.S. experienced two major blackouts which have caused large financial losses. The second factor to trigger the renovation of the grid is that in a near future, the imbalance between the growing demand and the diminishing fossil fuels, aging equipments, and lack of communications are foreseen to worsen the condition of the power grids. Growing demand is a result of growing population, as well as nations' becoming more dependent on electricity based services. The third factor that triggers the renovation, is the inefficiency of the existing grid. In (Lightner et al., 2010), the authors present that in the U.S. only, 50% of the generation capacity is used 100% of the time, annually, while over 90% capacity is only required for 5% of the time where the figures are similar for other countries. Moreover, more than half of the produced energy is wasted due to generation and transmission related inefficiencies (Lui et al., 2010). This means that the operation of the power grid is rather inefficient. In addition to those resilience and efficiency related problems, high amount of Green House Gases (GHG) emitted during the process of electricity generation need to be reduced as the Kyoto protocol is pressing the governments to reduce their emissions. The renovation targets to increase the penetration level of renewable energy resources, hence reduce the GHG emissions. Finally, the power grids are not well protected for malicious attacks and acts of terrorism. Physical components of the grid are easy to reach from outside and they can be compromised unless they are monitored well.

To address the above mentioned problems, the U.S., Canada, the E.U. and China have recently initiated the smart grid implementations. Smart grid aims to integrate the opportunities that have become available with the advances in Information and Communications Technology (ICT) to the grid technologies in order to modernize the operation and the components of the grid (Massoud-Amin & Wollenberg, 2005). The basic building blocks of the smart grid can be listed as; the assets, sensors used to monitor those assets, the control logic that realizes the desired operational status and finally communication among those blocks (Santacana et al., 2010). These layers are presented in Fig. 1.

The priorities of the governments in the implementation of the smart grid may be different. For instance, the U.S. focuses on energy-independence and security while the E.U. is more

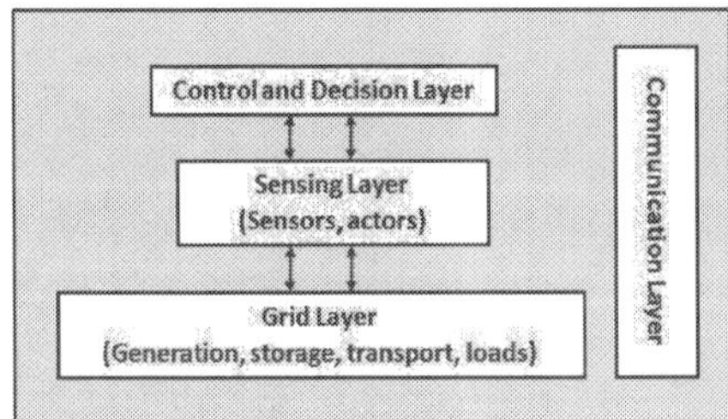

Fig. 1. Building blocks of the smart grid.

concerned about environmental issues and integrating renewable resources. On the other hand, China targets efficient transmission and delivery of electricity. The objectives that are set forward for smart grid implementation can be summarized as:

- Integrating renewable energy sources

- Enabling two-way flow of information and electricity

- Self-healing

- Being environment-friendly

- Enabling distributed energy storage

- Having efficient demand management

- Being secure

- Integrating Plug-in Hybrid Electric Vehicles (PHEV)

- Being future proof

An illustration of a city with smart grid is presented in Fig. 2. The illustration shows distributed renewable energy generation and storage, consumer energy management, integration of PHEVs, and communication between the utility and the parts of the grid.

Among the objectives of the smart grid, demand management will play a key role in increasing the efficiency of the grid (Medina et al., 2010). In the smart grid, demand management extends beyond controlling the loads on the demand-side. Controlling demand side load is known as Demand Response (DR), and it is already implemented in the traditional power grid for large-scale consumers although it is not fully automated yet. DR directly aims to control the load of the commercial and the industrial consumers during peak hours. Peak hours refer to the time of day when the consumption exceeds the capacity of the base power generation plants that are build to accommodate the base load. When the amount of load exceed the capacity of base power plants, they are accommodated by the peaker power plants. Commercial and industrial consumers can have a high impact on the overall load depending on their scale. Briefly, DR refers to those consumers' decreasing their demand following utility instructions and it is generally handled by the utility or an aggregator company. The subscribed consumers are notified by phone calls, for example, to turn off or to change the set point of their HVAC systems for a certain amount of time to reduce the load. In smart grid, Automated Demand Response (ADR) is being considered. In ADR programs, utilities send signals to buildings and industrial control systems to take a pre-programmed

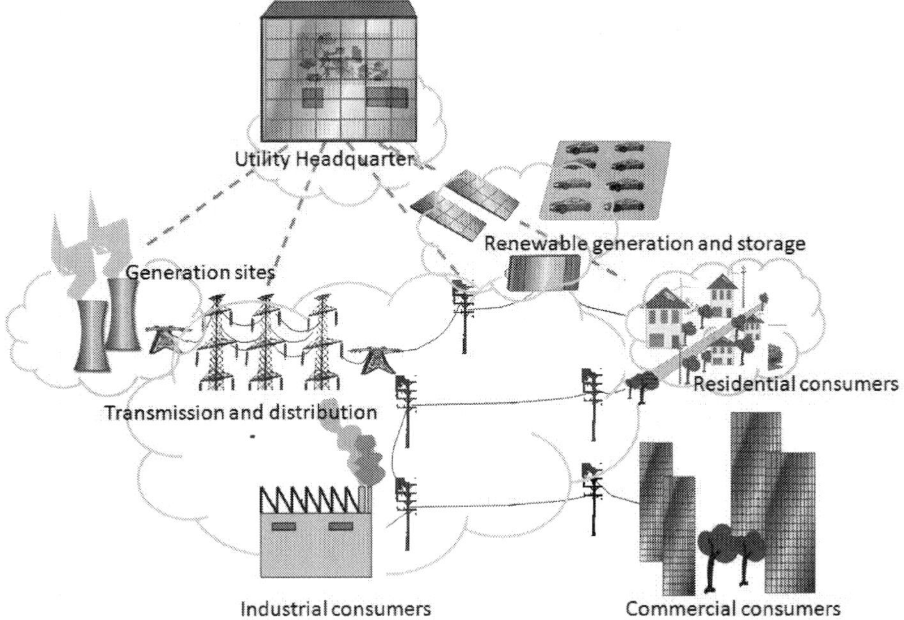

Fig. 2. Illustration of the smart grid with communications.

action based on the specific signal. Recently, OpenADR standard has been developed by the Lawrence Berkeley National Laboratory and the standard is being used in California (Piette et al, 2009). Another well-known data communication standard for Building Automation and Control network is the BACnet. BACnet has been initially developed by the American Society of Heating, Refrigerating and Air-Conditioning Engineers (ASHRAE) and later adopted by ANSI (Newman, 2010).

The traditional grid does not employ DR for residential consumers although demand-side management has been discussed since late 1990s (Newborough & Augood, 1999). Previously residential consumers used electricity without feedback about its availability and price (Ilic et al., 2010). In the smart grid, by the use of smart meters, consumers will have information about their consumption without waiting for their monthly or bi-monthly bills.

The smart grid provides vast opportunities in the DR field. The DR solutions target both peak load reduction and consumer expense reduction. Furthermore, in the smart grid, DR is extended to demand management since the consumers are also able to generate energy. Energy generation at the demand-side requires intelligent control and coordination algorithms. In addition to those, widespread adoption of the PHEVs will impose tight operation constraints for the power grids. PHEVs will be charged from the grid and their energy consumption rating may be as high as a households' daily consumption. The PHEV loads are anticipated to multiply the demand for electricity. For those reasons, demand management will become even more significant in the following years (Shao et a., 2010).

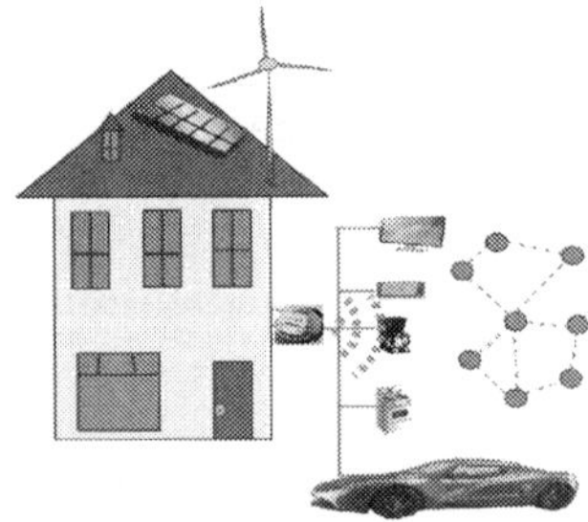

Fig. 3. Smart home with energy generation, WSN and a PHEV.

In the following sections, we will introduce the recent demand management schemes. One of the promising demand management techniques is employing Wireless Sensor Networks (WSNs) in demand management. A WSN is a group of small, low-cost devices that are able to sense some phenomena in their surroundings, perform limited processing on the data and transmit the data to a sink node by communicating with their peers using the wireless medium. The advances in the Micro-Electro-Mechanical Systems (MEMS) have made WSN technology feasible in the recent years, and WSNs find applications in diverse fields. Environmental monitoring and surveillance applications are the pioneering fields to utilize WSNs however following those successful applications, WSNs are today used in tele-health, intelligent transportation, disaster recovery and structure monitoring fields (Chong & Kumar, 2003). WSNs also provide vast opportunities for the smart grid (Erol-Kantarci & Mouftah, 2011a). Especially WSNs can have a large number of applications in demand management in the smart grid since they are able to provide pervasive communications and control capabilities at low cost. Furthermore, they can provide applications that comply with consumers' choices where leaving the consumer as the decision maker is stated as one of the desired properties of the smart grid demand management applications (Lui et al., 2010).

Briefly, there are a large number of opportunities that will become available with the new smart grid technologies however the implementation of the smart grid has several challenges. Regulations and standardization is one of the major challenges. Currently, various governmental agencies, alliances, committees and groups are working to provide standards so that smart grid implementations are effective, interoperable and future-proof. Security is another significant challenge since the grid is becoming digitized, integrating with the Internet, and generally using open media for data transfer. Smart grid may be vulnerable to physical and cyber attacks if security is not handled properly (Metke & Ekl, 2010). Furthermore, successful market penetration of demand management systems is important for the smart grid to achieve its goals. Last but not least, the load on the grid is expected to increase as PHEVs are plugged-in for charging. Unbalanced and uncoordinated charging may cause failures and the smart grid calls for novel coordinated PHEV charging mechanisms (Erol-Kantarci & Mouftah, 2011c). Moreover, as renewable resources become dominant and PHEVs are used as storage devices the intermittency of supply will require rethinking of the traditional planning, scheduling and dispatch practices of the grid operators (Rahimi & Ipakchi, 2010).

In the following sections, we first give a broad perspective on the possible utilization of WSNs in the smart grid. Then, we focus on demand management and introduce the recent demand management techniques which we group under communication-based, incentive-based, real-time and optimization-based demand management techniques. Demand management using WSNs falls under communication-based techniques and they are explained in detail in Section 3.1.

2. Smart grid and Wireless Sensor Networks

In this section, we will briefly summarize the literature on the use of WSNs in the power grid in order give a complete picture of the state of the art. The electrical power grid is a large network that can be partitioned into three main conceptual segments as energy generation, power transmission and electricity distribution, and consumption segments. In the smart grid, the traditional radial organization and this partitioning will change since the electricity will be also produced and used within a distribution system forming a microgrid.

In this section, we follow the organization of the traditional grid for the sake of increasing the understandability of the text. We start with electricity generation sites, continue with power transmission and electricity distribution and finally reach to consumption which is the last mile of the electricity delivery services. WSNs have broad range of applications in all of those segments.

2.1 WSNs for generation facilities

In the traditional power grid, energy generation facilities are generally monitored with wired sensors which are limited in amount and located only at a few critical places. This is due to the high cost of installation and maintenance of those sensors. WSNs offer low-cost sensors that can communicate via wireless links hence have flexible deployment opportunities. In fact, the utilization of WSNs becomes even more essential with the increasing number of renewable energy sites in the energy generation cycle. These renewable energy generation facilities can be in remote areas, and operate in harsh environments where fault-tolerance of WSNs makes them an ideal candidate for such applications. Furthermore, the output of the renewable energy resources is closely related with the ambient conditions such as wind velocity for wind power generation and cloudiness for solar panels. These varying ambient conditions cause intermittent power generation which makes renewable resources hard to integrate to the power grid. For instance, at high wind speeds, to avoid damage to the blades and gears inside the hub of the wind turbine, the turbines are shut off. This causes a steep reduction of output that has to be balanced with other resources (Ipakchi & Albuyeh, 2009). Prediction of such events will give opportunities for preparedness and fast restoration capabilities by the help of backup generators. This emphasizes the importance of ambient data collection. For those reasons, WSNs can offer solutions for renewable energy generation sites, such as solar (PV) farms or wind farms. Furthermore, wireless sensor and actor networks can take part in increasing the efficiency of the equipments.

In (Shen et al., 2008), the authors address the challenge of varying wind power output by employing prediction where WSNs are used to collect and communicate the wind speed prediction data to a central location. WSNs can also be used for condition monitoring of the wind turbines. Wind turbines are expensive equipments which may experience break downs in time due to wear. Early detection of malfunctioning components may increase the lifetime of the wind tribunes and reduce the time spared for maintenance which increases the efficiency of production. In (Al-Anbagi et al., 2011), the authors utilize WSNs for monitoring the condition of the bearings within the gearboxes where accelerometers are used to monitor wind turbine vibration. WSNs are used to provide early detection for bearing failures or other related problems. The authors address the issue of delay-sensitive data transmission in WSNs for a wind turbine by modifying the Medium Access Control (MAC) protocol of IEEE 802.15.4 standard in order to provide service differentiation for critical and non-critical data, and reduce the end-to-end delay for critical data.

A WSN-based energy evaluation and planning system for industrial plants have been introduced in (Lu et al., 2010). The authors have discussed the feasibility of using WSNs and the benefits of replacing the conventional wired sensor with WSNs. A similar WSN-based system can also be used for condition monitoring of power plants. Low-cost, ease of deployment, fault-tolerance, flexibility are among the advantages of the WSN-based systems.

2.2 WSNs for transmission and distribution assets

Transmission system consists of towers, overhead power lines, underground power lines, etc., that are responsible for transportation of electricity from the generation sites to the distribution system. In the traditional power grid, the voltage is stepped up in order to reduce the losses at the transportation, and then, it is step down at the distribution system. Distribution system consists of substations, transformers and wiring to the end-users. In the transmission and distribution segment, an equipment failure or breakdown may cause blackouts or it may even pose danger for public health. Moreover, these assets can be easily reached from outside, therefore they can be a target of terrorism. WSNs, once again, provide promising solutions for monitoring and securing the transmission and distribution segment.

In (Leon et al., 2007), the authors utilize WSNs for detection of mechanical failures in the transmission segment components such as conductor failure, tower collapses, hot spots, extreme mechanical conditions, etc. WSNs provide a complete physical and electrical picture of the power system in real time and ease diagnosing faults. Moreover, power grid operators are provided with appropriate control suggestions in order to reduce the down time of the system. The authors employ a two-level hierarchy where short-range sensor nodes collect data from a component and deliver the collected data to a gateway. This gateway is called as Local Data and Communications Processor (LDCP). The LDPC has the ability to aggregate the data from the sensors, besides it has a longer-range radio which it uses to reach the other LDPCs that are several hundreds of meters away. The mechanical status of the transmission system is processed and delivered to the substation by the LDPCs. This hierarchical deployment increases the scalability of the WSN which emerges as a necessity when the large geographical coverage of the transmission system is considered.

The use of an IEEE 802.15.4 based WSN in the substations has been discussed in (Ullo et al, 2010) and data link performance has been evaluated. The communication services provided by WSNs have been shown to be useful for automation and remote metering applications. Similarly in (Lim et al., 2010), the authors utilize WSNs in transmission and distribution system for power quality measurements. The authors proposed a data forwarding scheme between pole transformers and the substation using multi-hop WSNs. Power quality measurements include harmonics, voltage sags, swells, unbalanced voltage, etc. These measurements are communicated using the IEEE 802.15.4 standard.

Further potential applications of sensor networks in the power delivery system have been defined in (Yang et al., 2007) as:

- Temperature, sag and dynamic capacity measurements from overhead conductors
- Recloser, capacitor, and sectionalizer integrity monitoring
- Temperature and capacity measurements from underground cables
- Faulted circuit indication
- Padmount and underground network transformers
- Monitoring wildlife and vegetation contact
- Monitoring underground network components, e.g. transformers, switches, vaults, etc.

2.3 WSNs for demand-side applications

In the traditional power grid, power grid operators do not have services for the demand-side except the DR programs for large-scale consumers. However, in the smart grid, by using the smart meters and the utility Advanced Metering Infrastructure (AMI), it will be possible to communicate with the consumers. A smart meter and AMI interconnection using Zigbee has been considered in (Luan et al., 2009). Furthermore, energy generation at the consumer premises will be also available. In fact, energy generation by solar panels and wind turbines are already possible, even the locally generated energy can be sold to the grid operators. However, Distributed Generation (DG) is not fully implemented. DG refers to a subsystem that can intentionally island. There are several reasons why this has not been implemented yet. Power quality problems may occur in an islanded system, safety of power personnel may be endangered due to unintentionally energized lines and there might be synchronization problems. In this context, utilization of WSNs can provide efficient monitoring and control capabilities to increase the reliability of the DGs (Sood et al., 2009). WSN applications in the demand-side will be discussed in detail in Section 3.1.

3. Demand management in the smart grid

In the smart grid, it will be possible to communicate with the consumers for the purposes of monitoring and controlling their power consumption without disturbing their business or comfort. This will bring easier administration capabilities for the utilities. On the other hand, consumers will require more advanced home automation tools which can be implemented by using advanced sensor technologies. For instance, consumers may need to adapt their consumption according to the dynamically varying electricity prices which necessitates home automation tools. In the smart grid, time-differentiated billing schemes will be employed. For instance, very soon Time of Use (TOU) will be activated by most of the utilities in North America. TOU rates will be applied to the metering operations by the help of smart meters and the AMI.

TOU is a natural result of consumer activity. Consumer demands have seasonal, weekly and daily patterns. For instance, during overnight hours consumer activity decreases, or heating loads increase during cold days, or similarly cooling loads increase during hot days. The daily load pattern of a typical household on a winter weekday is illustrated in Fig. 4. Morning and evening peaks are visible from this plot. In Fig. 5, we present the accumulated loads of a large

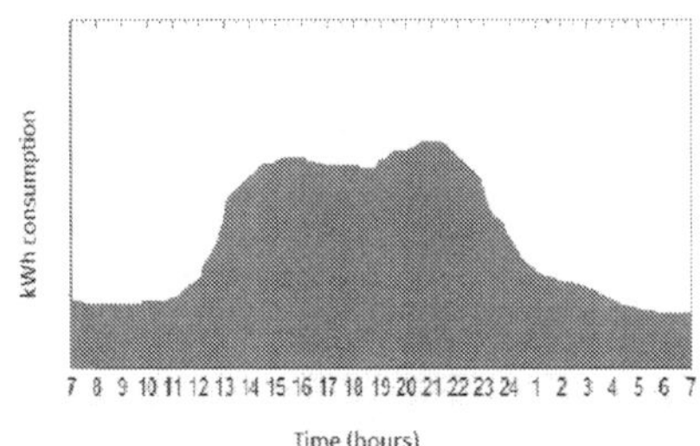

Fig. 4. Illustration of the daily load profile for a winter weekday.

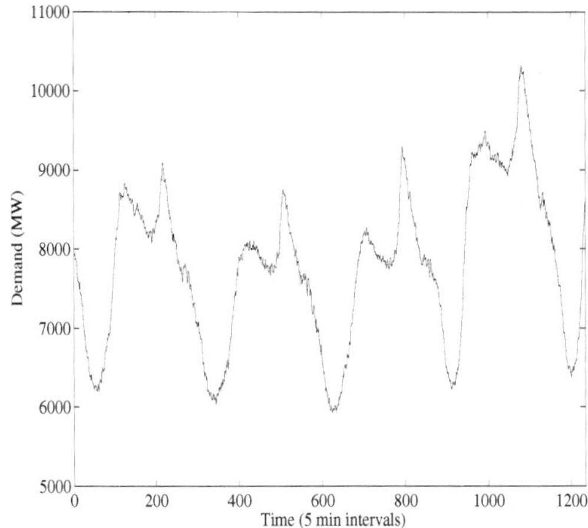

Fig. 5. Electricity load on the grid for four days in winter.

number of consumers collected by the Australian Independent System Operator (AISO). As seen from the figure the peaks become more significant as they are accumulated (Erol-Kantarci & Mouftah, 2010d). The hours of high consumer activity, i.e. high load durations, is called on-peak periods, while moderate and low load durations are called mid-peak and off-peak periods, respectively.

In TOU tariff, electricity is more expensive during peak hours because utilities handle peak load by bringing peaker plants online. Peaker plants have high maintenance costs and they use expensive fossil fuels. They burn coal, natural gas, or diesel which they have shorter response times. On the other hand, those fuels are fossil based and they incur higher CO_2 emissions (Erol-Kantarci &Mouftah, 2010b). They are also expensive fuels, therefore, the generation cost increases during peak hours. To compensate for these costs utilities apply block rates, i.e. TOU. Block rates are different than the conventional flat billing. The price of electricity is fixed during a block of consecutive hours, then it changes for another block of hours. The reason for varying rates are as follows. The length of the block of hours and

	Period	Time	Rate
Winter Weekdays	On-Peak	7:00am to 11:00am	9.3 cent/kWh
	Mid-Peak	11:00am to 5:00pm	8.0 cent/kWh
	On-Peak	5:00pm to 9:00pm	9.3 cent/kWh
	Off-Peak	9:00pm to 7:00am	4.4 cent/kWh
Summer Weekdays	Mid-Peak	7:00am to 11:00am	8.0 cent/kWh
	On-Peak	11:00am to 5:00pm	9.3 cent/kWh
	Mid-Peak	5:00pm to 9:00pm	8.0 cent/kWh
	Off-Peak	9:00pm to 7:00am	4.4 cent/kWh
Weekends	Off-peak	All day	4.4 cent/kWh

Table 1. TOU rates of an Ontario utility as of 2011.

the price for each block is determined by the utilities based on the consumption pattern and the raw market price of electricity. Electricity consumption during peak periods have higher price than consumption during off-peak periods as explained above. Furthermore, higher prices are employed to discourage consumers to use electricity during peak hours, and hence, avoid dangerous grid conditions. The rate chart of an Ontario-based utility is given in Table 1 (online:Hydro Ottawa, 2011) as an example of TOU rates. Note that, TOU hours and rates may vary from one utility to another based on the local load pattern and cost. For instance, cold weather conditions in northern countries increase heating demand throughout the winters whereas, southern countries may have less heating demand during the same period of the year.

In fact, residential demand control has been previously developed for the smart homes. Smart homes employ energy saving applications that can turn the lights off depending on the occupancy of the rooms, or dim the lights off based on outside light intensity and shutter positions, or adjust the thermostat based on the outside temperature and sensor measurements. etc. These type of comfort-focused energy management applications date back to 1990s (Brumitt et al., 2000; Lesser et al., 1999). However, smart home implementations have been rare. Today most of the residential premises do not have such energy management systems. Furthermore, smart home related techniques do not involve communication and coordination with the power grid. The smart grid introduces a number of opportunities for the home energy management and enables, communication-based, incentive-based, real-time demand management and optimization-based techniques which will be described in the following sections. Furthermore, smart grid and WSNs can enable consumers to have more control over their consumption. We will describe a WSN-based home energy management system in the following sections, as well.

3.1 Communication-based demand management
In this section, we introduce four communication-based demand management schemes, which are in-Home Energy Management, iPower, Energy Management Using Sensor Web Services and Whirlpool smart device network.

3.1.1 in-Home Energy Management (iHEM)
In (Erol-Kantarci & Mouftah, 2011b), the authors have used WSNs and smart appliances for residential demand management. This residential demand management scheme is called

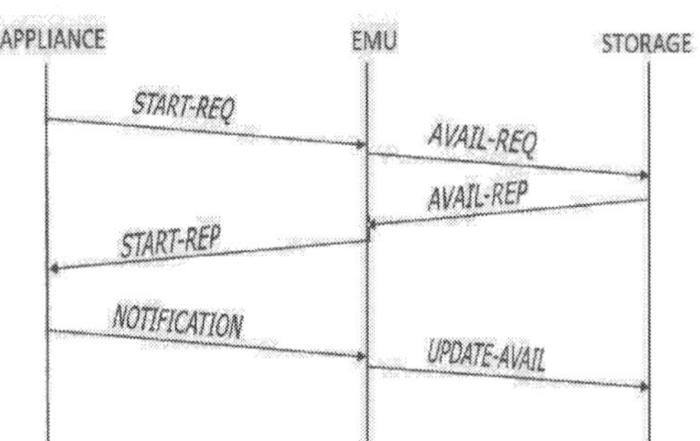

Fig. 6. Message flow for iHEM.

in-Home Energy Management (iHEM). iHEM employs a central Energy Management Unit (EMU) and appliances with communication capability. EMU and appliances communicate via wireless links where their packets are relayed by a WSN. iHEM is based on the appliance coordination scheme that was proposed in (Erol-Kantarci &Mouftah, 2010a;c). It attempts to shift consumer demands at times when electricity usage is less expensive according to the local TOU tariff.

The message flow of the iHEM application is given in Fig. 6. According to iHEM, when a consumer turns on an appliance, the appliance generates a START-REQ packet and sends it to EMU. EMU communicates with the smart meter regularly to receive the price updates of the TOU tariff applied by the grid operator. The authors assume that the smart home is also able to produce energy by solar panels or small wind turbines. Therefore, upon receiving a START-REQ packet, EMU communicates with the storage units of the local energy generators and retrieves the amount of the available energy by sending an AVAIL-REQ packet. Upon reception of AVAIL-REQ, the storage unit replies with an AVAIL-REP packet where the amount of available energy is sent to the EMU. After receiving the AVAIL-REP packet, EMU determines the convenient starting time of the appliance by using Algorithm 1. EMU computes the waiting time as the difference between the suggested and requested start time, and sends the waiting time in the START-REP packet to the appliance. The consumer decides whether to start the appliance right away or wait until the assigned timeslot depending on the waiting time. The decision of the consumer is sent back to the EMU with a NOTIFICATION packet. Afterwards, EMU sends an UPDATE-AVAIL packet to the storage unit to update the amount of available energy (unallocated) on the unit after receiving the consumer decision. This handshake protocol among the appliance and the EMU, ensures that EMU does not force an automated start time. We avoid this approach to increase the comfort of the consumers and to provide more flexibility. Furthermore, energy is allocated on the storage units as per request. Therefore, when the smart home exports electricity (sells), the amount of unallocated, hence available energy will be known.

The format of the iHEM packets are given in the figures below. START-REQ packet format is shown in Fig. 7. The first field of the packet is the Appliance ID. The sequence number field denotes the sequence number of the request generated by the appliance since the appliance may be turned on several times in one day. Start time is the timestamp given when the

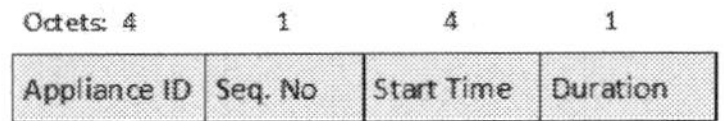

Fig. 7. START-REQ packet format.

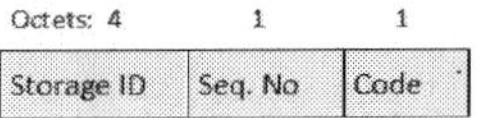

Fig. 8. AVAIL-REQ packet format.

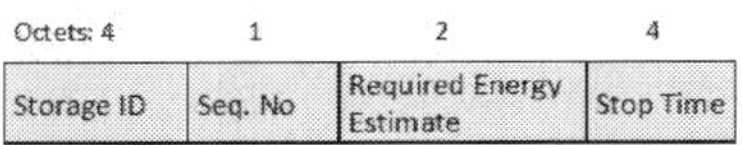

Fig. 9. UPDATE-AVAIL packet format.

consumer turns on the appliance. The duration field denotes the length of the appliance cycle. Each appliance has different cycle lengths. A cycle could be a washing cycle for a washer or the time required for the coffee maker to make the desired amount of coffee. This duration depends on the consumer preferences, i.e. the selected appliance program. The AVAIL-REQ packet format is given in Fig. 8. The storage ID field is the ID of the storage unit that is attached to the local energy generation unit. When the house is equipped with multiple energy generation devices such as solar panels and small wind tribunes, the amount of energy stored in their local storage units may have to be interrogated separately. The packet sequence number is used for the same purpose as described previously. Code field carries the controller command code. In iHEM, this field is used for inquiring the amount of available energy, hence it is a static value. However, other applications may also use this code field, e.g. to send a command to the storage unit to dispatch energy to the grid. Other code combinations have been reserved for future use. NOTIFICATION packet has the same format as the START-REQ packet. The start-time field of the NOTIFICATION packet denotes the negotiated running time of the appliance, i.e., it could be either the time when the appliance is turned on, or the start time suggested by the EMU. This information is required to allocate energy on the local storage unit when it is used as the energy source. As we mentioned before, since it is further possible to sell excess energy to the grid operators, the amount of energy that needs to be reserved for the appliances that will run with the local energy has to be known ahead. The format of the UPDATE-AVAIL packet is given in Fig. 9. Storage ID and the code fields are explained above. The required energy estimate field, is the power consumed by the appliance multiplied by the duration of a cycle. Stop time denotes the time when the appliance is scheduled to finish its cycle.

The algorithm of scheduling (Algorithm 1) works as follows. EMU first checks whether locally generated power is adequate for accommodating the demand. If this is the case, the appliance starts operating, otherwise the algorithm checks if the demand has arrived at a peak hour, based on the requested start time, St_i. If the demand corresponds to a peak hour, it is either shifted to off-peak hours or mid-peak hours as long as the waiting time does not exceed D_{max}, i.e. maximum delay. The computed delay, d_i is returned to the consumer as the waiting time. D_{max} parameter limits the delay, hence it guarantees a maximum delay for the

---

Algorithm 1 | Scheduling at the EMU

1: $\{D_{max}$: maximum allowable delay$\}$
2: $\{d_i$: delay of appliance $i\}$
3: $\{St_i$: requested start time of appliance $i\}$
4: **if** (stored energy available = TRUE) **then**
5:     *StartImmediately*()
6: **else**
7:     **if** (St_i is in peak) **then**
8:         $d_i \leftarrow ShiftToOffPeak()$
9:         **if** ($d_i > D_{max}$) **then**
10:             $d_i \leftarrow ShiftToMidPeak()$
11:             **if** ($d_i > D_{max}$) **then**
12:                 *StartImmediately*()
13:             **else**
14:                 *StartDelayed*()
15:             **end if**
16:         **else**
17:             *StartDelayed*()
18:         **end if**
19:     **else**
20:         **if** (St_i is in mid-peak) **then**
21:             $d_i \leftarrow ShiftToOffPeak()$
22:             **if** ($d_i > D_{max}$) **then**
23:                 *StartImmediately*()
24:             **else**
25:                 *StartDelayed*()
26:             **end if**
27:         **else**
28:             *StartImmediately*()
29:         **end if**
30:     **end if**
31: **end if**

---

consumers, and at the same time it prevents the requests to pile up at certain off-peak periods. *StartImmediately*() and *StartDelayed*() functions determine the scheduled time of operation. iHEM uses a WSN to relay the packets shown in Fig. 6. The same WSN may also be responsible for other smart home applications such as inhabitant health monitoring since installing a WSN for the sole purpose of iHEM would increase cost. The WSN uses the Zigbee protocol. In (Erol-Kantarci & Mouftah, 2011b), the authors show the impact of these underlying smart home applications on the performance of the WSN. They also demonstrate the savings achieved by the iHEM application. iHEM is shown to be able to reduce consumer expenses, appliance loads during peak hours and carbon emissions related with electricity usage during peak periods.

3.1.2 iPower

Intelligent and Personalized energy conservation system by wireless sensor networks (iPower) implements an energy conservation application for multi-dwelling homes and offices by using the context-awareness of WSNs (Yeh et al., 2009). iPower is similar to the energy management applications in the smart homes. It includes a WSN with sensor nodes and a gateway node, in addition to a control server, power-line control devices and user identification devices. Sensor nodes are deployed in each room and they monitor the rooms with light, sound and temperature sensors. When a sensor node detects that a measurement exceeds a certain threshold, it generates an event. Sensor nodes form a multi-hop WSN and they send their measurements to the gateway when an event occurs. The gateway node is able to communicate with the sensor nodes via wireless communications and it is also connected to the intelligent control server of the building. iPower uses Zigbee for WSN communication and X10 for power-line communications. Intelligent control server performs energy conservation actions based on sensor inputs and user profiles. The action of the server can be turning off an appliance or adjusting the electric appliances in a room according to the profiles of the users who are present in the room. Server requests are delivered to the appliances through their power-line controllers.

3.1.3 Energy management using sensor web services

Web services can invoke remote methods on other devices without the knowledge of the internal implementation details and enable machine-to-machine communications (Groba & Clarke, 2010). In (Asad et al., 2011), the authors consider a smart home that contains smart appliances with sensor modules that enable each appliance to join the WSN and communicate with its peers. The authors present three energy management applications that use sensor web services. The basic application enables users to learn the energy consumption of their appliances while they are away from home. The current drawn by each appliance is monitored by the sensors on board and this information is made available through a home gateway to the users. Users can access the gateway from their mobile devices using web services. Second application of (Asad et al., 2011) is a load shedding application for the utilities. Load shedding is applied to HVAC systems only during peak hours and when the load on the grid is critical. In addition to monitoring and load shedding applications, the third application focuses on a case when the energy generated and stored is either sold to the grid or consumed at home. The application enables the storage units to be controlled by the remote users.

3.1.4 Whirlpool Smart Device Network (WSDN)

Whirlpool Smart Device Network (WSDN) aims to provide simple smart grid participation options for the end-users (Lui et al., 2010). WSDN is based on machine-to-machine communications and it aims to minimize consumer interaction. WSDN consists of three networking domains which are the HAN, the Internet, and the smart meter network. WSDN utilizes several wired and wireless physical layer technologies together, e.g. Zigbee, Wi-Fi, Broadband Internet, Power Line Carrier (PLC). The Wi-Fi connects the smart appliances and forms the HAN. The ZigBee and the PLC connects the smart meters and the broadband Internet connects consumers to the Internet. Above the physical layer, there are the TCP and the IP layers. On top of those, Open Communication Protocol stack is placed which includes Extensible Markup Language (XML), Simple Authentication and Security Layer (SASL), Transport Layer Security (TLS), Extensible Messaging and Presence Protocol (XMPP) protocols. WSDN application is aimed to be easily downloadable from a smart phone. The WSDN also handles user authentication since security is a major concern for such a network.

Moreover, utilities are able to use WSDN and perform load shedding during critical peaks. All of the consumer or utility generated transactions are handled by the Whirlpool-Integrated Service Environment (WISE). Security objectives of WISE has been summarized as:

- Availability: the smart grid system is protected from denial-of-service attacks and always available

- Privacy: consumers have control over their own personal data

- Confidentiality: information is not disclosed unless authorized

- Integrity: data sent between the appliance and utility is not modified

3.2 Incentive-based demand management

In (Mohsenian-Rad et al., 2010a;b), the authors deploy an Energy Consumption Scheduling (ECS) mechanism for a local neighborhood. The ECS is assumed to be implemented in each smart meter. The smart meters communicate and interact in order to find an optimum consumption schedule for each subscriber in the neighborhood. ECS relies on a distributed algorithm. The objective of ECS is to reduce consumer expenses and reduce peak-to-average ratio in the load curve. ECS is an incentive-based scheme as the consumers are given incentives based on pricing which varies according to a game theoretic approach. The ECS does not reduce the overall consumption of the appliances, instead it shifts consumer demands to off-peak hours. This naturally reduces peak-to-average ratio since ECS basically does peak shaving and valley filling. Within a time horizon of $T = 24$ hours, the daily energy consumption of each consumer, $c \in C$, is formulated as:

$$\sum_{a \in A_c} E_{c,a}^t \quad t \in T \tag{1}$$

where $E_{c,a}^t$ is the hourly consumption of the appliances, a, in the appliance set of the cth consumer, A_c, (i.e. $a \in A_c$). When complete knowledge of the consumer demands are available and a central controller schedules the demands, it is possible to schedule demands by minimizing the $E_{c,a}^t$ of all A_c appliances that belongs to all C consumers during T hours. This can be formulated as:

$$minimize \sum_{t=1}^{T} \beta_t \sum_{c \in C} \sum_{a \in A_c} E_{c,a}^t \quad t \in H \tag{2}$$

where β denotes the cost function. The incentives are given regarding the billing of consumers. In the game theoretic approach, consumers select their consumption to minimize their payments to the utility. It has been shown in (Mohsenian-Rad et al., 2010b) that for increasing and strictly convex β, Nash equilibrium of the energy consumption game exists and is unique.

3.3 Real-time demand management

In (Mohsenian-Rad & Leon-Garcia, 2010), the authors propose the Residential Load Control (RLC) scheme considering a power grid that employs real-time pricing. According to real time pricing, the price of the electricity follows the raw market price of the electricity. The market price of electricity is generally determined by the regional independent system operator. The independent system operator arranges a settlement for the electricity prices of the next-day or

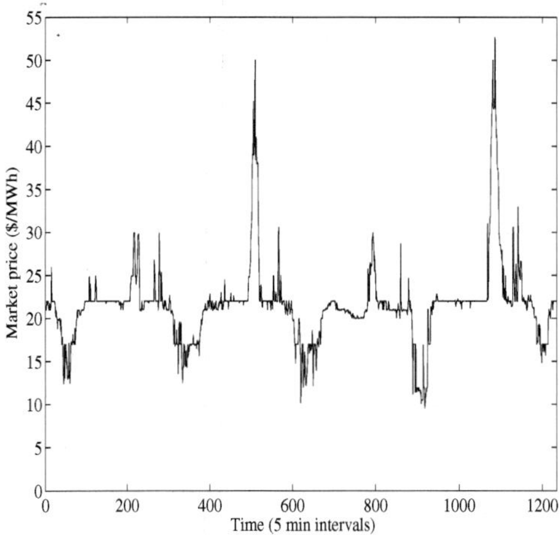

Fig. 10. Electricity price data from an ISO on four days.

next-hour, based on the load forecasts, supplier bids and importer bids. A typical price data is shown in Fig. 10 (Erol-Kantarci & Mouftah, 2010d).

(Mohsenian-Rad & Leon-Garcia, 2010) proposes an automated load control scheme that aims to minimize the consumer expenses as well as the waiting times of the delayed demands. The scheduling scheme is augmented with a price predictor in order to attain the prices of several hours ahead. This is necessary if the grid operator only announces the prices for the next one or two hours. In fact, load and price forecasting is widely studied in the literature. Load forecasts are essential for dispatchers, who are the commercial or governmental bodies responsible for dispatching electricity to the grid. Load forecasting provides tools for operation and planning of a dispatcher where decisions such as purchasing or generating power, bringing peaker plants online, load switching and infrastructure development can be made (Gross & Galiana, 1987). Electricity market regulators and dispatchers rely on forecasting tools that provide short, medium and long-term forecasts.

Short-term load forecasts cover hourly or daily forecasts where medium-term forecasts span a time interval from a week to a year and long-term forecasts cover several years. Forecasting techniques may differ according to this range. For short-term forecasting, the similar day approach searches the historical database of days to find a similar day with properties such as weather, day of the week, etc. (Feinberg & Genethliou, 2006). Regression is another widely used statistical technique for load forecasting. Regression methods aim to model the relationship of load and environmental factors, e.g. temperature (Charytoniuk et al., 1998). Time series methods try to fit a model to data. Previous studies have employed a wide variety of time series methods such as Autoregressive Moving Average (ARMA), Autoregressive Integrated Moving Average (ARIMA), Autoregressive Moving Average with eXogenous variables (ARMAX) and Autoregressive Integrated Moving Average with eXogenous variables (ARIMAX) methods. Neural networks, expert systems, support vector machines and fuzzy logic are among the recent forecasting techniques. The techniques proposed for load forecasting can be used for price forecasting, as well.

In (Mohsenian-Rad & Leon-Garcia, 2010), the authors use a simple AR process that uses the price values of previous two days and the same day of the last week. This is due to the weekly pattern of the consumption data.

3.4 Optimization-based demand management

In this section, we introduce two optimization-based demand management schemes, which are Decision-support Tool (DsT) and Domestic Optimization and Control techniques.

3.4.1 Decision-support Tool (DsT) for the smart home

In (Pedrasa et al., 2010), the authors propose a Decision-support Tool (DsT) for the smart homes. The DsT aims to help the household in making intelligent decisions when operating their appliances. The authors focus on appliances that have high energy consumption, e.g. PHEV, space heater, water heater and pool pump. The authors define an aggregate, must-have services such as lighting, cooking, refrigeration,etc., which exists beside the loads of space heating, water heating and pool pumping services and PHEV charging loads. The energy consumption properties such as duration, battery capacity, maximum charging rating are assumed to be determined by the consumer. In the initial phase of DsT, consumers assign values to those desired energy services. Moreover, DsT assumes the availability of generation via solar panels and the peak output of the PV is also set at the initial phase. Then, consumption is optimized by scheduling the available distributed generation, energy storage and controllable end-use loads which are called as distributed energy resources (DER). The scheduling algorithm attempts to maximize the net benefits for the consumer which is equal to the total energy service benefits minus the cost of energy. The cost of energy is based on a TOU tariff with critical pricing during several hours of a day. The must-run services are delivered regardless of cost and the other services are restricted to run only during defined hours. For instance, the pool pump is not allowed to work overnight due to noise issues.

The scheduling of the DER is formulated and solved via the particle swarm optimization (PSO) heuristic. PSO is a population-based optimization technique that enables to attain near-optimal schedules within manageable computation times.

In (Pedrasa et al., 2010), the communication among the DER and consumers has not been considered. However the authors emphasize the significance of coordinated scheduling using a centralized decision-maker that controls the operation of all the various DERs. The benefits of having a decision-maker that can access the dynamic prices of electricity as well as weather forecasts through the Internet, and that can communicate with the sensors have also been discussed in (Pedrasa et al., 2010).

In Table 2, we give a comparison of the presented demand management techniques that have similar objectives, i.e. iHEM, RLC, DsT and ECS.

3.4.2 Domestic optimization and control

In (Moldernik et al., 2009; 2010), the authors propose using domestic optimization and control scheme to achieve the following goals:

- optimize efficiency of power plants

- increase penetration of renewable resources

	Method	Pricing	Comm.	Coverage	Monthly cost reduction	Peak load reduction
iHEM (Erol-Kantarci & Mouftah, 2011b)	Interactive demand shifting	TOU	Yes	local	30%	40%
RLC (Mohsenian-Rad & Leon-Garcia, 2010)	Automated load control with LP-based optimization	Real-time pricing	No	local	10%-25%	22%
DsT * (Pedrasa et al., 2010)	Particle swarm optimization	TOU and Critical Peak Pricing (CPP)	No	local	16%-25%	N/A
ECS (Mohsenian-Rad et al., 2010a)	Game theoretic pricing and scheduling	Proportional to daily load and generation cost	No	neighbor-hood	37%	38%

* Using TOU tariff, no PHEV and no critical peak pricing scenario.

Table 2. Comparison of iHEM, RLC, DsT and ECS.

- optimize grid efficiency

Domestic optimization is based on predicting the demand and the day-ahead prices and optimize the resources accordingly. The authors use a neural network-based prediction approach to predict the next-day heat demand. The schedule of the Micro combined heat and power (micro-CHP) device is determined based on this prediction. CHP, also known as cogeneration, provides ability to simultaneously produce heat and electricity. Electricity is generated as a by-product of heating.

The neural network is trained such that a set of given inputs produce the desired outputs. In (Moldernik et al., 2010), the output of the neural network predictor is the heat demand per hour. The factors affecting the heat demand is assumed to be the behavior of the residents, the weather, and the characteristics of the house which are given as inputs to the prediction model. The data are derived from historical demand and consumer behavior databases.

Following the prediction step, planning of the runs of the microCHP is established. Thus, the times when the microCHP is switched on is planned. This planning is based on local decisions. However, a group of houses is considered to act as a virtual power plant where in the global planning phase, global production is optimized via iterative distributed dynamic programming. In the next step, the authors schedule the appliances in a single house based on the global planning decisions. Local appliances are controlled to optimize electricity import/export of home.

3.5 Summary and discussions

In this book chapter, we grouped the demand management schemes proposed for the smart grid under four categories as:

- Communication-based demand management
- Incentive-based demand management
- Real-time demand management
- Optimization-based demand management

Communication-based techniques have been studied in (Asad et al., 2011; Erol-Kantarci &Mouftah, 2010c; Erol-Kantarci & Mouftah, 2011b; Lui et al., 2010; Yeh et al., 2009). Demand management schemes that employ WSNs have been presented under communication-based techniques, as well. Communication-based techniques provide flexible solutions that can compromise between reducing the energy consumption of the consumers and accommodating their preferences.

Incentive-based techniques have been studied in (Mohsenian-Rad et al., 2010a;b). These schemes try to shift the consumer demands to off-peak hours, and in the meanwhile they provide incentives to the consumers by configuring the prices based on a game theoretic approach. Incentive-based schemes can shape consumer behavior according to the needs of the smart grid.

Real-time demand management has been studied in (Mohsenian-Rad & Leon-Garcia, 2010). In real-time demand management, scheduling makes use of the real-time price of the electricity. Based on the varying prices an automated load control scheme chooses the appliance schedules with the objective of minimizing the consumer expenses, as well as the waiting times of the delayed demands. Those schemes are suitable for the grids where the operators apply real-time pricing tariffs.

Optimization-based demand management has been studied in (Moldernik et al., 2009; 2010; Pedrasa et al., 2010). Optimization-based demand management assumes that the consumer demands are known ahead or at least they can be predicted. Local generation capacity of a house or group of houses is scheduled based the predicted demand profile. Optimization-based schemes may increase the efficiency of the demand management programs significantly.

4. Conclusion

Growing demand for energy, diminishing fossil fuels, desire to integrate renewable energy resources, efforts to reduce Green House Gases (GHG) emissions and resilience issues in the electrical power grid, have led to a common consensus on the necessity for renovating the power grid. The key to this renovation is the integration of the advances in the Information and Communication Technologies (ICTs) to the power grid. The new grid empowered by ICT is called smart grid.

Smart grid can employ ICT in almost every stage from generation to consumption, i.e. electricity generation, transport, delivery and consumption. ICT can increase the efficiency of the generation facilities, transmission and distribution assets and consumption at the demand-side. In this chapter, we reviewed the demand management schemes for the smart grid with a focus on the potential uses of Wireless Sensor Networks (WSN) in the building blocks of the smart grid. We first discussed the use of WSNs at the electricity generation sites. We, then, continued with power transmission and electricity distribution, and finally reached to demand-side which is the last mile of the delivery services. WSNs provide promising solutions for efficient integration of intermittent renewable energy resources, low-cost monitoring of traditional power plants and high-resolution monitoring of utility transport assets. Furthermore, WSNs offer vast variety of applications in the field of consumer demand management.

The ultimate aim of those demand management schemes is to schedule the appliance cycles so that the use of electricity from the grid during peak hours is reduced which consequently reduces the need for the power from the peaker plants and reduces the carbon footprint of the household. In addition, consumer expenses will be reduced as peak hour usage results in higher expenses. Moreover, the use of locally generated power is aimed to be maximized.

Demand management for the smart grid is still in its infancy. The demand management techniques introduced in this chapter have been recently proposed, and they need to be improved as the technology advances. For instance, consumer-in-the-loop or predicted demands can be mitigated by employing learning techniques from the Artificial Intelligence (AI) field. This would increase the consumer comfort and pervasiveness of the demand management applications. Furthermore, those schemes mostly consider conventional appliances, but in a close future, smart appliances will be commercially available. In this case, demand management schemes may be extended to allow sub-cycle scheduling. The availability of such appliances will enrich the opportunities that become available with the demand management applications of the smart grids.

5. References

Al-Anbagi, I., Mouftah, H. T., Erol-Kantarci, M., (2011). Design of a Delay-Sensitive WSN for Wind Generation Monitoring in The Smart Grid, IEEE Canadian Conference on Electrical and Computer Engineering (CCECE), Niagara Falls, ON, Canada.

Asad, O., Erol-Kantarci, M., Mouftah, H. T., (2011). Sensor Network Web Services for Demand-Side Energy Management Applications in the Smart Grid, IEEE Consumer Communications and Networking Conference (CCNC'11), Las Vegas, USA.

Brumitt, B., Meyers, B., Krumm, J., Kern, A., Shafer, S., (2000). EasyLiving: technologies for intelligent environments", in: Proc of Handheld Ubiquitous Computing, Bristol, U.K.

Charytoniuk, W., Chen, M. S., Van Olinda, P., (1998). Nonparametric Regression Based Short-Term Load Forecasting, IEEE Transactions on Power Systems, vol. 13, pp. 725-730.

Chong C., Kumar, S. P., (2003). Sensor networks: Evolution, opportunities, and challenges, In Proceedings of the IEEE, Volume 91 No 8, pp.1247-1256.

Erol-Kantarci, M., Mouftah, H. T., (2010).Using Wireless Sensor Networks for Energy-Aware Homes in Smart Grids, IEEE symposium on Computers and Communications (ISCC), Riccione, Italy.

Erol-Kantarci, M., Mouftah, H. T., (2010). The Impact of Smart Grid Residential Energy Management Schemes on the Carbon Footprint of the Household Electricity Consumption, IEEE Electrical Power and Energy Conference (EPEC), Halifax, NS, Canada.

Erol-Kantarci, M., Mouftah, H. T., (2010). TOU-Aware Energy Management and Wireless Sensor Networks for Reducing Peak Load in Smart Grids, Green Wireless Communications and Networks Workshop (GreeNet) in IEEE VTC2010-Fall, Ottawa, ON, Canada.

Erol-Kantarci, M., Mouftah, H. T., (2010).Prediction-Based Charging of PHEVs from the Smart Grid with Dynamic Pricing, First Workshop on Smart Grid Networking Infrastructure in IEEE LCN 2010, Denver, Colorado, U.S.A.

Erol-Kantarci, M., Mouftah, H. T., (2011). Wireless Multimedia Sensor and Actor Networks for the Next Generation Power Grid. Elsevier Ad Hoc Networks Journal, vol.9 no.4, pp. 542-511.

Erol-Kantarci, M., Mouftah, H. T., (2011). Wireless Sensor Networks for Cost-Efficient Residential Energy Management in the Smart Grid, accepted for publication in IEEE Transactions on Smart Grid.

Erol-Kantarci, M., Mouftah, H. T., Management of PHEV Batteries in the Smart Grid: Towards a Cyber-Physical Power Infrastructure, Workshop on Design, Modeling and Evaluation of Cyber Physical Systems (in IWCMC11), Istanbul, Turkey.

Feinberg, E. A.,Genethliou, D., (2006). Load Forecasting, Applied Mathematics for Restructured Electric Power Systems, Springer.

Groba, C., Clarke, S., (2010). Web services on embedded systems - a performance study, 8th IEEE International Conference on Pervasive Computing and Communications Workshops (PERCOM Workshops), pp.726-731.

Gross, G., Galiana, F. D., (1987). Short-term load forecasting, Proceedings of the IEEE, Vol. 75, no. 12, pp. 1558 - 1573.

Ilic, M. D.; Le X., Khan, U. A., Moura, J. M. F., (2010). Modeling of Future CyberÚPhysical Energy Systems for Distributed Sensing and Control, IEEE Transactions on Systems, Man and Cybernetics, Part A: Systems and Humans, vol.40, no.4, pp.825-838.

Ipakchi, A., Albuyeh, F., (2009). Grid of the future, IEEE Power and Energy Magazine, vol.7, no.2, pp.52-62.

Lightner, E. M., Widergren, S.E., (2010). An Orderly Transition to a Transformed Electricity System, IEEE Transactions on Smart Grid, vol.1, no.1, pp.3-10.

Leon, R. A., Vittal, V., Manimaran, G., (2007). Application of Sensor Network for Secure Electric Energy Infrastructure, IEEE Transactions on Power Delivery, vol.22, no.2, pp.1021-1028.

Lesser, V., Atighetchi, M., Benyo, B., Horling, B., Raja, A., Vincent, R., Wagner, T., Xuan, P., Zhang, S., (1999). The UMASS intelligent home project, in: Proc. of the 3rd Annual conference on Autonomous Agents, Seattle, Washington.

Lim Y., Kim H.-M., Kang S. A., (2010). Design of Wireless Sensor Networks for a Power Quality Monitoring System. Sensors. 2010; 10(11):9712-9725.

Lu, B., Habetler, T.G., Harley, R.G., Gutierrez, J.A., Durocher, D.B., (2007). Energy evaluation goes wireless, IEEE Industry Applications Magazine, vol.13, no.2, pp.17-23.

Luan, S.-W., Teng, J.-H., Chan, S.-Y, Hwang, L.-C.., (2009). Development of a smart power meter for AMI based on ZigBee communication, International Conference on Power Electronics and Drive Systems, pp.661-665.

Lui, T. J., Stirling, W., Marcy, H. O., (2010). Get Smart, IEEE Power and Energy Magazine, vol.8, no.3, pp.66-78.

Massoud Amin, S., Wollenberg, B.F., (2005). Toward a smart grid: power delivery for the 21st century, IEEE Power and Energy Magazine, vol.3, no.5, pp. 34- 41.

Medina, J., Muller, N., Roytelman, I., (2010). Demand Response and Distribution Grid Operations: Opportunities and Challenges, IEEE Transactions on Smart Grid, vol.1, no.2, pp.193-198.

Metke, A.R., Ekl, R.L., (2010). Security Technology for Smart Grid Networks, IEEE Transactions on Smart Grid, vol.1, no.1, pp.99-107.

Mohsenian-Rad, A.-H., Leon-Garcia, A., (2010). Optimal Residential Load Control With Price Prediction in Real-Time Electricity Pricing Environments, IEEE Transactions on Smart Grid, vol.1, no.2, pp.120-133.

Mohsenian-Rad, A.-H., Wong, V. W. S., Jatskevich, J., Schober, R., (2010). Optimal and autonomous incentive-based energy consumption scheduling algorithm for smart grid, Innovative Smart Grid Technologies (ISGT), pp.1-6.

Mohsenian-Rad, A.-H., Wong, V. W. S., Jatskevich, J., Schober, R., Leon-Garcia, A., (2010). Autonomous Demand-Side Management Based on Game-Theoretic Energy Consumption Scheduling for the Future Smart Grid, IEEE Transactions on Smart Grid, vol.1, no.3, pp.320-331.

Molderink, A., Bakker, V., Bosman, M. G. C., Hurink, J. L., Smit, G. J. M., (2009).Domestic energy management methodology for optimizing efficiency in Smart Grids, IEEE conference on Power Technology, Bucharest, Romania.

Molderink, A., Bakker, V., Bosman, M. G. C., Hurink, J. L.,Smit, G. J. M., (2010). Management and Control of Domestic Smart Grid Technology, IEEE Transactions on Smart Grid, vol.1, no.2, pp.109-119.

Newborough, M., Augood, P., (1999). Demand-side management opportunities for the UK domestic sector, IEE Proceedings Generation, Transmission and Distribution, vol.146, no.3, pp.283-293.

Newman, H. M., (2010). Broadcasting BACnet, BACnet Today, Supplement to ASHRAE Journal. Vol. 52, No. 11, pp. B8-12.

Piette, M. A., Ghatikar, G., Kiliccote, S., Koch, E., Hennage, D., Palensky, P., McParland, C., (2009). Open Automated Demand Response Communications Specification (Version 1.0). California Energy Commission, PIER Program. CEC-500-2009-063.

Pedrasa, M. A. A., Spooner, T. D., MacGill, I. F., (2010). Coordinated Scheduling of Residential Distributed Energy Resources to Optimize Smart Home Energy Services, IEEE Transactions on Smart Grid, vol.1, no.2, pp.134-143.

Rahimi, F., Ipakchi, A., (2010). Demand Response as a Market Resource Under the Smart Grid Paradigm, IEEE Transactions on Smart Grid, vol.1, no.1, pp.82-88.

Santacana, E., Rackliffe, G., Tang L., Feng, X. (2010). Getting Smart, IEEE Power and Energy Magazine, vol.8, no.2, pp.41-48.

Shao, S., Zhang, T., Pipattanasomporn, M., Rahman, S., (2010). Impact of TOU rates on distribution load shapes in a smart grid with PHEV penetration, IEEE PES Transmission and Distribution Conference and Exposition, pp.1-6.

Shen, L., Wang H., Duan, X., Li, X., (2008). Application of Wireless Sensor Networks in the Prediction of Wind Power Generation, 4th International Conference on Wireless Communications, Networking and Mobile Computing, vol., no., pp.1-4.

Sood, V. K., Fischer, D., Eklund, J. M., Brown, T., (2009). Developing a communication infrastructure for the Smart Grid, IEEE Electrical Power & Energy Conference (EPEC), pp.1-7.

Ullo, S., Vaccaro, A., Velotto, G. (2010). The role of pervasive and cooperative Sensor Networks in Smart Grids communication, 15th IEEE Mediterranean Electrotechnical Conference (MELECON), pp.443-447.

Yang, Y., Lambert, F., Divan, D., (2007).A Survey on Technologies for Implementing Sensor Networks for Power Delivery Systems,IEEE Power Engineering Society General Meeting, vol., no., pp.1-8.

Yeh, L., Wang, Y., Tseng, Y., (2009). iPower: an energy conservation system for intelligent buildings by wireless sensor networks, International Journal of Sensor Networks vol.5, no. 1, pp. 9712-9725.

TOU rates. http://www.hydroottawa.com.